FISH PHYSIOLOGY

VOLUME XI
The Physiology of Developing Fish

Part B
Viviparity and Posthatching Juveniles

CONTRIBUTORS

JEAN-GUY J. GODIN

BRYON D. GROVE

W. S. HOAR

C. C. LINDSEY

JULIAN LOMBARDI

DAVID L. G. NOAKES

JOHN P. WOURMS

JOHN H. YOUSON

FISH PHYSIOLOGY

Edited by

W. S. HOAR
DEPARTMENT OF ZOOLOGY
UNIVERSITY OF BRITISH COLUMBIA
VANCOUVER, BRITISH COLUMBIA, CANADA

D. J. RANDALL
DEPARTMENT OF ZOOLOGY
UNIVERSITY OF BRITISH COLUMBIA
VANCOUVER, BRITISH COLUMBIA, CANADA

VOLUME XI

The Physiology of Developing Fish

Part B

Viviparity and Posthatching Juveniles

ACADEMIC PRESS, INC.
Harcourt Brace Jovanovich, Publishers
San Diego New York Berkeley Boston
London Sydney Tokyo Toronto

ACADEMIC PRESS, INC.
1250 Sixth Avenue, San Diego, California 92101

United Kingdom Edition published by
ACADEMIC PRESS INC. (LONDON) LTD.
24–28 Oval Road, London NW1 7DX

Library of Congress Cataloging in Publication Data
(Revised for vol. 11)

Hoar, William Stewart, Date
    Fish physiology.

    Vols. 8-   edited by W. S. Hoar [et al.]
    Includes bibliographies and indexes.
    Contents: v. 1. Excretion, ionic regulation, and
metabolism — v. 2. The endocrine system —   — v. 11.
The physiology of developing fish. pt. A. Eggs and
larvae. pt. B. Viviparity and posthatching juveniles
(2 v.)
    1. Fishes—Physiology—Collected works.
I. Randall, D. J.   II. Conte, Frank P., Date
III. Title.
QL639.1.H6   597'.01   76-84233
ISBN 0–12–350434–1 (v. 11, pt. B) (alk. paper)

PRINTED IN THE UNITED STATES OF AMERICA

88 89 90 91   9 8 7 6 5 4 3 2 1

CONTENTS

1. The Maternal–Embryonic Relationship in Viviparous Fishes
John P. Wourms, Bryon D. Grove, and Julian Lombardi

2. First Metamorphosis
John H. Youson

3. Factors Controlling Meristic Variation
C. C. Lindsey

4.   The Physiology of Smolting Salmonids
    W. S. Hoar

5.   Ontogeny of Behavior and Concurrent Developmental
    Changes in Sensory Systems in Teleost Fishes
    David L. G. Noakes and Jean-Guy J. Godin

CONTRIBUTORS

Numbers in parentheses indicate the pages on which the authors' contributions begin.

JEAN-GUY J. GODIN *(345), Department of Biology, Mount Allison University, Sackville, New Brunswick, Canada E0A 3C0*

BRYON D. GROVE *(1), Department of Anatomy, University of British Columbia, Vancouver, British Columbia, Canada V6T 1W5*

W. S. HOAR *(275), Department of Zoology, University of British Columbia, Vancouver, British Columbia, Canada V6T 2A9*

C. C. LINDSEY *(197), Department of Zoology, University of British Columbia, Vancouver, British Columbia, Canada V6T 2A9*

JULIAN LOMBARDI *(1), Department of Biology, The University of North Carolina at Greensboro, Greensboro, North Carolina 27412*

DAVID L. G. NOAKES *(345), Department of Zoology, College of Biological Science, University of Guelph, Guelph, Ontario, Canada N1G 2W1*

JOHN P. WOURMS *(1), Department of Biological Sciences, Clemson University, Clemson, South Carolina 29634*

JOHN H. YOUSON *(135), Departments of Zoology and Anatomy, and Scarborough Campus, University of Toronto, Toronto, Ontario, Canada M5S 1A8*

PREFACE

Dramatic changes occur in the physiology of most animals during their development. Among the vertebrates, birds are entirely oviparous, live for variable periods in a cleidoic egg, and show fundamental alterations in excretion, nutrition, and respiration at the time of hatching. In contrast, the eutherian mammals are all viviparous, depending on the maternal circulation and a specialized placenta to provide food, exchange gases, and discharge wastes. The physiology of both mother and fetus is highly specialized during gestation and changes fundamentally at the time of birth. Fishes exemplify both the oviparous and the viviparous modes of development, with some examples that are intermediate between the two. In these two volumes, we present reviews of many, but not all, aspects of development. The chapters in Part A relate to the physiology of eggs and larvae: different patterns of larval development, osmotic and ionic regulation, gas exchange, effects of pollutants, vitellogenesis, the absorption of yolk, and the mechanisms of hatching. Chapters in Part B deal with maternal–fetal relations, meristic variation, smolting salmonids, the ontogeny of behavior, and the development of sensory systems.

The editors wish to thank the authors for their cooperation and dedication to this project and also to express their deep appreciation to the many reviewers whose careful readings and constructive criticisms have greatly improved the final presentations.

W. S. HOAR
D. J. RANDALL

CONTENTS OF OTHER VOLUMES

THE MATERNAL–EMBRYONIC RELATIONSHIP IN VIVIPAROUS FISHES

JOHN P. WOURMS

Department of Biological Sciences
Clemson University
Clemson, South Carolina 29634

BRYON D. GROVE

Department of Anatomy
University of British Columbia
Vancouver, British Columbia Canada V6 T1W5

JULIAN LOMBARDI

Department of Biology
The University of North Carolina at Greensboro
Greensboro, North Carolina 27412

  I. Introduction
      A. Viviparity in Fishes
      B. Overview of Maternal–Embryonic Relationships
      C. Trophic Patterns in Viviparous Fishes
      D. Trophic Transfer Sites
 II. Trophic Specializations for Uterine Gestation
      A. Introduction
      B. Embryonic Specializations
      C. Maternal Specializations
III. Trophic Specializations for Ovarian Gestation
      A. Intralumenal Gestation
      B. Intrafollicular Gestation
 IV. Conclusions
     References

FISH PHYSIOLOGY, VOL. XIB

I. INTRODUCTION

Viviparity is a highly successful mode of reproduction that has evolved independently many times and with many variations in widely separated taxonomic groups. It occurs in all classes of vertebrates except birds and among many different groups of invertebrates. In the vertebrate line, viviparity first evolved among the fishes, and it is among the fishes that the diversity of maternal and embryonic adaptations for viviparity is most pronounced.

The great majority of animals reproduces by laying eggs, and viviparous forms seldom, if ever, are found among the primitive species of a taxonomic group. Oviparity, therefore, is considered to be the unspecialized, primitive mode of reproduction, and viviparity must have evolved from oviparity. During the transition from oviparity to viviparity, profound changes occurred in the egg, embryo, and maternal organism. Diverse structural and physiological specializations for internal embryonic development evolved repeatedly and independently in diverse taxonomic groups through convergence and parallelism.

Initial steps in the evolution of viviparity involved a shift from external to internal fertilization and the retention of fertilized eggs in the female reproductive system. Simple retention of the developing embryo to term is the most primitive form of viviparity. In organisms limited to this developmental pattern, the embryo is afforded protection during a vulnerable phase of its life history but little else. Such embryos tend to be metabolically autonomous. Even in its most primitive form, however, viviparity is characterized by a variety of new maternal—embryonic interactions involving respiratory, osmoregulatory, endocrinological, immunological, and trophic relationships. Initially, the extent of these relationships was minimal, but once established, they have tended to evolve from the simple to the complex. A shift toward increased maternal dependency with a concomitant loss of the embryo's metabolic autonomy presumably enhanced the survival of offspring and conferred a selective advantage. The increase in maternal dependency was accomplished by the structural and functional specialization of maternal and embryonic tissues. As a result, not only were developing embryos afforded protection, but rate of development was no longer limited by their ability to regulate their environment and size at term was no longer limited by the supply of yolk. With a shift to maternal dependency, larger and more mature offspring could be produced. These features are presumed advantageous. Thus, the trophic relationship emerges as a dominant theme in the evolution of viviparity. Nevertheless, the importance of the hormonal control of viviparity, the immunological status of the fetus as a graft, and the physiological mechanisms involved in embryonic maintenance cannot be discounted. Unfortunately, these areas have not been well studied in fishes. Consequently, based both on its importance and on the relative availability of information, the maternal–embryonic trophic relationship is the focus of this review.

A. Viviparity in Fishes

Fishes provide a key to understanding vertebrate viviparity inasmuch as the first viviparous vertebrates were fishes and they display the most diverse maternal–fetal relationships of all the live-bearing vertebrates. Viviparity occurs in three major fish groups: the chondrichthyans, the teleosts, and the actinistians. Although widespread, it represents the dominant mode of reproduction only among the sharks and rays. Approximately 420 of the estimated 600–800 species of chondrichthyan fishes are viviparous (Wourms, 1977, 1981; Dodd, 1983). In contrast, only 510 of the estimated 20,000 species of teleost fishes are viviparous (Hoar, 1969; Wourms, 1981). The coelacanth (*Latimeria chalumnae*), the only extant species of actinistian, is viviparous (Smith *et al.*, 1975).

Among the various groups of fishes, there is an array of species that forms an almost continuous progression from primitively viviparous species, whose embryos are essentially metabolically autonomous, to highly specialized viviparous species with a high degree of maternal dependency (Amoroso, 1960; Hoar, 1969; Wourms, 1981). All of the known structural adaptations for maternal–fetal metabolic transfer, such as the yolk sac, yolk-sac placenta, amniochorion-like structures, placental analogs, and follicular and uterine secretions, have evolved in several different fish lineages. Evolutionary sequences with intermediate stages exist among living species in some instances (vide infra). In emphasizing the fundamental importance of fishes in the study of vertebrate viviparity, our operating hypothesis has been that the critical evolutionary step was the innovation in fishes of the structural and functional characteristics of the vertebrate female reproductive system, that is, (1) the vertebrate pattern of genital tract development and sex differentiation, (2) the vertebrate reproductive neuroendocrine system, and (3) the vertebrate ovary and oviduct with its analogs. Inasmuch as these features are critical for reproductive success, once they were established, they imposed constraints on possible pathways of further successful evolutionary innovation. Fishes, which are the oldest, most numerous, and most diverse group of vertebrates, in effect have served as an evolutionary laboratory for the testing of viviparity and the various modes of the maternal–fetal relationship. Virtually every known form of vertebrate viviparity and possible maternal–fetal relationship may be found in fishes. No matter how unusual or bizarre these adaptations appear, they are all modifications of preexisting cell types and tissues. What has evolved in other vertebrates represents little more than variations on a diverse basic theme. Some adaptations appear only in fishes while others, such as the placenta, which presumably are the more successfully advantageous ones, have evolved repeatedly. Any inquiry into the physiology of piscine viviparity not only is a legitimate end in itself, but also is fundamental to the general study of vertebrate viviparity.

Viviparity in fishes has been the subject of several reviews. Early studies on the physiology of development, especially the pioneering research of Ranzi (1932, 1934), were synthesized by Needham (1942). Turner (1947) summarized his own extensive studies of teleost viviparity, and Matthews (1955) considered various live-bearing fishes as indicators of possible paths that the evolution of viviparity may have taken. Since then there have been the following noteworthy reviews: (1) chondrichthyan viviparity (Budker, 1958); (2) teleostean viviparity (Bertin, 1958); (3) viviparity in fishes (Amoroso, 1960; Wourms, 1981); (4) modes of fish reproduction including viviparity (Breder and Rosen, 1966); (5) fish reproduction and development (Hoar, 1969); (6) reproductive guilds of fishes (Balon, 1975, 1981, 1984); (7) a popular compendium of viviparous teleost fishes (Jacobs, 1971); (8) an overview of viviparity including fishes (Hogarth, 1976); (9) chondrichthyan reproduction (Wourms, 1977; Dodd, 1983; Dodd *et al.*, 1983; Dodd and Dodd, 1986; Gilbert, 1981); and (10) prenatal nutrient absorptive structures in selachians (Hamlett, 1986).

Viviparity in general and viviparity in amphibians and reptiles have also been the subject of recent reviews (Blackburn, 1982a; Packard *et al.*, 1977; Wake, 1977, 1982, 1985). This review will build on the previous reviews, especially those of Hoar (1969), Dodd (1983), Dodd and Dodd, (1986), and Wourms (1977, 1981). Inasmuch as Wourms (1981) provides an extensive systematic account of the occurrence of viviparity, that information will not be repeated except in general terms and by way of up-dating. Our emphasis will be on the structural and functional basis of the maternal–embryonic trophic relationship. In conclusion, it is important to realize that the study of viviparity and embryonic nutrition, especially in terms of ultrastructure and physiology, is still in its infancy. For many groups of viviparous fishes, even the most basic morphological information is lacking. Unexpected discoveries are still being made, such as viviparity in *Latimeria*, the living coelacanth (Smith *et al.*, 1975), and trophotaeniae in embryos of the pile perch *Rhacochilus*, a member of the embiotocids, a group whose development was supposedly well known (Wourms and Lombardi, 1985).

1. CHONDRICHTHYAN VIVIPARITY

The chondrichthyan or cartilaginous fishes include the sharks, skates, rays, and chimeras. Contemporary cartilaginous fishes are assigned to two subclasses of very unequal size, Elasmobranchii and Holocephali. The holocephalans are a small group (six extant genera and about 30 species) that inhabit deep, cool, marine waters. The elasmobranchs are the dominant group of cartilaginous fishes, numbering 98 genera and about 700 species (Nelson, 1984). Elasmobranchs have evolved along two general lines: sharks and rays (Nelson, 1984; Compagno, 1984).

Reproductive patterns in chondrichthyan fishes have been reviewed recently by Dodd (1983) and Wourms (1977, 1981). Dodd (1983) stated that virtually nothing is known about the reproductive biology of about two-thirds of the estimated 700 species of elasmobranchs and that the information about the remaining third is fragmentary, descriptive, and highly selective.

All recent chondrichthyan fishes employ internal fertilization. With few exceptions, they produce a relatively small number of heavily yolked eggs whether they are oviparous or viviparous. Oviparity is confined to the three extant families of chimaeras, all four families of skates, and 10 of 22 families of sharks. Viviparity is widespread (approximately 453 of 700 species or about 65% of the chondrichthyans are viviparous). Viviparity is characteristic of sharks (253 of 367 extant species or about 69%), all rays, and a fossil holocephalan (cf. Table I). Eggs are retained in viviparous species and development is normally completed in utero. Within some genera, such as *Galeus* and *Halaelurus*, there appear to be transitionally viviparous species. Viviparous species are characterized by (1) lecithotrophy, (2) oophagy or adelphophagy, (3) trophodermy, and (4) placentotrophy.

There are contemporary species of sharks whose reproductive modes may represent stages in the sequential evolution of elasmobranch viviparity (Smedley, 1927; Breder and Rosen, 1966; Nakaya, 1975; Wourms, 1977; Teshima, 1981; Dodd and Dodd, 1986). Transi-

Table I

Families of Viviparous Chondrichthyan Fishes[a]

Class Chondrichthyes (after Compagno, 1973, 1984)

I. Subclass Elasmobranchii (sharks, rays, skates)
   Order Hexanchiformes (frilled and cow sharks)
   1. Chlamydoselachidae (frilled shark)
   2. Hexanchidae (cow sharks)
   Order Squaliformes (dogfish sharks)
   3. Echinorhinidae (bramble sharks)
   4. Squalidae (dogfish sharks)
   5. Oxynotidae (rough sharks)
   Order Pristiophoriformes (saw sharks)
   6. Fam. Pristiophoridae (saw sharks)
   Order Squatiniformes (angel sharks)
   7. Squatinidae (angel sharks)
   Order Orectolobiformes (carpet sharks)
   8. Brachaeluridae (blind sharks)
   9. Orectolobidae (wobbegons)
   10. Ginglymostomatidae (nurse sharks)
   Order Lamniformes (mackerel sharks)
   11. Odontaspididae–0 (sand tiger sharks)
   12. Pseudocarchariidae–0 (crocodile shark)
   13. Mitsukurinidae[b] (goblin shark)
   14. Alopiidae–0 (thresher sharks)
   15. Cetorhinidae–0 (basking shark)
   16. Lamnidae–0 (mackerel and white sharks)
   Order Carcharhiniformes (ground sharks)
   17. Scyliorhinidae (catsharks)
   18. Psuedotriakidae–0(?) (false catsharks)
   19. Proscylliidae (finback catsharks)
   20. Leptochariidae–P (barbeled hound shark)
   21. Triakidae–P (houndsharks)
   22. Hemigaleidae–P (weasel sharks)
   23. Carcharhinidae–P (requiem sharks)
   24. Sphyrnidae–P (hammerhead shark)
   Order Rajiformes (guitar rays)
   25. Rhinidae (shark ray)
   26. Rhynchobatidae (shovelnose ray)
   27. Rhinobatidae (guitarfishes)
   28. Platyrhinidae (thornbacks)
   Order Pristiformes (sawfishes)
   29. Pristidae (sawfishes)
   Order Torpediniformes (electric rays)
   30. Torpedinidae (torpedos)
   31. Hypnidae (numbfish)
   32. Narcinidae (electric rays)
   33. Narkidae (electric rays)
   Order Myliobatiformes (rays)
   34. Dasyatidae (stingrays)
   35. Potamotrygonidae (river stingrays)
   36. Urolophidae (round rays)
   37. Gymnuridae (butterfly rays)
   38. Myliobatidae (eagle rays)
   39. Rhinopteridae (cow-nosed rays)
   40. Mobulidae (manta rays)

II. Subclass Holocephali[c] (chimeras or rat-fishes)

---

[a] O, Oophagous or adelphophagous; P, placental.
[b] Presumed viviparous
[c] One-fossil species, −0.

---

tional species are mostly found within the Orectolobiformes (carpet sharks) and the Scyliorhinidae (catsharks), a family within the Carcharhiniformes. Transitional species are characterized by (1) the retention of the developing egg in the oviduct for long periods, (2) a tendency toward multiple oviparity (Nakaya, 1975), (3) a reduction in the thickness of the egg case, and (4) a loss of the surface ornamentation used for egg case attachment. Dodd and Dodd (1986) present a sequence illustrating a possible oviparous–viviparous transition. It is derived from Nakaya (1975), and we have expanded upon it using information contained in Bass *et al.* (1975), Springer (1979), and Compagno (1984). Nakaya subdivided oviparity into single and multiple oviparity. Single oviparity includes sharks such as *Scyliorhinus canicula*, *S. stellaris*, *S. torazame*, and *Proscyllium haberari* in which each oviduct contains but a single egg at a time. In *S. canicula* and *S. stellaris*, the egg is laid after a short (1–2 days) stay in the oviduct, usually in the blastoderm stage, and development is completed during the course of an 8- to 9-month incubation. In both *S. torazame* and *P. haberari*, however, the embryos develop to an advanced stage prior to oviposition.

Multiple oviparity is exemplified by five species within the scyliorhinid genus *Halaelurus*. In these species, several (4–12) eggs are present in the oviduct of spawning females. Each egg case contains an embryo at a stage of development corresponding to its place in the temporal sequence in which the eggs were ovulated, the so-called "conveyor belt model" (terminology of Myagkov and Kondyurin, 1978). There are four eggs per oviduct in *H. burgeri* and *H. boesmani* and 6–11 eggcases per oviduct in *H. natalensis*. Egg cases of *H. burgeri* are retained in utero for 6–8 months, during which the embryos develop to an advanced (70 mm) state (Kudo, 1959). Egg cases of *H. boesmani* and *H. natalensis* are assumed to be retained for a prolonged period, during which their embryos develop to an advanced (45 mm) state. In *H. lineatus* there may be up to eight egg cases per oviduct, and these are retained until the embryos are in a well-advanced state (30–40 mm). Under aquarium conditions, egg cases have been deposited and embryos hatched in 23–36 days at a length of 80 mm. Since egg cases have not been reported in nature, it is possible that the egg cases are retained for the entire duration of development and that hatching either coincides with or precedes parturition. The transition appears to have progressed even further in *H. lutarius*, which Bass *et al.* (1975) have described as "ovoviviparous." Its embryos are enclosed in thin-walled, fragile, bag-like egg cases that would offer little protection in the external environment. Encapsulated embryos, 11 cm in length with external yolk sacs 10 × 15 mm, have been reported *in utero*. This species probably is viviparous and displays a pattern of development similar to *Squalus acanthias* in which the embryos hatch from egg cases at a fairly advanced stage and complete their development in utero.

We have been able to construct another sequence for the genus *Galeus* based on information summarized in Compagno (1984). *Ga-*

leus eastmani is oviparous. One egg case at a time transverses the
oviduct prior to oviposition. *Galeus melastomus* is also oviparous, and
up to 13 egg cases have been found within the oviduct. *Galeus polli* is
the only known viviparous species in the genus. Embryos develop
within egg cases, which are retained until the embryonic development is well advanced. Embryos then hatch and complete their development in utero (Cadenat, 1959). Compagno (1984) suggested that the
island subspecies of *G. arae* is oviparous while the continental subspecies is viviparous.

As long ago as 1927, Smedley proposed a transitional series from
typically oviparous species with long-tendriled egg cases, through
oviparous species with unanchored egg cases, and finally to viviparous species. His model was based on the observation of three Ma-
layan species of sharks. Since his model is based on species within
three separate genera, it is not as convincing as models based on
species within a single genus (cf. Breder and Rosen, 1966, for a summary). A key factor in Smedley's series is the progressive increase in
the duration of egg retention. *Chiloscyllium* is oviparous, extruding
pairs of eggs that hatch after a short period of $2\frac{1}{2}$–3 months. *Stegos-
toma fasciatus* has up to four egg cases per oviduct, a reduced egg
size, and lacks all anchoring tendrils. Finally, in *Nebrius ferrugineus,*
four egg cases are retained until they hatch. Parturition coincides with
hatching.

There are other "transitional" species of interest and potential
importance that deserve more detailed study. The nurse shark *Ginglymostoma cirratum* carries as many as 21 huge, thick-shelled eggs in
each oviduct. Embryos develop to an advanced state and then hatch.
It is uncertain how closely hatching and parturition coincide (Gudger,
1940; Breder and Rosen, 1966). In the frilled shark *Chlamydoselachus,* there are up to 12 huge (80–90 mm) eggs per oviduct. Embryos
develop to an advanced state, hatch, and complete development in
utero (Gudger, 1940). The genus *Apristurus* is large (25 species) but
not well known biologically. Insofar as is known, all of its species are
oviparous except for *A. saldanha* (=*A. saldancha*), which is both viviparous and displays a "conveyor belt" mode of development
(Myagkov and Kondyurin, 1978). Three to five nonencapsulated embryos were found in each oviduct. Embryos were of different sizes,
6.8–9.2 cm in one female and 5.9–8.7 cm in another. The largest
embryos were close to term size (11–12 cm). Difference in size is
interpreted as difference in gestational age and corresponds to the
sequence of ovulation. In oviparous species such as *A. brunneus* there

is only a single egg per oviduct. Incubation requires about 1 year (Compagno, 1984).

Recognition of transitional stages in the evolution of viviparity may provide insight into the reproduction of the whale shark *Rhincodon*. The species was originally thought to be oviparous based on the finding of a single egg case that contained a late-stage, 36-cm embryo (Baughman, 1955). Wolfson (1983) has suggested an alternate explanation. The whale shark may either exhibit a simple form of viviparity, such as is found in *Ginglymostoma*, in which egg cases are retained, embryos develop to an advanced condition, hatch, and complete development in utero prior to parturition, or else the egg cases may be retained in utero for the major portion of development and then be deposited. The egg case described by Baughman (1955) may have been aborted. Wolfson's conclusions are based on the extreme rarity of egg cases in nature, the extreme thinness of the egg case, the presence of embryonic yolk reserves, the partially developed state of the gill sieves, and the presence of yolk-stalk scars on free-living, 55-cm specimens. It is interesting to note that female whale sharks have been reported to contain as many as 16 egg cases in utero.

There is considerable variability in the organization of the female reproductive system, a probable correlate of the wide range of reproductive patterns, especially viviparity. The unspecialized, basic pattern of the female reproductive system consists of a pair of ovaries and oviducts. These organs begin the embryonic development as paired structures, but often only one develops and differentiates, producing asymmetry in adults. This modification is prevalent in viviparous species. In most chondrichthyans, the ovary is in the gymnovarian condition. Mature ova are discharged from its outer surface into the coelom and then enter the oviduct where they are fertilized. The oviduct originates as a simple Müllerian duct that differentiates into four regions: (1) anterior ostium tubae, (2) shell or nidamental gland, (3) connecting isthmus, and (4) posterior uterus. The shell gland is the site of sperm storage and egg-case formation. It tends to be reduced in size or vestigial in viviparous species. The posterior region of the oviduct is highly developed in viviparous species, in which it is termed the uterus, and may display various modifications for viviparity. In all viviparous chondrichthyans, fertilized eggs are retained and develop to term in the uterus. [For further information and reference on the organization of the female reproductive system, refer to Dodd (1983), Dodd *et al.* (1983), Wourms (1977, 1981), and earlier papers cited in these references.]

2. Osteichthyan Viviparity

The Osteichthyes or bony fishes are divided into two subclasses, the Sarcopterygii and Actinopterygii (Rosen *et al.*, 1981). The former contains six living species of lungfishes, the Dipnoi, and the Actinistia with a single living species, the coelacanth *Latimeria*. The Actinopterygii contain virtually all of the 20,000 species of living bony fishes (Nelson, 1984). Reproduction is generally oviparous and fertilization external. Fertilization is internal in some species, the great majority of which are viviparous. Viviparity, although widespread, is the exception rather than the rule. It occurs in 14 or 15 of the estimated 424 families of living bony fishes (Nelson, 1984) and is distributed among seven orders with a total of 123 genera and about 510 species (cf. Table II). Viviparity evolved independently in the coelacanth and two groups of actinopterygeans: over 500 species of teleosts and a single fossil chondrostean (cf. Wourms, 1981). The degree of maternal dependence in viviparous osteichthyans varies as greatly as it does in the chondrichthyans, a fact that is reflected in the diversity of the functional and structural relationships between mother and offspring.

The female reproductive system of the coelacanth consists of a single ovary and an oviduct, while in teleosts it consists of either a single ovary or pair of ovaries and a single gonoduct. The ovaries begin development as paired structures; subsequently, (1) they develop into paired structures, (2) they develop and differentiate unilaterally, or (3) the two primordia fuse to form a single compound ovary. The entire system exhibits considerable variation in structure. [For details refer to Amoroso (1960), Amoroso *et al.* (1979), Dodd (1977), Hoar (1969), Wourms (1981), and Nagahama (1983).]

In the coelacanth *Latimeria*, only the right ovary, which is a solid ovary, is functional. Mature eggs are shed into the coelom and enter the oviduct, a modified Müllerian duct. Only the right oviduct is functionally differentiated. Gestation is uterine, and embryos develop to term in compartmentalized regions of the oviduct (Millot *et al.*, 1978; Smith *et al.*, 1975).

In contrast, viviparous teleosts are characterized by a cystovarian ovary, that is, a hollow ovary in which the germinal epithelium lines the interior surface of the ovarian cavity. Almost all viviparous teleosts have a single, fused, median ovary. The ovary is enclosed by a capsule formed by the peritoneal folds during development. Beneath the capsule there are one or more layers of heavily vascularized connective tissue and smooth muscle. The germinal epithelium with its ovarian follicles and associated connective tissue undergoes complex folding

Table II

Families of Viviparous Osteichthyan Fishes

Class Osteichthyes (modified from Wourms, 1981)[a]

I. Subclass Sarcopterygii
   Order Coelacanthiformes
   1. Latimeridae (coelacanth)
II. Subclass Actinopterygii
   Order Gadiformes
   2. Zoarcidae (eelpouts)
   3. Parabrotulidae (parabrotulids)
   Order Ophidiiformes
   4. Bythitidae (Suborder Bythitoidei) ⎫ viviparous
   5. Aphyonidae (Suborder Bythitoidei) ⎭ brotulids
   Order Atheriniformes
   6. Hemiramphidae (halfbeaks)
   Order Cyprinodontiformes
   7. Goodeidae (goodeids)
   8. Anablepidae (including *Jenynsia*) (four-eyed fishes)
   9. Poeciliidae (live-bearing tooth carps)
   Order Scorpaeniformes
   10. Scorpaenidae (rockfishes)
   11. Comephoridae (Baikal oilfishes)
   Order Perciformes
   12. Embiotocidae (surfperches)
   13. Clinidae (Suborder Blennioidei) (clinids)
   14. Labrisomidae (Suborder Blennioidei) (labrisomids)

[a] Systematic references to the groups listed here are given in Wourms (1981). We have adhered to his arrangement although alternative arrangements are also accepted. Some systematists (e.g., Nelson, 1984) place the Zoarcidae (eelpouts) among the Perciformes. Moreover, the Family Parabrotulidae is not universally recognized; instead, some systematists (Nelson, 1984) include its members within the Zoarcidae. It is also worth noting that the Bythitidae and Aphyonidae are joined together in the Suborder Bythitoidei and the Clinidae and Labrisomidae are similarly joined in the Suborder Blennioidei. In both cases, inclusion of the subordinal relationship indicates that the viviparity expressed by these groups may be the result of parallelism rather than convergence.

to form ovigerous folds or sheets that project into the ovarian lumen. In ovulation, the ova of teleosts are released into the lumen of the ovary, whereas in most other vertebrates ova are released into the coelom. In viviparous teleosts, the wall of the ovary is continuous with a gonoduct that extends posteriorly and opens to the exterior at the genital pore. The gonoduct is formed by the posterior growth of the ovarian tunic and its associated elastic connective tissue and smooth muscle layers. Even though it serves similar functions, the teleost gonoduct is not homologous with the oviduct of chondrichthyan fishes or the coelacanth (Amoroso *et al.*, 1979; Dodd, 1977; Hoar, 1969; Wourms, 1981, Nagahama, 1983).

Gestation in viviparous teleosts occurs either in the ovarian lumen (intralumenal gestation) or in the ovarian follicle (intrafollicular gestation). The ovary of viviparous teleosts is unique among vertebrates, since it is the site of both egg production and gestation. The ovarian lumen is the most common site of embryonic development. Nevertheless, intrafollicular fertilization is the general rule in viviparous teleosts, and embryonic development commences in the ovarian follicle in most teleosts with intralumenal gestation (Wourms, 1981). Embryos of embiotocids, goodeids, and some ophidioids leave the follicle early in development at the cleavage or blastula stages (Eigenmann, 1892; Turner, 1940a,b, 1947; Hoar, 1969; Wourms, 1981). Mendoza (1940, 1962) stated that in at least some embiotocids and goodeids, fertilization and ovulation occur so closely together that it is difficult to determine the sequence. In the comephorid *Comephorus baicalensis*, ovulation either just precedes or coincides with fertilization, gestation is intralumenal, and hatching coincides with parturition (Chernyayev, 1974). In the anablepid genus *Jenynsia*, some ophidioids (*Dinematichthys*, Wourms and Bayne, 1973), and some hemiramphids (*Nomorhamphus hageni*, Mohr, 1936), however, intrafollicular gestation is prolonged to an embryonic stage; hatching and development is completed within the ovarian lumen. Only in the viviparous zoarcids and scorpaenids is ovulation known to precede fertilization (Stuhlmann, 1887; Bretschneider and Duyvene de Wit, 1947; Moser, 1967a,b; Kristofferson *et al.*, 1973). Development in parabrotulids is intralumenal, but the site of fertilization is unknown (Turner, 1936). Gestation is exclusively intrafollicular in the cyprinidontiform families, Anablepidae (in *Anableps* but not *Jenynsia*) and Poeciliidae (Turner, 1940a,b, 1942), and in the perciform families, Clinidae and Labrisomidae (Penrith, 1969; Veith, 1979a,b, 1980).

Recently, Wake (1985) has attempted to homologize teleost viviparity with that of the other vertebrates. She points out that in the past there has been a general tendency to emphasize the differences in teleost viviparity as a result of the failure to analyze the evolution of teleost viviparity in terms of homologous and nonhomologous developmental patterns and structural and physiological convergences and constraints. According to Wake, two major constraints have been ignored: (1) teleosts never develop Müllerian ducts, so they lack the "true" oviduct, which is the organ utilized for embryonic mainte-nance in other vertebrates, and (2) teleost ovaries develop differently from other vertebrate ovaries. Wake has speculated that teleost vivi-parity evolved by using the ovary as the only available reproductive structure that was competent to differentiate cellular mechanisms for embryonic maintenance. Intraovarian gestation is a direct result of the developmental constraint imposed by internal ovulation, that is, ovu-lation into the lumen of the ovary instead of into the coelom. Wake suggests that

in the absence of an oviduct of Müllerian duct origin with its competency for hypervas-cularization and secretion, the vascularization and secretory epithelia of the follicle and ovary lumen are available for sperm maintenance and then for embryonic maintenance. Relatively slight modification of maternal physiology under endocrine control would facilitate longer-term retention and nutrition of embryos. The gonoduct, being a perito-neal sheath or peritoneal–ovarian muscular derivative, and often an interrupted struc-ture, lacks the competence and the responses that would facilitate maintenance. (p. 432)

B. Overview of Maternal–Embryonic Relationships

1. INTRODUCTION

Although the major emphasis of this review is on the maternal–fetal trophic relationship, other relationships fundamental to the evo-lution of viviparity and the survival of viviparous embryos cannot be ignored. Viviparity presents several special endocrinological prob-lems even though the basic endocrinology of reproduction undoubt-edly has remained similar in both oviparous and viviparous fishes. Once viviparity is achieved, the developing embryo depends on the female for the control of its environment—especially gas exchange–respiration and osmoregulation–excretion. Since the embryo pos-sesses and expresses both maternal and paternal genes, it can be re-garded as an allograft in its maternal host. This raises the question of immunological relationships. Finally, viviparity, as a reproductive strategy of frequent evolutionary occurrence, needs to be considered in terms of its ecological implications.

2. ENDOCRINE RELATIONSHIPS

According to Amoroso *et al.* (1979), "hormones are believed to hold a key position in the evolution of viviparity." Hogarth (1976), nevertheless, thought it unlikely that any major endocrine innovations occurred during the acquisition of viviparity. Instead, new tissue responses to preexisting categories of hormones evolved and new factors, such as embryonic influences, have been added to the regulation of basic hormonal interactions. In truth, there are few data to support either view.

In 1973 Atz stated that "the key problems associated with the hormonal regulation of viviparity to which Chieffi and Bern (1966) called attention . . . are still largely untouched." This opinion was echoed by Dodd (1975), who declared that "virtually nothing is known of the endocrinology of reproduction in viviparous teleosts . . . and the subject remains virtually unexplored in elasmobranchs." Eight years later, Dodd (1983) reiterated his opinion, stating that "nothing is known of possible endocrine involvements in gestation in either aplacental or placental elasmobranchs." Today there is still a dearth of pertinent information.

During the past 25 years, a number of reviews have appeared on the subject of the reproductive endocrinology of fishes and related topics. They show that what is needed is information specifically dealing with endocrine influences on gestation, copulation, mother–young relations, the development of placenta-like structures, superfetation, parturition, etc. Unfortunately, there are few pertinent original papers and reviews. Pickford and Atz (1957, pp. 216–222) reviewed the evidence for pituitary control over viviparity in fishes up to 1956. Dodd (1960) specifically considered the role of gonadal and gonadotrophic hormones in the viviparity of elasmobranchs and bony fishes. Hoar (1969) dealt with the effects of hypophysectomy on gestation, and Liley (1969) considered the effects of pituitary hormones on parturition. In 1973, Browning reviewed the evolutionary history of the corpus luteum. Chambolle (1973) addressed the question of hormonal regulation of gestation in the poeciliids. Overviews have been provided by Donaldson (1973) and Dodd (1975). Recently, Amoroso *et al.* (1979) and Amoroso (1981) considered the role of hormones in the evolution of viviparity in fishes and other vertebrates. More recently, the endocrine control of reproduction in chondrichthyan fishes has been thoroughly reviewed by Dodd (1983) and Dodd *et al.* (1983). Dodd (1983) concluded that his "summary exposes only some of the many gaps in our knowledge of the reproductive physiology of elasmobranchs and emphasizes the danger in generalizing."

Hogarth (1976) postulated that once internal fertilization had been achieved in fishes, the transition from oviparity to viviparity involved the interpolation into the reproductive cycle of a prolonged delay between fertilization and deposition of the fertilized eggs. Alterations in endocrine events normally associated with egg development and deposition could account for this transition. Such changes in endocrine regulation may have included alterations in the (1) regulation of follicle development through feedback mechanisms, (2) timing of ovulation, (3) control of cycles in the female reproductive tract, (4) regulation of the maternal–embryonic trophic relationship (e.g., morphological associations, osmoregulation of embryonic environment, trophic relationships), and (5) control of parturition.

Despite the limited amount known about the endocrine control of viviparity in fishes, it appears that the changes in the endocrine regulation of viviparity in elasmobranchs and teleosts have not been the same. The pituitary of both oviparous and viviparous elasmobranchs is believed to maintain gonadal integrity and stimulate steriod production by the ovary through gonadtropin secretion (Dodd, 1960; Amoroso *et al.*, 1979; Amoroso, 1981), but its role in gestation remains uncertain. Indeed, Amoroso (1981) stated that "the transition from oviparity to viviparity in this group of fishes cannot . . . be definitely linked to fundamental changes in endocrine regulation." Hisaw and Abramowitz (1938, 1939) had reported that when *Mustelus canis* females were hypophysectomized at the beginning of pregnancy, their embryos continued to develop. However, recent studies suggest that at least some endocrine events associated with gestation and parturition in elasmobranchs may be similar to those in viviparous amniotes. Tsang and Callard (1983, 1984) found that progesterone levels in *Squalus acanthias* remained high throughout pregnancy until near term and that estrogen levels were low during this period. In vitro experiments suggested that the corpus luteum may be the primary site of progesterone secretion. Relaxin, an ovarian polypeptide postulated to play a role in parturition in mammals (Sherwood and Downing, 1983), also has been identified in elasmobranchs (Schwabe *et al.*, 1978; Gowan *et al.*, 1981). Shark relaxin is biochemically similar to mammalian relaxin, cross-reacts with antibodies to mammalian relaxin, and induces pubic symphysis relaxation in guinea pigs (Reinig *et al.*, 1981). Conversely, mammalian relaxin has been shown to relax the caudal portion of the oviduct in near-term *S. acanthias* (Koob *et al.*, 1984).

In the teleosts, some of the reproductive events associated with viviparity suggest that changes in endocrine regulation may have been important in the evolution of viviparity, as Turner (1942) recognized. Fertilization in many viviparous teleosts is intrafollicular instead of occurring after ovulation. In some of these species, zygotes are released from the follicle into the ovarian lumen, where they complete development; in others, embryos develop in the follicle and are released only at the conclusion of gestation. Turner (1942) postulated that the retention or release of zygotes or embryos from follicular gestation sites and the retention or expulsion of embryos from the ovarian lumen are homologous to equivalent events in the reproduction of oviparous teleost fishes, namely, ovulation and egg-laying. Presumably, both sets of events are under the same or similar hormonal control. It would appear that the timing of the endocrine events normally associated with ovulation and egg-laying in oviparous species has been altered in viviparous species. Turner's conclusions were based on the following observations (Turner, 1942). He demonstrated that rupture of the follicle and release of the embryo was morphologically identical to ovulation. In species with superfetation, only the oldest embryos were released from the follicle, implying a differential response on the part of the follicular tissue. In *Gambusia,* where gestation to term is intrafollicular, release of the embryos from the follicles into the ovarian lumen coincides with birth, that is, expulsion of the embryos from the ovarian lumen. Birth or parturition is accomplished by the initiation of contraction of the muscular walls of the ovary. Embryos are quiescent and do not have any mechanical role in follicular rupture or parturition. (There is no information on the possible role of embryos in production of chemical or hormonal signals.) In one series of experiments, destruction of the brain by pithing initiated release of embryos from the follicles, either by removal of a central neural inhibition or by release of pituitary hormones. Evidence from other sources (Ishii, 1961, 1963) indicates that hormones can induce the premature release of embryos. Turner (1942) concludes that the overall control of gestation as well as differences in the duration of intrafollicular and intralumenal gestation among viviparous fishes can be accounted for by temporarily suppressing and subsequently reactivating production of the hormones that regulate follicular rupture and ovarian contraction and by the temporal separation of the latter two events. In light of contemporary thought, Turner's original model of the regulation of gestation by temporal differences in hormone production can be enhanced by including temporal differences in the hormonal competence or responsiveness of target tissues, that is, follicles and ovarian muscles.

At present, our understanding of the endocrine control of reproduction in viviparous teleosts is limited to the general role of the pituitary and steriod hormones in gestation and parturition. A few experimental and morphological studies suggest that the many hypophyseal hormones, among them, the gonadotropins and thyroid-stimulating hormone, may not be essential for the maintenance of pregnancy in poeciliids (Ball, 1962; Young and Ball, 1982). However, Chambolle (1964, 1966, 1967) found that the pituitary in *Gambusia* may play an important role in osmoregulation of the embryonic environment. Several studies also have implicated neurohypophyseal peptides, ACTH secretion, and corticosteriods in poeciliid parturition (Ishii, 1963; Heller, 1972; Kujala, 1978; Young, 1980; Young and Ball, 1983a,b). The role of estrogen and progesterone in gestation and parturition, however, is still problematic. Ishii (1961, 1963) found that estradiol induced premature release of *Gambusia* embryos during late pregnancy, but increased embryo mortality when administered during early pregnancy. Progesterone, on the other hand, had no effect (Ishii, 1961). In contrast, Korsgaard and Petersen (1979) reported that in *Zoarces viviparus*, treatment with progesterone or progesterone and estradiol together during late pregnancy induced parturition, but that estradiol treatment alone had no effect. In the embiotocid *Neodimetra*, short-term estrone treatment early in gestation accelerated thickening of the ovarian lining, suggesting that estrogen may play a role in preparing female tissues for pregnancy (Ishii, 1960). However, long-term estrone treatment inhibited embryonic development and suppressed histological changes in the ovarian lining. Recently, de Vlaming *et al.* (1983) reported that estradiol and progesterone levels were low during gestation in two embiotocids, *Cymatogaster* and *Hysterocarpus.*

In her review of viviparity among the vertebrates, Wake (1985) declared that "an endocrine control system that mediates ovarian or oviducal vascularization and secretion, and in some species aids in prolonging gestation, has also evolved a number of times, and must be structurally and physiologically homologous." These views appear to be unfounded or, at best, premature. Even if one assumes that such endocrine systems do, in fact, exist and that it is possible to homologize their physiological functions, in spite of the absence of any morphological correlates, evidence to support Wake's assertions is hard to find. Wake adduces none, referring instead to admittedly convergent biochemical and morphological similarities among the elasmobranchs, teleosts, and tetrapods.

3. RESPIRATION AND GAS EXCHANGE

During the development of viviparous fishes, two critical phases require special physiological adaptations of the ovary or oviduct as well as the developing embryos. The first phase occurs when the demand for additional nutrients has to be met after the yolk reserves have been absorbed completely. The second phase occurs "when the respiratory requirements exceed the capacity of the unspecialized exchange surfaces" (Webb and Brett, 1972a). The subject of respiration and gas exchange in viviparous fishes has been reviewed by Gulidov (1963) and Soin (1971). Most of the available information is contained in anatomical and morphological studies in which function is deduced from structure. It is fortunate that these deductions are probably valid, because only a handful of experimental studies have been carried out to test them.

Since most of the available information is derived from morphological studies, it is appropriate to consider them first. The morphology of embryonic and ovarian or oviducal structures of viviparous fishes has been described in detail (Amoroso, 1960; Jasinski, 1966; Hoar, 1969; Dodd, 1977; Wourms, 1981). Gas exchange and respiration are achieved by the modification of both maternal and embryonic structures. Ultimately, growth is limited by the capacity of the maternal ovary or oviduct to supply the needs of the developing embryos. The maternal structures of interest are the uterine villi and trophonemata, ovarian vascularization, the branchial placenta, the ovigerous folds, and the follicular placenta. The uterus of viviparous sharks and rays undergoes increased vascularization during gestation. The organization of the inner surface of the uterine wall ranges from an unspecialized, smooth epithelium (e.g., in the spiny dogfish *S. acanthias*) to complex uterine villi or trophonemata, (e.g., in the butterfly ray *Gymnura micrura* (Ranzi, 1932, 1934; Needham, 1942). Trophonemata may actually enter the branchial chamber. In some sharks, such as *M. canis,* closer approximation between the uterine wall and the embryo has been achieved by a gestation-induced hypertrophy of the uterine wall to form compartments around the embryo. The formation of compartments usually is associated with species having a yolk-sac placenta (Ranzi, 1932, 1934; Schlernitzauer and Gilbert, 1966). Gas exchange is assumed to be effected through the close apposition of highly vascularized maternal and fetal tissues. The efficiency of exchange may be enhanced by the presence of uterine fluid. In some fishes, such as the rockfishes *Sebastes,* the ovarian vascularization has been modified to supply the respiratory demands of large numbers (2.5×10^6) embryos. In *Sebastes,* this involves an arterial loop formed by the confluence of branches of the anterior and posterior ovarian arteries. Among bony fishes, the branchial placenta consists of a modification of the ovarian epithelium, the structure, and probably function, of which parallels that of the uterine trophonemata of elasmobranchs. In *Jenynsia*, processes of the ovarian epithelium enter the embryonic gill cavity (Turner, 1940d). In goodeids and embiotocids, the inner ovarian epithelium is thrown into a series of lamellae or sheets called the ovigerous folds. Turner (1938b), and subsequently Gardiner (1978) and McMenamin (1979), reported that the ovigerous folds of *Cymatogaster aggregata* serve a dual function. During early gestation, they are primarily secretory. During late gestation their vascularization increases and they serve in respiration. Turner (1938b) also described unusually thick-walled arterial vessels in these ovigerous folds. Webb and Brett (1972a) suggested that the sheathed arteries prevent premature oxygen diffusion between the arterial blood and both the venous blood and ovarian fluid while ensuring a uniform supply of oxygenated blood to the distal capillary beds. In *Anableps* and the Poeciliidae, a follicular placenta is formed during gestation. Vascularized villi of the follicular epithelium of *Anableps* are most numerous in regions opposite the portal circulation of the yolk sac and pericardial sac (Knight *et al.*, 1985). In poeciliids, where the maternal and embryonic tissues are only closely apposed, the intervening follicular fluid serves as a transport medium. On theoretical grounds, Webb and Brett (1972a) deduced that the system for oxygenation of the follicular fluid is more efficient than that for the ovarian fluid. The functional morphology of the maternal portion of the follicular placenta is similar to that of the uterine villi and the ovarian flaps of the branchial placenta.

Webb and Brett (1972a) distinguished between primary and secondary embryonic exchange surfaces. The former category includes the yolk sac, the pericardial sacs and their derivatives, the pericardial chorion and amnion, portal blood networks, gill filaments, and the general surface epithelium. Secondary exchange surfaces are found in the vertical fin system of embryonic surf perches, the trophotaeniae of goodeids, ophidioids, and parabrotulids, and the yolk-sac placenta and umbilical-cord appendiculae of sharks. These structures will be discussed at length under the section on trophic relationships. At this time, it can be said that all of them seem to incorporate a basic principle of design: the possession of a thin surface epithelium, often amplified, that is in contact with the capillary webs of a hypertrophied vascular supply. In discussing what are presumed to be respiratory adaptations, it should be borne in mind that many of the ovarian and fetal structures in question may serve a dual function. Structures that are thought primarily to play a role in nutrient secretion or absorption, such as trophonemata and trophotaeniae, may also participate in gas exchange. In the absence of physiological or ultrastructural evidence, it would be imprudent to exclude the possibility of such a functional duality. In this respect, the multiple functions of the mammalian placenta should be recalled (Miller *et al.*, 1976). Moreover, structures that were originally evolved for one function are not infrequently coopted to serve additional ones (Simpson, 1950).

Physiological studies on respiration and gas exchange in embryos of viviparous fishes are few. To date, maternal–embryonic gas exchange has been studied most extensively in the embiotocids *Rhacochilus (Damalichthys) vacca* and *Embiotoca lateralis.* Surfperch embryos develop in the ovarian cavity in close association with the ovigerous folds (cf. Section III,A). During midgestation, there is a progressive increase in the area of the ovigerous folds, the thickness of the ovarian epithelium, the capillary density of the ovarian surface, and the amount of ovarian fluid. The embryonic yolk sac is the first exchange surface to be elaborated. During late gestation, ovigerous folds penetrate the embryonic gill cavity and come in close contact with the gills (Webb and Brett, 1972a,b). During gestation the gill area increases to a maximum of 30% of the entire exchange surface at parturition. The most conspicuous feature of the embryos is the hypertrophy of its median and caudal fins (cf. Plate 1, Webb and Brett, 1972a). They become greatly expanded to represent 60% of the effective exchange surface at term, are highly vascularized, and terminate in vascularized, spatulate extensions. The paired fins are also vascularized. Presumably, gas exchange takes place between the ovigerous folds and the gills and fins.

Using a number of parameters, Webb and Brett (1972b) estimated the oxygen transfer characteristics of the brood/ovary system, maximum oxygen consumption, and ovarian blood flow. They found that during gestation, the respiratory requirements of the young increased 2.5 times, attaining a rate of 222 mg O_2 kg^{-1} h^{-1} at parturition. As gestation progressed, however, the capacity of the system to meet these needs decreased. The oxygen tension of the ovarian fluid reached a minimum value of 13.7 mm Hg just prior to parturition. The estimated flow of ovarian blood, however, was highest at parturition (714 ml h^{-1}) and was estimated to represent 12% of the cardiac output, a value similar to that of the mammalian placenta. Webb and Brett (1972b) concluded that the transfer of oxygen to the young is controlled and limited by the rate of flow of ovarian blood rather than by any structural characteristics. For comparison, Veith (1979b) determined that the oxygen consumption of 20-mm prepartum embryos of the clinid *Clinus superciliosus* was 473.5 ± 155 μl g^{-1} h^{-1} or 219 μl embryo^{-1} h^{-1}, a value twice that reported by Webb and Brett (1972b) for the surfperch at similar temperatures. Moser (1967a) found that oxygen consumption for stage-28 and -33 embryos of the rockfish *Sebastes eos,* and stage-34 (full-term) embryos of *S. rhodocholaris* was 335, 1750, and 2013 μl (g dry weight)$^{-1}$ h^{-1}, respectively. Again, these rates are considerably higher than those reported for embiotocids or clinids. Boehlert and Yoklavich (1984) reported that the oxygen uptake of *S. melanops* continually increased during gestation from an initial low value of 0.0078 μl O$_2$ embryo^{-1} h^{-1} in early embryos (i.e., stage 9, 2.4 days) to 0.087 μl O$_2$ embryo^{-1} h^{-1} in near-term embryos, (i.e., stage 30, 27.4 days). Oxygen uptake has been found to be linearly correlated with body weight in embryos of the eelpout Z. *viviparus.* Korsgaard (1986) showed that oxygen consumption was low at the time of hatching, 2.36 μmol O$_2$ (g wet weight)$^{-1}$ h^{-1} at 11°C, but doubled within 1 week to an average value of 5.67 μmol g^{-1} h^{-1} and remained fairly constant during yolk-sac absorption. When expressed per whole embryo, oxygen consumption increased with increased weight of the embryo. In a previous study, Korsgaard and Andersen (1985) found that oxygen uptake per unit weight of embryos subsequently decreased with increasing body weight of the embryos during the period after the yolk sac had been absorbed. The average rate of oxygen uptake was 4.20 μmol g^{-1} h^{-1} for embryos with an average wet weight of 254 mg.

Differences in embryonic and maternal blood oxygen affinity may also be important in facilitating maternal—embryonic gas exchange in viviparous fishes. Ingermann *et al.* (1984) reported that in the embiotocid *E. lateralis,* oxygen affinity of whole blood is higher in embryos than in adults and that embryonic blood oxygen affinity decreases from midgestation to term. This difference between embryonic and maternal blood is believed to be primarily due to differences between embryonic and adult hemoglobin (Ingermann and Terwilliger, 1981a; Ingermann *et al.,* 1984), although differences in the concentration of erythrocyte hemoglobin and organic phosphate have also been implicated (Ingermann and Terwilliger, 1981b, 1982). Using low pH, disc gel electrophoresis, amino acid analysis, and peptide mapping, Ingermann and Terwilliger (1981a) found that embryonic hemoglobin differs biochemically from adult hemoglobin. They also found that hemoglobin from midgestation embryos has a higher oxygen affinity than adult hemoglobin, while hemoglobin from late-gestation embryos has an oxygen affinity intermediate between those of midgestation embryos and adults. Ingermann *et al.* (1984) suggested that replacement of embryonic hemoglobin with adult hemoglobin may account for the decrease in blood oxygen affinity during the last half of gestation in *E. lateralis*.

Biochemical and functional differences between embryonic and adult hemoglobins also have been identified in other viviparous fishes. Manwell (1958, 1963) demonstrated that embryonic hemoglobin from *Squalus suckleyi* has a higher oxygen affinity than adult hemoglobin. Electrophoretic patterns of proteolytic digests of embryonic and adult hemoglobins also differed, indicating biochemical differences. Similarly, Hjorth (1974) analyzed embryonic and adult hemoglobins from *Z. viviparus* using starch gel electrophoresis and reported differences in their electrophoretic profiles.

4. OSMOREGULATION AND EXCRETION

Recently, osmoregulation and the removal of metabolic wastes from the fetal environment by the mother have been recognized as a fundamental aspect of fish viviparity. In 1967, Price and Daiber suggested that the inability of elasmobranch embryos to regulate urea content and osmotic pressure during early development constituted a selective disadvantage that led to the evolution of viviparity. Read's (1968) demonstration of urea-cycle enzymes at all stages of development appeared to contradict this. Pang *et al.* (1977) have reviewed the subject but they did not resolve the issue. More recently, Evans and Mansberger (1979) demonstrated that late-term *Squalus acanthias* embryos were able to osmoregulate, while Evans and Oikari (1980) showed that mechanisms of ionic and urea regulation are resident in early *S. acanthias* embryos. The rectal gland seems to be the major site of Na^+ regulation in this species. Extra-rectal-gland mechanisms, possibly branchial chloride cells, may play the major role in Cl^- balance. In all likelihood, the ability to osmoregulate increases with embryonic age and the functional differentiation of osmoregulatory tissues. Osmoregulation of early embryos can be accomplished more efficiently and with less expenditure of embryonic energy in a maternally controlled uterine environment, but as development progresses to term, the embryos presumably acquire an increasing degree of osmoregulatory independence.

Available evidence suggests that maternal regulation of the osmotic and chemical environment of the embryo also confers a selective advantage on viviparous teleosts. Triplett and Barrymore (1960b) found that the ability of the embryos of the surfperch (*Cymatogaster*) to osmoregulate is proportional to their stage of development. The youngest, midterm embryos (18–31 mm, as compared with 41-mm full-term embryos) were able to survive only in media equivalent to 25–36% seawater. Near-term embryos (32–38 mm) were able to regulate in salinities between 25 and 75% seawater and above. The ovarian fluid is hypotonic, osmotically equivalent to 36% seawater. Preliminary $^{39}Cl^-$ tracer experiments suggested that the branchial salt-secreting mechanisms, present in adults, were absent or operating with less efficiency in the embryos. It was also found that advanced embryos drink the surrounding medium, thus creating an osmotic gradient between the embryonic body fluids and the gut lumen. Veith (1979a) determined the osmolarity of the follicular fluid in *Clinus superciliosus* and found it 16% higher than that of the maternal plasma (377 ± 29 mosmol kg^{-1} versus 324 ± 21 mosmol kg^{-1}). He postulated that the embryos osmoconform to the follicular fluid and therefore do not need to expend energy on maintaining a positive water balance. At parturition, the full-term embryo abruptly leaves the hypoosmotic follicular environment (377 mosmol kg^{-1}) and is released into seawater with an osmolarity of 1000 mosmol kg^{-1}. The result is an increased osmoregulatory demand. As Webb and Brett (1972a,b) have pointed out, absorptive and exchange structures that might have been advantageous during gestation now have become an osmotic liability. This problem can be solved by reducing the surface/volume relationship of absorptive surfaces and by rendering embryonic exchange surfaces nonfunctional at birth. In embiotocids, this is accomplished by shunting blood away from the capillary web of the hypertrophied vertical fins (Webb and Brett, 1972a,b). To test the hypothesis that embryonic exchange surfaces are disadvantageous outside of the ovary, Veith (1980) compared the osmoregulatory ability of 6- to 8-mm "small" and 20-mm full-term clinid (*C. superciliosus*) embryos. Full-term embryos survived all concentrations of seawater from 40 to 100%. In contrast, the "small" embryos were not effected by 40–50% seawater, but had a reduced (80%) survival rate in 60% seawater, and any further increase in salinity was fatal. The time required for 50% mortality sharply decreased as salinity increased. Veith concluded that the "small" embryos are capable of some degree of osmoregulation, since 40–50% seawater differs from ovarian fluid in osmolarity and ion content, but that water and ion fluxes across their large exchange surfaces are too great to be regulated at higher salinities. In the eelpout (*Zoarces*), the concentration of inorganic ions in the histotrophe and plasma is about the same and differs strikingly from the surrounding brackish water, which is hypotonic to the plasma. No information is available on the ability of eelpout embryos to osmoregulate. The histotrophe provides a medium for the development of embryos that is more favorable to them from an osmoregulatory point of view than is the surrounding water (Kristoffersson *et al.*, 1973; Korsgaard, 1983). Dépêche (1976) has confirmed and extended these findings in the guppy, *Poecilia reticulata*, a freshwater fish. During the first part of gestation, the embryo is dependent on maternal control of its osmotic and ionic environment. Young embryos produce and accumulate urea which may have an osmoregulatory role. Ouabain-sensitive active-transport systems of the embryo are most sensitive during the first half of gestation. As development proceeds, the embryo becomes progressively more capable of independent osmoregulation (Dépêche, 1976). Ultrastructural examination of the yolk sac and the embryonic pericardial sac surface of *P. reticulata* revealed that "chloride cells" were common (Dépêche, 1973). Since the chloride cells altered their morphology in response to osmotic stress, they have been implicated in osmoregulation. The role of the pituitary and interrenal organ in controlling osmoregulation during gestation of the poeciliid *Gambusia* was determined by Chambolle (1973).

There is little useful information on the maternal regulation of fetal metabolic wastes. A logical assumption, made without benefit of experimental evidence, is that waste products are transported across the follicular, ovarian, or uterine epithelium and subsequently removed via the maternal vascular system. Nevertheless, waste storage sites analogous in function to the allantois may occur in some fishes. The hypertrophied urinary bladder of *Heterandria* (Scrimshaw, 1944b) and *Poecilia* (Kunz, 1971) and the hypertrophied hind gut of *Anableps* (Turner 1940b, 1947) may perform this function.

5. IMMUNOLOGICAL RELATIONSHIP

Medawar (1953) has stated that the evolution of viviparity poses special immunological difficulties for the fetus. The mammalian embryo, and probably the embryos of most other viviparous vertebrates as well, constitutes an allograft in the maternal host. Except in highly inbred matings, the embryo inherits paternal histocompatibility antigens that are foreign to the mother but fails to inherit all of the maternal histocompatibility antigens (Billingham and Beer, 1985). Four hypotheses have been put forth to account for the absence of rejection of the fetal allograft: (1) antigenic immaturity of the fetus, (2) immunological indolence of the pregnant female, (3) the existence of anatomical structures that act as maternal–fetal immunological barriers, and, more specifically, (4) the uterus as an immunologically privileged site in which immune transplantation reactions are suppressed. Other immunological topics important to consider in connection with viviparity are spermatozoan antigenicity and the transfer of passive immunity to the fetus.

Until recently, the major research emphasis has been on mammalian systems (Brambell, 1970; Edidin, 1976). Triplett and Barrymore (1960a) and Hogarth (1968) were the first to draw attention to the maternal—embryonic immune relationship of viviparous fishes. Using the surfperch *Cymatogaster aggregata*, Triplett and Barrymore (1960a) investigated the rejection rate of scale homografts in newly born young of females that were sensitized to homografts during gestation. Homograft reaction time was considerably shortened in the young of sensitized mothers. It was concluded that pregnant females can transfer homograft sensitivity to intraovarian embryos and hypothesized that the transfer is effected by a circulating antibody passed to the embryo via the ovarian fluid and absorbed in the embryonic hindgut. Recently, Tamura *et al.* (1981) found that differentiation of the embryonic thymus, along with lymphocyte proliferation and differentiation, begins at approximately midgestation in another surfperch, *Dimetra temmincki*. Interestingly, the thymus of pregnant females in this species remains atrophic until late gestation, when it increases tremendously in volume with a simultaneous increase in the volume of the thymus of the developing embryo.

Hogarth (1968) demonstrated that developing embryos of the poeciliid *Xiphophorus helleri* elicited an immune response in gravid females—that is, intraperitoneal transplants of embryonic tissue were destroyed. The mother acted like an immune-competent host and did not become tolerant of embryonic antigens. Hogarth (1972a) further demonstrated that midterm *X. helleri* embryos carried active adult histocompatability genes before birth. The possibility remained that the ovary lacked a functional afferent lymphatic drainage and as a result homografts were not rejected. Hogarth (1972b) tested this by showing that allografts placed within the ovary were rejected as rapidly as those located elsewhere, and he concluded that the ovary is not a privileged site for allograft survival. Moreover, removal of one embryo from the ovary and its implantation elsewhere resulted in that embryo's destruction but did not affect the development of the remaining members of the brood (Hogarth, 1973). These studies also confirmed that the developing embryo is encased in an acellular egg envelope. More recently, Hogarth (1976) has suggested that the egg envelope acts as an immune barrier by preventing the access of embryonic antigens to the mother and thus avoiding sensitization or protecting the embryo from immune rejection by the sensitized mother. Grove (1985) recently described tight junctions in the follicular epithelium of the poeciliid *Heterandria formosa*. Tight junctions have been demonstrated to establish immunological barriers in other tissue systems, such as the mammalian retina and testis (Fawcett, 1986), and their presence in the follicular epithelium of species with intrafollicular gestation may establish an additional immune barrier. In species with intralumenal gestation (cf. Section III,A), embryos are not encapsulated by an egg envelope, and tight junctions in the internal ovarian epithelium may be important in protecting embryos from immunological rejection.

Another area also in need of study concerns the possible passive transfer of immunity from mother to fetus (Brambell, 1970). Fishes are not only known to have circulating antibodies, but uterine and ovarian fluids of fishes have been reported to contain leukocytes and serum proteins. Ranzi (1932, 1934) reported leukocytes in the uterine fluid of elasmobranchs, and Kristofferson *et al.* (1973) also found them in the ovarian fluid of *Zoarces*. Chambolle (1973) found that the intraovarian fluid of the poeciliid *Gambusia affinis* contained nine bands of serum proteins, which displayed the same electrophoretic mobility as nine protein bands in the maternal plasma. Veith (1979a) demonstrated that the follicular fluid of the clinid *Clinus superciliosus* has the same five protein fractions present in the maternal plasma. In some species of surfperches (Embiotocidae), ovarian fluid contains proteins identical with those of maternal serum (de Vlaming *et al.*, 1983). Wourms and Bodine (1983, 1984) reported that the histotrophe and maternal serum of the butterfly ray *G. micrura* contained proteins of identical electrophoretic mobility and that these proteins appear to correspond to immunoglobulin M (Ig M), macroglobulin, and serum albumin. Fetal absorption of proteins in goodeids has been experimentally demonstrated by Wourms and Lombardi (1979b), Lombardi and Wourms (1979, 1985c), and Lombardi (1983). Exogenous proteins are hydrolyzed within the lysosomal system. The unanswered questions are whether some histotrophe proteins are immunoglobulins, and if they are, whether they are selectively absorbed and transported intact by the fetus.

A final point concerns sperm antigenicity. Hogarth (1972c) presented evidence that the sperm of *Poecilia* is antigenic. This again poses a problem of allograft rejection. Following insemination, viable spermatozoa can be stored for periods of at least 10 months in the ovary of poeciliid fishes. Why is there no immune reaction against them?

6. Ecological Considerations

The relationship between viviparity and the ecology of viviparous organisms is not well understood. Although reproductive strategies have often been explained in terms of general life history strategy models (reviewed by Stearns, 1976), attempts to account for viviparity in this way are inadequate. Current models of life history strategies predict that viviparity, a reproductive strategy characterized by the production of a few well-developed offspring, is adaptive when the environment is stable and competition is high (K-selection) or when a fluctuating environment leads to increased juvenile mortality (bet-hedging) (Stearns, 1976). However, it is difficult to find clear-cut examples in which this is the case.

As a reproductive strategy, viviparity has several advantages. It not only affords developing embryos protection from predation and other hazards, but offspring are released to the environment later in development when they are better able to cope with predation and competition. The acquisition of specializations for maternal—embryonic nutrient transfer may further enhance offspring survival by extending the duration of parental care or accelerating embryonic growth. A number of advantages accrue to large-sized neonates of viviparous species (Wourms, 1977). Viviparity also may be advantageous in environments with changing local conditions. Live bearers' broods are portable, and they are easily moved away from place where conditions are deteriorating (Baylis, 1981). Viviparity may facilitate dispersal or recolonization, inasmuch as a single pregnant female has the potential to colonize a new environment. Finally, viviparity offers the advantages of parental care with none of the constraints on mobility so often associated with the care of offspring. Consequently, adults can be more opportunistic in escaping predation and competition or can exploit environments unsuitable for embryonic development (e.g., pelagic sharks). Given these advantages, viviparity is considered to be a highly effective form of parental care, despite obvious tradeoffs (Williams, 1966; Shine and Bull, 1979). In fishes, the evolution of viviparity appears to have been limited primarily by the acquisition of internal fertilization (Wourms, 1981; Gross and Shine, 1981; Gross and Sargent, 1985). Approximately 3% of all the teleost families have viviparous members, but 57% of those with internal fertilization are viviparous (Wourms, 1981; Gross and Sargent, 1985). Similarly, 65% of the chondrichthyan species, all of which display internal fertilization, are viviparous (cf. Section I,A,1).

Attempts to identify ecological conditions that favor viviparity in fishes have led to some general observations regarding the relationship between ecology and viviparity and the maternal–embryonic relationship. Tortonese (1950) called attention to a possible relationship between reproductive strategies and habitat in sharks. He noted that small, benthic or littoral sharks tend to be oviparous, while viviparous species are more diverse in habitat. Similarly, oviparity in skates and chimeras is associated with benthic or littoral habitats, while viviparous rays are benthic or pelagic (e.g. Mobulidae) (Wourms, 1977). The association of viviparity with diverse habitats suggests that viviparity may place fewer constraints on lifestyle. Barlow (1981) suggested that viviparity in chondrichthyans may facilitate dispersal and noted that small, benthic sharks (often oviparous) are conspicuously absent from coral reefs.

An association between viviparity and feeding ecology has also been noted (Wourms, 1977). Large predatory sharks and sawfishes are viviparous, while skates and smaller sharks (e.g., Heterodontidae, Scyliorhinidae), which feed on benthic invertebrates and small fish, are oviparous. However, this relationship is not clear-cut. Torpedos, stingrays and eagle rays, which have a similar feeding ecology to that of the smaller oviparous sharks, are viviparous. In addition, some large macroplanktivores are oviparous (e.g., whale shark) while others are viviparous (e.g., *Cetorhinus;* Mobulidae).

Embryonic nutrition in viviparous sharks appears to be related to lifestyle and feeding ecology as well. Species displaying limited or no maternal–embryonic nutrient transfer tend to be small and feed on smaller prey. In contrast, embryos of many of the large predatory sharks are supplied continuously with nutrients from the female (Wourms, 1981). In several species (e.g., *Lamna, Odontaspis, Alopias*), embryos obtain nutrients through oophagy and uterine cannabalism, while in others, maternal-embryonic nutrient transfer occurs across a well-developed placenta. Continuous transfer of nutrients to embryos during gestation presumably enhances embryonic growth and results in larger offspring at birth. Large offspring not only face less predation and competition, but are efficient and active predators at birth, a necessity for pelagic predatory species. Moreover, large offspring generally attain a large adult size. Oophagy and uterine cannibalism in particular are well suited to a predatory lifestyle. Embryonic growth is enormous, with embryonic dry weight increasing as much as 12,000 times during gestation (Wourms, 1981). Often dentition develops precociously, and generally only a few offspring survive to term. Embryos of these species face intense selection for rapid growth and aggressiveness and must be experienced predators at birth. Interestingly, matrotrophy is well developed in many of the rays (Wourms, 1981), but does not appear to be related to feeding ecology. Other explanations, however, may be in order, such as the advantage accruing to large neonates. Moreover, matrotrophy, which usually results in large embryos but small broods, may shorten the gestation period and thus permit the production of more broods per year.

Viviparous teleosts have diverse lifestyles and occur in a wide variety of environments. The fact that it is difficult to correlate viviparity in this group with specific ecological parameters suggests that viviparity is a flexible teleostean reproductive strategy. In contrast, the maternal-embryonic relationship in teleosts may be subject to environmental constraints. Thibault and Schultz (1978) examined the reproductive strategies of species of *Poeciliopsis* that inhabit very different aquatic environments. They concluded that lecithotrophy is successful in diverse, unpredictable environments, but that maternal–embryonic nutrient transfer requires a predictable food supply. Because lecithotrophic embryos are nourished by yolk stored in the egg, embryonic growth, and hence offspring size, is not dependent on food availability once vitellogenesis has been completed. On the other hand, the growth of embryos that obtain their nutrients from the female during gestation may be seriously affected by changes in food availability. The energetics of viviparity in *Poeciliopsis occidentalis* have been considered by Constantz (1980). Scrimshaw (1944a) measured embryonic growth in *Heterandria formosa,* a species that displays substantial maternal–embryonic nutrient transfer, and suggested that food supply may limit embryonic growth in this species. The findings of Cheong *et al.* (1984) also suggest that the amount of nutrients transferred to *H. formosa* embryos, which in turn affects offspring size, is a function of food availability. They found that when food was not limiting, offspring size increased in consecutive broods. Moreover, offspring size was positively correlated with female size, a phenomenon that may be due, in part, to larger females being better at foraging for food.

Superfetation, a condition in which multiple broods develop simultaneously in the female reproductive system, is common in viviparous teleosts that display maternal–embryonic nutrient transfer (e.g., *Heterandria, Poeciliopsis, Clinus*). The adaptiveness of superfetation, which appears to be subject to environmental constraints, has been the subject of debate (Downhower and Brown, 1975); Thibault, 1974, 1975). On theoretical grounds, Downhower and Brown (1975) argued that fishes with superfetation are favored in transient environments characterized by low levels of predation and competition and, consequently, high adult survivorship. Thibault (1975), however, reported that superfetation is most pronounced in the species of *Poeciliopsis* that inhabit stable environments. In addition, Thibault and Schultz (1978) reported that in *P. monacha*, a species that inhabits an unstable environment, superfetation is abandoned when conditions deteriorate and predation and competition increase.

C. Trophic Patterns in Viviparous Fishes

Fishes are either oviparous or viviparous. Oviparity is considered to be primitive and less specialized, while viviparity has repeatedly evolved from it among taxonomically divergent groups. In the older literature, a third category, ovoviviparity, has been used to describe a condition in which developing embryos are retained within the maternal organism but not provided with any additional nutrients. The category is artificial and difficult to apply, since the dependency on a continuing supply of maternal nutrients varies from almost nothing to complete and cannot be assessed without laboratory study, preferably of living material (Wourms, 1981; Dodd, 1983; Blackburn *et al.*, 1985). The term ovoviviparity has been rejected since the only operational distinction is that between oviparity and viviparity (Wourms, 1981; Balon, 1981). As Blackburn *et al.* (1985) have pointed out, the simple dichotomy between oviparity and viviparity, which was first proposed by Aristotle, has several features to recommend it, namely, (1) the classification is operational because it is based on easily observable phenomena; (2) it is applicalbe to all vertebrates; (3) it represents an important and meaningful biological distinction; and (4) by adopting the literal meanings of the two terms, present and future confusion can be minimized. The subsequent discussion of oviparity and viviparity is linked to the sequence in which four critical events of early life history occur: ovulation, fertilization, hatching, and parturition or oviposition (cf. Section I,A and Wourms, 1981).

Oviparity, literally egg-laying, presents no conceptual problems as it consists of the deposition of eggs enclosed in some form of egg envelope such as a shell or egg case in the environment external to the female's body. The egg, which may be unfertilized, in the first stages of development, or embryonated, completes its developmental program and a young organism hatches from the egg envelope. From the viewpoints of development, physiology, and evolution, it is useful to recognize subgroups within the category of oviparity (cf. Table III). The need for this should be readily apparent from the preceding dis-

Table III

Reproductive and Trophic Patterns in Viviparous Fishes

I. Oviparity
   A. Ovuliparity
   B. Zygoparity
   C. Embryoparity
II. Viviparity
   A. Facultative viviparity
   B. Obligate viviparity
      1. Lecithotrophy
      2. Matrotrophy
         a. Oophagy, adelphophagy, and matrophagy
         b. Trophodermy
         c. Placentotrophy
            i. Buccal and branchial placenta
            ii. Yolk-sac placenta
            iii. Follicular placenta
            iv. Trophotaenial placenta

cussion of transitional stages in the evolution of elasmobranch viviparity (Section I,A,1). Bertin (1958) coined the term "ovuliparity" to categorize the oviposition of unfertilized eggs. This is the reproductive mode encountered in most organisms with external fertilization, such as most oviparous teleost fishes. Blackburn (1981, 1982b, and personal communication) introduced the term "zygoparity" to describe reproductive modes that involve the oviposition of fertilized eggs, that is, zygotes, the product of egg–sperm fusion. Eggs are often in an early, that is, cleavage, phase of development. The original definition of zygoparity should be extended to include self-fertilizing hermaphrodites, such as *Rivulus marmoratus*, and other egg-laying fishes that reproduce by gynogenesis, e.g. gynogenetic carp and amazon molly. In the latter instance, developing eggs are not zygotes. Internal fertilization is a prerequisite in gonochoristic species. Zygoparity is the reproductive mode encountered in all skates, and some sharks such as *Scyliorhinus* and *Heterodontus*. Zygoparity is equivalent to "ovi-ovoviviparity" of Balon (1975) but has the advantage of being far less cumbersome and confusing. A difficult problem is posed by the existence of oviparous species in which encapsulated eggs are retained for prolonged periods and in which there is some or moderate embryonic development prior to oviposition. Reproductive patterns such as this are intermediate between oviparity and viviparity. This pattern seems to be more frequently encountered among elasmobranchs than teleosts (cf. Section I,A,1 for example and discussion). Blackburn (1981, 1982b, and personal communication) considers this a type of zygoparity in which fertilization long precedes oviposition. There seems to be a preferable option. We coin the term "embryoparity" to categorize a pattern of oviparous reproduction in which eggs are retained for long periods during which a definitive embryo is formed that may develop to an advanced state prior to oviposition. Completion of development and hatching (i.e., eclosion from egg envelopes) is external. The extreme limits of embryoparity would overlap facultative viviparity (vide infra).

Viviparity, also called "internal bearing" (Balon, 1981), has been defined by Wourms (1981) as "a process in which eggs are fertilized internally and are retained within the maternal reproductive system for a significant period of time during which they develop to an advanced state and are then released." We amend the definition as follows: Viviparity is a process in which eggs are fertilized internally and are retained and undergo development in the maternal reproductive system. Hatching (that is, eclosion from an egg envelope if one is present) precedes or coincides with parturition, and the result is a free-living fish. Previous disagreements on formal definitions of oviparity and viviparity have resulted from unsuccessful attempts to deal conceptually with the occurrence of intermediate or transitional phases in the evolution of viviparity. This difficulty, however, is part of the problem of establishing discrete boundaries or categories in essentially continuous processes either of ontogenetic development or evolution, regardless of whether one espouses gradualism or punctuated equilibria. Elsewhere, one of us has attempted to deal with viviparity in terms of either facultative or obligate viviparity (Wourms, 1981).

Viviparity in most fishes is obligate—that is, during reproduction gravid females bear living young at the end of a temporally defined gestation period that coincides with the developmental program of the embryo. In contrast, facultative viviparity is encountered, usually sporadically, among species that are normally oviparous. Internal fertilization is effected either routinely or accidentally, or else eggs are parthenogenetically or gynogenetically activated. Fertilized or otherwise activated eggs are retained within the female reproductive tract, where they undergo embryonic development (cf. Wourms, 1981, for examples). Transitional forms of viviparity, such as occur in the sharks *Halaelurus lineatus, H. lutarius, Galeus cirratum*, and *N. ferrugineus* (cf. Section I,A,1) represent the culmination of embryoparous oviparity and are best treated as an extreme form of facultative viviparity in which egg retention has become routine rather than sporadic. The phenomena of facultative viviparity with internal fertilization and embryoparous oviparity are of considerable interest, since they probably constitute the initial stages in the evolution of obligate viviparity (Rosen, 1962).

Viviparity can be categorized on the basis of trophic relationships (cf. Table III). Embryonic nutrition ranges from strict lecithotrophy to extreme matrotrophy (Wourms, 1981). Lecithotrophic embryos derive their nutrition solely from yolk reserves, and oviparous fishes, of course, are lecithotrophic by definition. This is considered the primitive trophic situation (Amoroso, 1960; Hoar, 1969; Wourms, 1981). On the other hand, in matrotrophic species, the reserve of stored yolk is not adequate to meet the matabolic requirements of embryonic development, so the developing embryo depends on a continuous supply of maternal nutrients during gestation (Ranzi, 1932, 1934; Needham, 1942; Hoar, 1969; Wourms, 1981).

Lecithotrophy, essentially the retention of the developing egg to term, is the most primitive form of obligate viviparity and is trophically unspecialized. Strictly lecithotrophic, viviparous species undergo an embryonic dry weight decrease of roughly 35% during development, the decrease ranging from 25 to 55% (cf. Table IV). This decrease is comparable to the decrease in embryonic dry weight observed in oviparous fishes during development (Gray, 1928; Smith, 1957; Paffenhofer and Rosenthal 1968; Terner, 1979). Presumably, the lost biomass is made available and consumed to provide metabolic energy for embryonic development and growth.

In contrast some matrotrophs undergo enormous increases in dry weight, such as 238,300–842,900% in *Anableps* (Knight *et al.*, 1985) and $1.2 \times 10^6\%$ in *Eugomphodus taurus* (Wourms *et al.*, 1981), a most substantial maternal—embryonic nutrient transfer thus being indicated (cf. Table IV). In matrotrophy, the maternal contribution to the offspring takes place during two distinct periods. During oogenesis, female biomass is converted and stored as high-energy yolk (Ng and Idler, 1983; Wallace, 1985). During embryonic development, additional nutrients are supplied, including amino acids, fatty acids, proteins, and lipids.

When interpreting data on the embryonic energetics of viviparous fishes, an important caveat must be heeded. To ascertain whether a particular species is lecithotrophic or matrotrophic, the dry weight or total organic weight of the egg is compared with that of the full-term embryo (cf. Ranzi, 1932, 1934; Needham, 1942; Scrimshaw, 1945; Amoroso, 1960). Based on these data, it is possible to calculate either

Table IV

Developmental Mass Changes in Viviparous Fishes

Species	Egg		Embryo		Change in mass (%)	Nutritional mode
I. Chondrichthyes						
1. *Squalus acanthias*, Atlantic spiny dogfish	8.6	g	5.2	g	−40	Viviparous, lecithotrophy
2. *Squalus blainvillei*, Mediterranean spiny dogfish	10.7	g	10.8	g	+1	Uterine villi and secretion
3. *Torpedo ocellata*, electric ray	3.78	g	2.91	g	−23	Uterine villi and secretion
4. *Centrophorus granulosus*, European brown dogfish	162	g	74	g	−54	Uterine villi and secretion
5. *Dasyatis violacea*, pelagic stingray	0.9	g	16	g	+1680	Trophonemata and uterine milk
6. *Myliobatis bovina*, bat ray	1.9	g	61	g	+3120	Trophonemata and uterine milk
7. *Gymnura micrura*, butterfly ray	0.2	g	10	g	+4900	Trophonemata and uterine milk
8. *Prionace glauca*, blue shark	3.4	g	32	g	+840	Placenta
9. *Mustelus canis*, smooth dogfish	2.8	g	32	g	+1050	Placenta
10. *Eugomphodus taurus*, sand tiger shark	0.1626 g		1920	g	+1,200,000	Oophagy, adelophophagy
11. *Scoliodon laticaudus*, spadenose shark	0.06–0.07 mg		0.6–0.9	g	+1,000,000	Placenta

Taxon	Egg weight	Newborn weight	% change	Mode
II. Osteichthyes				
A. Latimeridae				
12. *Latimeria chalumnae*, coelacanth	184 g	180–228 g(?)	−2 to +23(?)	Oophagy (?)
B. Scorpaenidae (rockfishes)				
13. *Sebastes marinus*	3 mg	2 mg	−34	Lecithotrophic
14. *S. melanops*	0.071 mg	0.0667 mg	−10	Trophodermy (gut)
C. Poeciliidae (live-bearing toothcarps)				
15,16. *Gambusia*; *Poecilia*	1–3 mg	1–3 mg	0	Lecithotrophic
17. *Belonesox*	9.9 mg	6.9 mg	−30	Lecithotrophic
18. *Poeciliopsis monacha*	2 mg	1.26 mg	−38	Lecithotrophic
19. *Poeciliopsis turneri*	0.18 mg	3.39 mg	+1840	Follicular placenta
20. *Heterandria formosa*	0.017 mg	0.68 mg	+3900	Follicular placenta
D. Zoarcidae (eelpouts)				
21. *Zoarces viviparus*, eelpout	20 mg	240 mg	+1100	Hypertrophied hindgut
E. Goodeidae (goodeids)				
22. *Goodea atripinnis*	0.245 mg	3.15 mg	+1100	Trophotaeniae and gut
23. *Chapalichthys encaustus*	0.12 mg	3.38 mg	+2700	Trophotaeniae and gut
24. *Ameca splendens*	0.21 mg	31.7 mg	+15,000	Trophotaeniae and gut
F. Embiotocidae (surfperches)				
25. *Embiotoca lateralis*	4.42 mg	910 mg	+20,400	Gut, branchial placenta, fins
G. Clinidae (clinids)				
26. *Clinus superciliosus*, clinid, klipfish	0.047 mg	16.2 mg	+34,370(?)	Body surface, gut
H. Anablepidae (four-eyed fishes and allies)				
27. *Anableps anableps*	0.049 mg	149 mg	238,300	Follicular placenta and hypertrophied gut
28. *A. dowi*	0.108 mg	910 mg	842,900	Follicular placenta and hypertrophied gut
29. *Jenynsia lineata*	0.024 mg	5.8 mg	24,000	Branchial placenta

the percent weight change during gestation or the plastic efficiency coefficient, that is, the ratio of dry weight of the developed embryo to that of the fertilized egg (Gray, 1928). A net loss in weight during gestation on the order of 25–35% indicates a lack of nutrient transfer, while a net increase, stasis, or even a slight loss (about 5–10%) indicates maternal nutrient transfer. This approach, however, is useful only as a first approximation and for comparative purposes, especially when preserved material is all that is available for study. Weight comparisons are static and do not provide information on the changing energetics of embryonic metabolism. In an open system, static measurements that do not take into account embryonic catabolism lead to a serious underestimate of both the embryonic mass converted into energy and the maternal nutrients transferred to the developing embryo. One method of compensating for this deficiency is to assume that the embryonic mass converted into energy in matrotrophic, viviparous species is equivalent to the catabolic loss in mass in more of less closely related oviparous species and to introduce this figure (about 35%) as a correction factor. In some viviparous sharks (e.g., *Squalus acanthias*) and certain rays (e.g., *Torpedo*), this is indeed the case. More useful, however, are firsthand measurements of metabolism such as those carried out by Boehlert and Yoklavich (1984) and Boehlert *et al.* (1986) on three species of rockfish *Sebastes*. For example, embryonic catabolism in *S. melanops,* when measured by respirometry, required 64% of the original energy of the egg, but when measured calorimetrically, the embryo at birth contained 81% of the initial energy of the egg. thus, the total energy required for development from fertilization to birth was about 1.45 times the initial endogenous energy supply. About 70% of the energy of catabolism during gestation was of maternal origin. In contrast, weight change was only −11%. For *S. schlegeli,* values of 88% and 93% of the initial energy of the egg, respectively, were obtained for embryonic catabolism and embryonic biomass, making the total energy expenditure during development 1.8 times the initial endogenous supply. In this case, weight change was +22%. Similarly, values of 66%, 39%, 1.05 times and −21% were obtained for *S. caurinus.* These studies of Boehlert *et al.* (1986) present an unresolved paradox. There is a marked discrepancy when the total loss of organic weight (approximately 25–35% during development in oviparous species such as the trout and herring), presumed to represent the mass converted to energy, is compared with the catabolic metabolism of *Sebastes* embryos, about 64–88% of the original egg energy. Are these differences real or artifactual? Could they be accounted for on the basis of a different methodology and different assumptions used to calculate caloric content and catabolism? Alternatively, is the viviparous mode of reproduction energetically more demanding than oviparity? It would be informative and prudent in future studies on the energetics of viviparous development to employ a combination of measurements in weight change, respirometry, and direct calorimetry.

D. Trophic Transfer Sites

1. PATTERNS OF TROPHIC TRANSFER

To facilitate nutrient transfer in matrotrophic fishes, a number of different developmental adaptations appear to have evolved. In most instances, both maternal and embryonic tissues have been modified. Such trophic adaptations must have evolved independently in taxonomically unrelated groups. They typically differ both in the structure of the component tissues and in the extent of nutrient transfer. All of these adaptations appear to be modifications of preexisting cell types and tissues, albeit sometimes strikingly exaggerated. Evolution of trophic adaptations seems to represent a prime example of Jacob's (1977) "evolution by tinkering." Maternal tissues involved are limited to the oviduct/uterus, the follicle, and the ovarian epithelum, while embryonic tissues are limited to the epithelium of the body surface or the gut and its derivatives. Given these constraints on the availability of evolutionary substrates as well as an apparent high degree of plasticity in the extent of modification, it is not surprising that trophic adaptations display much convergence and parallelism.

Several systems have been proposed to define and categorize the different types of trophic specializations. They differ in approach but agree in substance. Wourms (1977, 1981) stated that three major classes of trophic adaptations are responsible for nutrient transfer: (1) oophagy and adelphophagy, (2) placental analogs, and (3) yolk-sac placenta. Balon (1981) considered matrotrophic fishes to be either matrotrophous oophages or viviparous trophoderms, the latter category referring to embryos that obtain nutrients by transport across an epithelial surface and including both the yolk-sac placenta and placental analogs of Wourms. Blackburn *et al.* (1985) proposed a more elaborate scheme, namely, (1) oophagy/adelphophagy, in which embryos feed upon sibling ova or developing siblings, (2) histophagy, in which embryos ingest maternal secretions, (3) histotrophy, in which embryos absorb maternal secretions, and (4) placentotrophy, in which nutrient transfer is accomplished via a chorioallantoic or yolk-sac placenta. The first category of oophagy was derived from Wourms (1977) and due to its usefulness will be retained. The distinction between histophagy and histotrophy seems blurred, and the terminology is confusing. How does the ingestion of histotrophe and its subsequent absorption by gut epithelial cells differ from the absorption of histotrophe by epithelia of the embryo's external body surface? Placentotrophy is a useful term, but it seems amniocentric to confine the term placenta to only the yolk-sac and chorioallantoic placenta (cf. Mossman, 1937).

Here, we propose a system that incorporates and modifies elements of Blackburn's, Balon's and Wourm's systems (cf. Table III). It recognizes three types of matrotrophy In the first category, we retain oophagy and adelphophagy as originally defined by Wourms (1981) and include matrophagy, the eating of maternal tissues by the embryo within it. Although matrophagy occurs in viviparous caecelians (Wake, 1977), its existence among fishes is problematic; ingestion of exfoliated maternal cells together with histotrophe is probably not uncommon, but instances in which piscine embryos actively attack and devour maternal tissues, although suspected, have never been proven (e.g., the buccal placenta of *Ogilbia*) (Suarez, 1975).

Trophodermy, the second category, is derived from Balon's (1981) guild of viviparous trophoderms. The latter group displays an extreme form of matrotrophy in which the absorptive and secretory structures involved in maternal–embryonic nutrient transfer consist of "some kind of modifications involving an epithelium: intestinal, uterine, pericardial, gill or fin." The key feature of Balon's concept is that maternal–embryonic nutrient transfer takes place across epithelial surfaces. Balon included both the placental analog and the yolk-sac placenta within the trophoderms. Placental analog was originally used by Wourms (1977, 1981) to describe a number of specialized embryonic or maternal tissues that facilitated maternal–embryonic nutrient transfer. Although these structures performed placental functions, they cannot be equated to classical placental types of structures such as choriovitelline, chorioallantoic, or yolk-sac placentas. We propose to restrict the term "trophodermy" to those types of matrotrophy in which maternal nutrients are transferred from their epithelial site of origin across intratissue spaces to distally located embryonic epithelial absorptive sites that are not in intimate association with maternal tissues. Several structural associations that traditionally have been considered to be placental analogs, such as the buccal/branchial, follicular, and trophotaeniael placentas, which were included in Balon's category of trophodermy, are excluded and assigned to a third category, placentotrophy. We establish the category of placentotrophy to emphasize the functional and evolutionary uniqueness of the placental relationship. Consequently, the term placental analog is abolished. Within trophodermy, the epithelium of the oviduct/uterus, follicle, and ovarian lumen, as well as embryonic structures such as the general body surface, fin epithelum, gill epithelium, and gut, may be involved in maternal–embryonic nutrient transfer. Some examples include ingestion and absorption of ovarian secretions through the gut of embiotocid embryos and absorption of nutrients through finfolds and the general body surface of embryos of the clinid *Clinus superciliosus*.

Placentotrophy, our third category, is the transfer of maternal nutrients to the embryo via a placenta. Placenta is used in its modern sense, as proposed by Mossman (1937, p. 156) and subsequently adopted by other authorities, among them Amoroso (1952, 1960): "an animal placenta is an intimate apposition or fusion of the fetal organs to the maternal (or paternal) tissues for physiological exchange." Mossman's definition of the placenta is of particular significance in the comparative study of viviparity, since it avoids the pitfall of amniocentrism. It shifts the placental concept away from criteria that depend on the stereotyped patterns of extraembryonic membranes in the amniotes, especially the mammals, and toward criteria based on the functional role of parental and embryonic tissues in physiological exchange. By emphasizing functional morphology, the placental concept can be extended to include many but not necessarily all of the wide variety of parental–embryonic exchange systems of viviparous invertebrates and poikilothermous vertebrates. It should be noted that in some cases, a suite of maternal and embryonic specializations for nutrient transfer, fitting into more than one category, may, more or less simultaneously, be employed—for example, the follicular placenta and the hypertrophied, trophodermic hindgut of *Anableps*.

At this time, we recognize four placental relationships among viviparous fishes, although it is possible that further research on some forms of trophodermy (e.g., finfolds and epaulettes) will result in their being reinterpreted as placentas. The first type of placenta is the yolksac placenta, confined to about 68 species from five families of sharks. It consists of the embryonic yolk sac and umbilical stalk, which may or may not have its surface area amplified by the differentiation of appendiculae. The maternal component consists of the uterine-wall attachment site. The second type is the follicular placenta, which consists of an intimate association between follicle wall and embryonic surfaces, such as the general body surface, finfolds, yolk sac, pericardial trophoderm, and pericardial amniochorion. Intrafollicular gestation and the follicular placenta are known to occur in the clinids, some labrisosmids, the poeciliids, and the anablepid *Anableps*. The third type includes the buccal and branchial placentae. The buccal placenta occurs in the ophidioid *Ogilbia* and involves the buccal investment of ovigerous bulbs, that is, projections of the internal ovarian epithelum, by the embryo (Suarez, 1975). (It is possible that this might also be actual or incipient matrophagy.) Branchial placenta have been reported in some rays in which trophonemata, long villiform, secretory processes of the uterine wall, enter the spiracles of the embryo and apparently release nutrients that enter the foregut. In the anablepid *Jenynsia*, as well as in some goodeids and embiotocids, there is an intimate association between regions of the internal ovarian epithelium and the gill filaments and pharyngeal epithelium. The fourth type consists of the trophotaenial placentas. These have evolved in four different orders of teleosts. They have been found in all species of goodeids except one, in one parabrotulid, in some ophidioids, and, most recently, have been discovered in the embiotocid *Rhacochilus*. The maternal portion of the trophotaenial placenta consists of the ovigerous folds of the internal ovarian epithelium, while the embryonic component consists of trophotaeniae, external rosettes or ribbonlike projections derived from the embryonic gut.

2. MATERNAL SPECIALIZATIONS

Maternal tissues specialized for nutrient transfer are confined to the oviduct and uterus, follicle, and internal ovarian epithelum. The oviduct and uterus are involved in the uterine gestation of chondrichthyans and the coelacanth. The lumenal epithelium of the oviduct–uterus exhibits sequential stages in the elaboration of structural and functional modifications for the transport or secretion of histotrophe (i.e., uterine fluid) (Wourms, 1981). The sequence reaches its zenith in the long villous trophonemata of some rays. In placental sharks, part of the uterine wall becomes the placental attachment site. The shell gland and its secretory product, the egg envelope, which is common to all egg-laying chondrichthyans, may be much reduced or even absent in some viviparous species. The primary specialization of the follicle is to form the maternal portion of the follicular placenta. In some instances, such as *Anableps* (Knight *et al.*, 1985), the follicular epithelium may form villous processes that lie more or less free or are compressed and interdigitate with the pericardial trophoderm of the embryo. In the eelpout *Zoarces* (Kristofferson *et al.*, 1973), gestation occurs in the ovarian lumen, and postovulatory follicles located at the tips of villiform processes extending from the ovarian wall become "nutrices calyces," structures involved in physiological exchange. Modifications of the internal ovarian epithelium are most commonly encountered in instances of gestation within the ovarian lumen. The epithelium usually becomes specialized for exchange, transport, or secretion. Modified regions of the ovarian epithelium form the maternal component of the trophotaenial, buccal, and branchial placentas (vide infra and Wourms, 1981).

3. EMBRYONIC SPECIALIZATIONS

The passage of maternal nutrients into the embryo takes place across epithelial surfaces. Because two major classes of embryonic epithelial surfaces—integument and gut—participate in the transfer processes, we have categorized these processes as dermotrophic and enterotrophic, respectively (Wourms and Lombardi, 1985). Dermotrophic transfer takes place across the epithelium of the general body surface and its derivatives. Transfer sites include general body surface, gill filaments, buccal epithelium, finfolds and epaulettes, yolk sac, pericaridal sac, pericardial amniochorion, and pericardial trophoderm. Enterotrophic transfer takes place across the epithelum of the gut and gut derivatives. Enterotrophic transfer sites include gut, branchial portion of the branchial placenta, and trophotaeniae. Enterotrophic transfer primarily occurs in oophagy and trophodermy, whereas dermotrophic transfer is confined to trophodermy and placentotrophy.

II. TROPHIC SPECIALIZATIONS FOR UTERINE GESTATION

A. Introduction

Among extant viviparous fishes, uterine gestation is confined to sharks, rays, and the coelacanth *Latimeria*. Embryos of some fossil viviparous holocephalans probably developed in the uterus (Lund, 1980). The uterus in all these fishes consists of an expanded area of the posterior region of the oviduct that may have transitory or permanent modifications for the maintenance of developing embryos (Wourms,

1977, 1981; Dodd, 1983; Hoar, 1969). Uterine gestation involves all four of the major embryonic trophic relationships: (1) lecithotrophy, (2) oophagy, (3) trophodermy, and (4) placentotrophy.

B. Embryonic Specializations

1. Yolk Utilization and Lecithotrophy

The utilization of stored yolk is the initial source of nutrients in almost all fishes, and for some fishes it is the primary source during development (Terner, 1979). Thus, yolk reserves are utilized during the entire development of oviparous and viviparous lecithotrophic species and during the early development of most matrotrophic species, except those that develop from yolk-deficient eggs, such as *Gymnura* and *Scoliodon*. Lecithotrophic viviparous elasmobranchs such as the electric rays *Torpedo marmorata* and *T. ocellata*, the European brown dogfish *Centrophorus granulosus,* and the spiny dogfish *Squalus acanthias* undergo a considerable (15–55%) loss of organic weight during development (Table IV) (Ranzi, 1932, 1934; Needham, 1942). Yolk utilization appears to be similar in both oviparous and viviparous lecithotrophic species. Although yolk utilization has not been studied in matrotrophic species, the events are probably similar (Hamlett and Wourms, 1984), with the possible exception that the intestinal absorptive stage is absent in embryos with yolk-deficient eggs, such as *Scoliodon* (Mahadevan, 1940). Transfer of stored nutrients from yolk to embryo is accomplished in three sequential stages. In the first and earliest phase, yolk is phagocytized by blastoderm cells and digested intracellularly. During the next phase, yolk is digested extracellularly by the peripheral syncytial cytoplasm and endodermal epithelium of the yolk sac. Digestion products are absorbed into the vitelline circulation. During later stages, yolk platelets are moved from the external yolk sac up the yolk stalk and into the intestine. In those species, such as *Squalus acanthias*, in which an internal yolk sac is present, yolk first passes into the internal yolk sac and then into the intestine (Te Winkel, 1943; Jollie and Jollie, 1967a; Hamlett and Wourms, 1984). Absorption of yolk from the gigantic (9 cm diameter, 185 g dry weight) eggs of the coelacanth *Latimeria* appears to follow a similar sequence (Smith *et al.,* 1975; J. P. Wourms and J. W. Atz, unpublished observations).

2. Oophagy, Adelphophagy, and Matrophagy

Oophagy and adelphophagy, which are forms of embryonic cannibalism, occur in both cartilaginous and bony fishes. They are variations on the tactic of siblicide (Mock, 1984) as a reproductive strategy. In oophagy, embryos feed upon sibling eggs. Ovulation may continue throughout part or all of gestation, thus providing a continuous food supply. Adelphophagy, literally eating one's brother, is also known in sharks as intrauterine embryonic cannibalism (Wourms, 1977, 1981). It occurs when the dominant embryos in a brood prey on their siblings while still in the uterus. Oophagy is a relatively primitive specialization that probably evolved to take advantage of the gamete wastage that is not uncommon in viviparous animals with large broods. It may have had its origins in the ingestion of moribund eggs and embryos by the surviving members. At present, there is no information about the occurence of matrophagy as an active form of embryonic nutrition.

Oophagy and adelphophagy are characteristic reproductive modes in sharks of the order Lamniformes (Compagno, 1984). They are either known or suspected to occur in five families—Odontaspididae, Pseudocarchariidae, Alopiidae, Cetorhinidae, and Lamnidae, which include the genera *Eugomphodus*, *Odontaspis*, *Pseudocarcharias*, *Alopias*, *Cetorhinus*, *Isurus*, *Lamna*, and *Carcharodon*. At least 14 species are involved. The frequency with which each mode occurs is not fully known. It is likely that members of the monotypic families, Mitsukurinidae and Megachasmidae are oophagous. The occurrence of oophagy in the carcharhinid *Pseudotriakis microdon* remains unproven, with conflicting accounts by Forster *et al.* (1970) and Taniuchi *et al.* (1984). Lund (1980) presented evidence of oophagy in a fossil holocephalan. Recent findings by Wourms *et al.* (1980) also suggest that the coelacanth is oophagous (vide infra).

Oophagy was first described in the porbeagle shark *Lamna* (cf. Wourms, 1977, 1981, for historical references), but the phenomenon has been best studied in the sand tiger shark *Eugomphodus taurus*. Springer (1948) discovered oophagy and presented evidence for adelphophagy in this species. Unlike the case for other viviparous sharks, he found only one embryo in each oviduct. Full-term embryos attained a length of slightly more than one meter and a weight of at least 6.4 kg. Living midterm embryos, about 260 mm long, were exceedingly active in utero and displayed aggressive predatory behavior. There were 60–70 egg capsules per oviduct in addition to the single embryo. Intact ova and yolk were found in the stomachs of embryos. It was assumed that the dominant embryo ate its siblings and ingested eggs. Subsequently, in an extensive study of *E. taurus*, Gilmore *et al.* (1983) found that there were six phases of embryonic growth and nutrition. We have added a seventh:

1. During the earliest phases of development, embryos are assumed to use yolk by absorbing solubilized yolk through the yolk sac as do other early shark embryos.

2. By 13.5–18.5 mm (total length), a definitive "amphibian-like" embryo has formed, and this was the earliest stage observed by Gilmore *et al.* A yolk stalk joins the external yolk sac to the distended abdomen, but a membrane appears to isolate the contents of the yolk sac from the abdominal cavity. The coelomic cavity, cardiac stomach, valvular intestine, and pericardial cavity contain yolk. At this stage, the embryos appear to derive nutrients from yolk in the internal coelom, not the yolk sac.

3. In the third phase, 18.5–51 mm, embryos derive nutrition from albumen and other encapsulated ova.

4. Hatching occurs between 49 and 63 mm, after which the embryos absorb uterine fluid and yolk.

5. The period of intrauterine cannibalism starts at about 100 mm, when the embryo begins to hunt and consume other embryos.

6. After consuming all the smaller embryos and reaching a size of 300–400 mm, the surviving embryo begins to consume egg capsules, each of which contains 7–23 unfertilized ova. During this phase, the embryo acquires its characteristically bloated yolk stomach and grows rapidly.

7. In the preparturition phase, the maternal ovary is reduced in size, few egg capsules are found in the uterus, embryonic yolk consumption is reduced, and the distension of the embryonic stomach declines. The embryonic liver increases in relative and absolute size, attaining 6.4% of the total body weight, a value comparable with that of adult fish. It would appear that metabolic reserves formerly stored as yolk are stored in liver cells at later stages. Birth occurs after a 9- to 12-month gestation period when the pups are 100–120 cm long and weigh more than 7–10 kg. During gestation, the female shark produces six distinct egg capsule types due to variation in ovulation rate and shell gland activity. Production of specific capsule types appears to be correlated with patterns of embryonic nutrition (cf. Section II,C,2).

Weight determinations of yolk utilization and scanning electron microscopic observations by Wourms *et al.* (1981) complement and extend the report by Gilmore *et al.* (1983). At the onset of gestation, gravid females may have 50–60 egg capsules per oviduct. Most of the capsules contained fertilized eggs or small embryos. In one instance, seven embryos, 20–95 mm, were recovered from one oviduct, the larger, 60- to 95-mm embryos being free in the oviduct. Weight determinations revealed that the relatively small (9 mm diameter) eggs had an average dry weight of 162 mg and a caloric content of 940 cal. Full-term embryos were assigned a dry weight of 2 kg, based on a 6-kg wet weight and the wet weight/dry weight ratios for shark embryos given in Needham (1942). On this basis, term embryos have undergone a $1.2 \times 10^6\%$ increase in dry weight during gestation. Assuming a 70% efficiency of yolk utilization, based on values for oviparous sharks (Needham, 1942), it is estimated that about 17,000 eggs (about 1000–1700 egg cases) with a caloric content of 16,000 kcal would be ingested by a typical fetus. The former figure may need to be readjusted in light of the finding that empty egg cases and egg cases with only 1–3 eggs are also ingested (Gilmore *et al.*, 1983). By sampling eggs and embryos at early stages, it was found that there is a marked loss in dry weight, 60–70%, from the fertilized egg through the 40-mm stage. Weight became stabilized at the 45- to 60-mm stages and rapidly increased in the 60- to 80-mm stages, eventually exceeding the initial egg weight by almost 100%. Weight changes could be correlated with changes in structure and behavior. Embryos at 22 mm had large yolk sacs, an open mouth but no jaws, and branchial filaments. Jaws had formed by 27 mm, but teeth were still absent. An extensive lateral-line system appeared by 30 mm and assumed a typical juvenile pattern by 35 mm. Precocious development of this system probably aids in prey detection. Functional dentition was acquired by the 40- to 45-mm stage and became pronounced by 60 mm. Development of dentition was precocious, occurring at 4% of term-embryo length in *E. taurus*, compared with 60–100% of term-embryo length in nonoophagous, viviparous sharks. The yolk sac underwent a marked reduction in size, beginning at 40 mm, and yolk depletion was completed by 60 mm. Ingestion of eggs appeared to begin at 50 mm and adelphophagy at about 80–100 mm. The implications of oophagy and adelphophagy have been discussed by Wourms (1977, 1981). They are considered simple but efficient strategies for attaining neonatal gigantism, a feature considered to have considerable survival value.

Oophagy, and to a lesser extent adelphophagy, have been documented in other species of sharks. As might be expected, details vary somewhat. A characteristic feature of some species is that, unlike the situation in *E. taurus,* more than one embryo per oviduct routinely survives to term. How is this accomplished? Is adelphophagy regulated by size relationships or by other behavioral mechanisms that protect some embryos from potentially fatal encounters with their sibs? Adelphophagy of small embryos does occur, so adelphophagy is not entirely surpressed.

Oophagy was first described in the porbeagle shark *L. nasa.* Each oviduct may contain two embryos, which can attain a length in excess of 1 m and a weight of more than 9 kg (Shann, 1923; Wourms, 1981). In *Pseudocarcharias kamoharai,* there are usually four pups to a litter, two in each uterus. Fujita (1981) observed a gravid female that had two 38- to 41-mm embryos and many egg capsules with two to nine ova in each oviduct. It appears that the developmental program specifies an initial development of only two embryos in each uterus or that only two ova are fertilized. At term, two large embryos, 400–430 mm total length (TL), with distended abdomens occurred in each oviduct. Each embryo weighted about 300 g; the liver accounted for about 25% of the body weight and ingested yolk from the stomach another 25%. Gilmore (1983) reported one embryo per uterus in the bigeye thresher shark *Alopias superciliosus,* confirming an earlier report of Gruber and Compagno (1981). Term embryos were 53 cm (SL) or 105 cm (TL). (The considerable difference between SL and TL is due to the greatly elongated dorsal lobe of the caudal fin.) Embryos weighed 2.5–3.0 kg and had distended cardiac stomachs filled with yolk. The liver was large, about 22% of standard length (SL). The same situation prevails in *A. pelagicus* (Otake and Mizue, 1981) and in *A. vulpinus* (Gubanov, 1972). The shortfin mako shark *Isurus oxyrinchus* may have up to 16 pups per brood, with up to eight embryos developing in each uterus (Stevens, 1983). At birth, they are about 70 cm long. Embryos are oophagous. Adelphophagy, at least at early stages, also is likely, inasmuch as Gilmore (1983) observed blastodiscs on many encapsulated uterine ova (Gilmore, 1983; Gubanov, 1972; Stevens, 1983). In the longfin mako shark *I. paucus,* only one embryo per uterus develops to term. Term embryos attain a length of at least 97 cm and a weight of 5.2 kg, being substantially larger than those of *I. oxyrinchus* (Gilmore, 1983).

Several other lamniform sharks (e.g., the goblin shark *Mitsukurina,* the great white shark *Carcharodon,* and the basking shark *Cetorhinus*) have been considered oophagous, even though definitive reproductive information has been lacking. The case for oophagy in

Cetorhinus is now much stronger, although its embryonic development is still enigmatic. Matthews (1950) had shown that the ovary in this species could produce large quantities of small eggs approximately 5 mm in diameter. The uterine region of the oviduct in nongravid females was a meter long. Nothing was known of development except for one 200-year-old report and a recent poorly substantiated report of very large embryos. Information on development from the Scandanavian literature has been called to our attention by Professor J. M. Dodd (cf. Sund, 1943; Aasen, 1966). In 1943, Sund reported that the process of birth in a basking shark had been observed in Norwegian waters in late August 1936. The mother was estimated to have weighed between 1525 and 2250 kg, and after being harpooned and while being towed, she gave birth to five living pups and one stillborn. The pups were estimated to be 1.5–2 m long. On the basis of liver size and length, they were considered to weigh about 20 kg. The living pups began swimming immediately after birth with their mouths open in the characteristic manner of basking sharks grazing on plankton. Neither yolk sac nor umbilical cord was observed. One can safely conclude that these pups were full-term. On the basis of Matthews's (1950) observations on egg size and number, it is reasonable to conclude that the reproductive pattern of *Cetorhinus* is basically similar to that of *Odontaspis*, with a large number of small eggs being ovulated and ingested throughout gestation. It is not possible, however, to say whether any adelphophagy occurs during early gestation. The minute size of the teeth in adults would mitigate against this, unless a special set of embryonic dentition was present. Moreover, it appears that at least three embryos per uterus developed to term, de facto evidence against late-stage adelphophagy.

The coelacanth *Latimeria chalumnae* is the only extant osteichthyan fish with uterine gestation. Information on its development is very limited (Wourms, 1981; Balon, 1984). It was not until 1975 that it was discovered to be viviparous. In that year, Smith *et al.* (1975) reported that dissection of a large female had revealed five advanced young, averaging 318 mm in length, each of which had a large yolk sac not connected with the uterine wall. Originally, it was suggested that coelacanths exhibit a simple lecithotrophic form of viviparity. It is now obvious that during the early and middle phases of development the embryos must be lecithotrophic. More recently, Wourms *et al.* (1980, and unpublished) have suggested that the developing embryos are oophagous. One of the five near-term embryos weighed 547 g (wet) (Wourms *et al.*, 1980) and the large (9 cm) eggs weighed 185 g (dry) Devys *et al.* (1972). Using these values and extrapolating to the dry weight of the embryo, Wourms *et al.* (1980) estimated that there is a −2% to +23% change in weight during a gestation period considered to last 10–13 months (Smith *et al.*, 1975). These data indicate a limited transfer of maternal nutrients from mother to offspring. If the smallest free-living coelacanth (420 mm, 800 g wet weight) (Anthony and Robineau, 1976) is not a juvenile but a newborn, then the estimated increase in its weight of +9 to +43% indicates a more extensive transfer of nutrients. Histological and scanning electron-microscopic studies of the yolk sac and lumenal epithelum of the oviduct did not reveal any trophic adaptations for viviparity. The uterine epithelium contained tubular glands that produce metachromatic secretions. The number of glands and the amount of secretion was insignificant, however, and no uterine fluid was present. The large, flaccid, highly vascularized yolk sac of one of the embryos was nearly devoid of yolk. The well-developed embryonic gut contained brown, amorphous material. Uterine secretions may be one source of additional maternal nutrients, but there may be another more important source. Anthony and Millot (1972) reported on a female that had 19 mature but unfertilized eggs lying free in the body cavity and apparently about to enter the oviduct. Due to spatial constraints in the one functional oviduct, it is highly unlikely that all 19 eggs could have developed to term. Wourms *et al.* (1980) postulated that some of these eggs would have developed while the rest eventually would have been ingested. They also have suggested (Wourms *et al.*, 1980, and personal communication) that the ingestion of excess eggs or their breakdown products may be the major supplemental source of maternal nutrients in developing *Latimeria* embryos.

3. External Gill Filaments

Some chondrichthyan embryos develop numerous long, external gill filaments that give the branchial region a bushy appearance. Similar structures also occur in a few larval osteichthyans, such as mormyrids. These are transitory structures that are replaced by the internal gill filaments that characterize most adult fishes. Chondrichthyan external gill filaments arise from the posterior surface of the gill arches and thus differ from the "true" external gills formed in amphibian larvae (Goodrich, 1930; Nelsen, 1953). The surface epithelium of the filaments is ectodermal in orgin, derived from the surface of the gill clefts. (Goodrich, 1930; Hughes, 1984). External gill filaments serve as a dermotrophic transport route.

Gill filaments in adult fishes are considered to function in gas exchange and ion transport (Laurent, 1984). Less attention has been given to the role played by external gill filaments in embryonic chondrichthyans. In oviparous species, such as skates, the gill filaments are bathed by protein-rich "albuminous" fluids within the egg case, while in viviparous species they are bathed by histotrophe. There is a consensus that the external gill filaments of embryos also function in gas and ion exchange. In viviparous sharks and rays, Kryvi (1976) suggested that hypertrophy of the external gill filaments is an adaption to increase the surface area available for exchange in embryos that develop in uterine fluids with a negligible circulation. This explanation also applies to oviparous species. Inasmuch as the structural organization of external gill filaments is well suited for exchange and transport, early workers (Goodrich, 1930; Ranzi, 1932, 1934) speculated that they also served as sites for nutrient absorption. Recent studies tend to confirm this view.

Each external gill filament consists of an epithelium that encloses a single vascular loop, which passes from the afferent to the efferent branchial vessel (Goodrich, 1930; Hughes, 1984, Laurent, 1984). Recently, the ultrastructure of embryonic gill filaments has been examined in the velvet belly shark *Etmopterus spinax* (Kryvi, 1976) and the Atlantic sharpnose shark *Rhizoprionodon terraenovae* (Hamlett *et al.*, 1985d). In the latter species, the surface epithelium is bilaminar and squamous and is separated from the underlying vascular endothelium by a collagenous stroma. In 4.5-cm, preimplantation embryos of this placental shark, the epithelium possesses a luminal glycocalyx, microvilli with smooth-walled vesicles at their bases, prominent tubular and vesicular elements, coated vesicles, lipid-like inclusions, rough endoplasmic reticulum (RER), Golgi complexes, and flattened nuclei. The endothelium lacks a basal lamina and exhibits many micropinocytotic vesicles on both its ad- and abluminal surfaces. The endothelial cytoplasm also contains RER, Golgi, mitochondria, and coated vesicles. Marked differences occur in the gill filaments of 10-cm embryos. Within epithelial cells there are numerous cytoplasmic filaments and a dense terminal web has formed. The epithelial cytoplasm contains fewer vesicles, tubules, mitochondria, and a less extensive RER and Golgi complex. The endothelium is unchanged. The amount of collagen within the stroma has increased and fibroblasts are more pronounced. These changes are associated with an apparent increase in mechanical strength of the gill filaments. Subsequently, as the yolk-sac placenta becomes functionally differentiated, the external gill filaments are resorbed.

Recently, Hamlett *et al.* (1985d) demonstrated that the external gill

filaments of the Atlantic sharpnose shark *R. terraenovae* are able to
take up the macromolecular tracer horseradish peroxidase (HRP). Using relatively high concentrations of HRP (10 mg of type IV per 50 ml),
they found that after 10 min of exposure, HRP uptake was most intense in the surface epithelial cells—to the extent that the reaction
product "nearly occludes the cytoplasm." Deeper epithelial cells contained reaction product in smooth-walled endothelial cells. Unfortunately, no information was provided about the possible passage of
HRP into the circulation. The finding of uptake is not surprising, since
it has been known for some time that juvenile teleosts are able to
absorb protein via their gills. Amend and Fender (1976) demonstrated
that juvenile rainbow trout could take up a 2% solution of bovine
serum albumin via the gills and lateral-line system during a three-
minute exposure. The rapid passage of the low-molecular-weight (molecular weight 261) organic compound tricaine methane sulfonate, a
derivative of amino benzoic acid, has been known for some time,
since it is widely used as a piscine anesthetic. It is reasonable to
assume that small nutrient molecules may also enter embryos via the
gill epithelium.

In the butterfly ray *Gymnura micrura*, and possibly other rays, a
branchial placenta is formed. This structure forms when uterine
trophonemata enter the branchial chamber of the embryo and establish close contact with the internal gill epithelium (Wood-Mason and
Alcock, 1891). A dual function was attributed to the branchial placenta. The intimate association of the highly vascularized trophonemata with the gill epithelium was considered primarily to have a
respiratory function. In addition, histotrophe, either transported
across or secreted by the trophonematal epithelium, presumably
passes through the pharynx and directly enters the gut, where it
serves as a nutrient substrate. A branchial placenta does not cocur in
Rhinoptera, but the overall frequency with which this structure occurs is not known (Hamlett *et al.*, 1985e).

4. Gut

There is a limited amount of information available on the embryological development of the gut and associated digestive organs [cf.
Wourms (1977) for an introduction to the general literature on development, e.g., Ziegler and Balfour and also Scammon's monograph on
Squalus]. The situation is even less satisfactory as far as the role of the
gut in the nutrition of viviparous embryos is concerned. Te Winkel
(1943) wrote that "there seems to be almost no general knowledge of

the role of the fetal intestine in developing elasmobranchs." This state of affairs has hardly improved during the intervening forty-three years.

Three patterns of embryonic nutrition involve the gut: (1) utilization of yolk platelets transported from yolk sac to intestine during the lecithotrophic phase of development, (2) ingestion and utilization of histotrophe–uterine fluid, and (3) oophagy, adelphophagy, and matrophagy. The latter two processes, when they do occur, usually attain maximal function after the lecithotrophic phase. Hence, patterns of digestion and absorption that served to metabolize yolk would subsequently be available for other nutrient substrates. Of the three processes, lecithotrophy and trophodermy (i.e., histotrophe absorption) appear to be strictly embryonic processes that can also be performed by tissues other than the gut, such as by the yolk sac and gill filaments. Endocytosis and intracellular digestion (i.e., lysosomal degradation) would be expected to be important in embryonic tissues. Adelphophagy and matrophagy appear to be qualitatively different inasmuch as they involve patterns of digestion more closely akin to postnatal and adult nutrition.

The role of the gut in trophodermy and oophagy has been deduced primarily from descriptive reports (Wourms, 1977, 1981). In species where the embryonic gut is suspected to play an absorptive role, embryos increase in weight, indicating nutrient transfer. Histotrophe or ova are released into the uterus by trophonemata and ovary, respec tively, and the embryonic gut becomes well developed. In some instances, histotrophe or ova could be removed from the embryo's stomach; in others, material found in the intestine appeared to be partially digested histotrophe or yolk. Unfortunately, details are lacking (Gudger, 1912; Shann, 1923; Gilmore et al., 1983). Trophodermic development of the tiger shark Galeocerdo poses a particularly interesting and puzzling problem. Embryos develop to term encased in an egg envelope that contains copious amounts of periembryonic fluid. Nutrient molecules released into the histotrophe by the uterine trophonemata must be small enough to traverse the egg envelope at a sufficiently high rate to bring about growth of the large (60–75 cm) embryos. Presumably, the embryo continuously ingests the periembryonic fluid, absorbs and subsequently digests the nutrients in the gut, and voids the nutrient-deficient fluid (J. P. Wourms and J. I. Castro, unpublished). Ingestion and recycling of perivitelline fluid has been reported in embryos of an oviparous teleost (Moskal'kova, 1985). It is tempting to speculate whether this same pattern may serve as a primary nutrient pathway in some aplacental sharks and rays and

as a secondary nutrient pathway in those placental sharks that develop
to term within an egg envelope.

In the lecithotrophic embryos of *Squalus acanthias*, intestinal
function is initiated when the embryo is approximately 65–70 mm in
length (Te Winkel, 1943). Yolk platelets are moved from the external
yolk sac up the yolk stalk into the internal yolk sac and from there into
the intestine. Yolk digestion and absorption increase and continue
even after birth, which occurs at 250–300 mm. At the 65- to 70-mm
stage, yolk platelets in the intestine display only a slight amount of the
degradation that is indicative of the onset of function. Since the pancreas contains zymogen granules at this stage, it is considered to be
functionally differentiated. More striking evidence of digestion is
found in the 150-mm embryos. Yolk platelets are found in all stages of
degradation. Breakdown is most complete in the lumen in the central
portion of the intestinal spiral valve and also in the internal yolk sac.
At 150 mm, intestinal epithelial cells were found to contain large
quantities of glycogen and many lipid droplets. Embryonic liver cells
are also distended with fat and glycogen. Te Winkel (1943) concluded
that the intestine, aided by the liver and pancreas, is the most important embryonic digestive organ during the remaining three-fourths of
gestation. It is reasonable to assume that a similar developmental
sequence occurs during the lecithtrophic phase of matrotrophic sharks
and rays. In point of fact, absorption of particulate material, namely,
India ink, in the gut has been demonstrated. Moreover, according to
Ranzi (1934) and Needham (1942), digestive glands in the stomach
become functional at an early stage. Recent electron microscopic studies of *Mustelus manazo* and *M. griseus* embryos indicate that their
intestinal epithelial cells are engaged in fluid phase endocytosis
(Okano *et al.*, 1981).

5. Yolk-Sac Placenta and Appendiculae

a. Introduction. The yolk-sac placenta is formed by the apposition
of a modified embryonic yolk sac to the uterine mucosa. Among fishes,
with the possible exception of the coelacanth *Latimeria*, the yolk-sac
placenta occurs only in sharks, in a diverse number of which it appears to have evolved independently. Based on data contained in
Compagno (1984), 68 of the 253 species of viviparous sharks, or about
27%, are known to be placental. Placental species occur in 17 genera
within five families: (1) Leptochariidae, one species of *Leptocharias*;
(2) Triakidae, 13 species in the genera *Hypogaleus*, *Iago*, and *Mustelus*; (3) Hemigaleidae, four species in the genera *Chaenogaleus*,

Hemigaleus, Hemipristis, and *Paragaleus;* (4) Carcharhinidae, 21–29 species in the genera *Carcharhinus, Isogomphodon, Loxodon, Nasolamia, Prionace, Rhizoprionodon,* and *Scoliodon;* and (5) Sphyrnidae, nine species in the genera *Eusphyra* and *Sphyrna.* Within a single genus such as *Mustelus,* 10 species are placental (e.g., *M. canis*) and 10 species aplacental (e.g., *M. antarcticus*) (cf. Teshima, 1981). Thus, it would appear that placental viviparity is far more widespread than generally thought and that the yolk-sac placenta has evolved on a number of different occasions. The latter conclusion suggests that the shark yolk sac is developmentally plastic and can easily evolve the more differentiated pattern of the yolk-sac placenta. Figure 1 illustrates these patterns.

The coelacanth *Latimeria* presents a problem. Does the yolk sac constitute a yolk-sac placenta in the late-term embryos (Smith *et al.,* 1975), as suggested by Amoroso (1981)? After reconsideration of the anatomical evidence and in accordance with Mossman's (1937) definition of a placenta, we conclude that the coelacanth yolk sac *is* a placenta. Amoroso's insight was correct but for the wrong reasons, a situation that also affected his interpretations. In the only known brood, five large, presumably late- or near-term, coelacanth embryos were developing within constricted, compartment-like regions of the oviduct. Embryos possessed large, flaccid, heavily vascularized yolk sacs that were nearly devoid of yolk. The yolk sac was in very close contact with the oviducal wall, but there was no evidence of a stable connection with the wall—that is, yolk-sac tissue could easily be translocated. The oviducal tissue conformed to the approximate shape of the yolk sac, thus forming a distinct zone of contact, but this could be an artifact of fixation. It is reasonable to conclude that there is very close contact between yolk sac and oviducal tissue and that there may even be a distinct zone of contact. Retention of a large yolk sac in a late- or near-term embryo is a specialized condition. In lecithotrophic sharks (e.g., *Squalus acanthias*) there is a progressive diminution and loss of the yolk sac during gestation. The same process occurs during the initial lecithotrophic phase in the development of nonplacental matrotrophic rays (Wourms, 1981; Hamlett *et al.,* 1985e). By contrast, it is only in the placental sharks that the yolk sac is retained after partial or near-complete depletion of the yolk reserves and prior to implantation. In the coelacanth, the latter condition is approximated. We regard the coelacanth yolk sac as the fetal portion of the yolk-sac placenta. The maternal portion consists either of a specialized zone of contact or of the entire oviducal mucosa within the embryo's compartment. Based on the extensive vascularity of maternal and fetal tissues

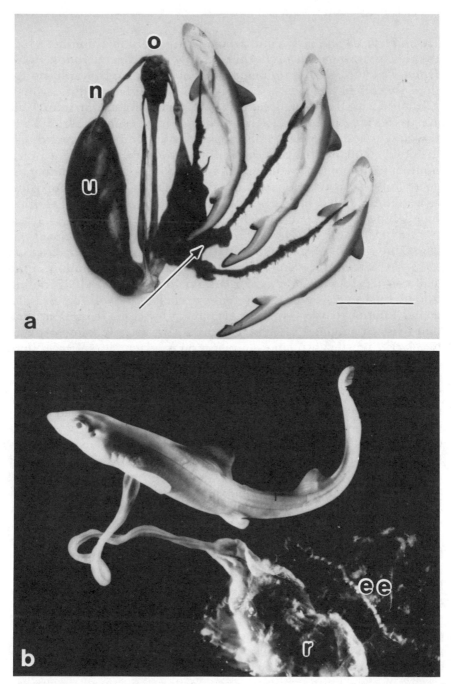

Fig. 1. (a) The reproductive system dissected from a gravid, full-term placental shark *Rhizoprionodon porosus* showing the ovary (o) with mature eggs, the shell gland (n), and on the left side an intact uterus (u) containing three embryos. On the right side,

and in the absence of obvious trophic modifications (Wourms *et al.*, 1980; J. P. Wourms and J. W. Atz, unpublished), we conclude that the primary functions of the coelacanth yolk-sac placenta are gas exchange and possibly the exchange of low-molecular-weight constituents, such as either metabolic wastes or metabolites. Our interpretation differs from that of Amoroso (1981), who concluded that there was a stable configuration between maternal and fetal tissues such as occurs in the yolk-sac placenta of the shark *M. laevis*.

Investigation of the structure and physiology of the selachian yolk sac placenta is an emerging field (Wourms, 1981). Light-microscopic descriptions of placental structure are limited to a handful of species, such as *M. canis* (= *laevis*) (Ranzi, 1934; Te Winkel, 1963; Graham, 1967), *M. griseus* (Teshima, 1975, 1981); *Carcharhinus falciformis* (Gilbert and Schlernitzauer, 1966), *C. dussmieri* (Teshima and Mizue, 1972; Teshima, 1973), *Scoliodon laticaudus* (= *S. sorrakowah*) (Mahadevan, 1940; Teshima *et al.*, 1978), and *Sphyrna tiburo* (Gilbert and Schlernitzauer, 1966). Ultrastructural studies have been confined to the sandbar shark *C. plumbeus* (Hamlett *et al.*, 1985a,b,c), and the blue shark *Prionace glauca* (Otake and Mizue, 1985; Wourms and Hamlett, 1978). Investigation of placental function is even more limited, to glucose tracer studies (Graham, 1967) and horseradish peroxidase tracer studies (Hamlett *et al.*, 1985a). Based on relatively few studies, there is a rather surprising diversity of placental structure and function, specifically; (1) early versus late implantation, (2) superficial versus interdigitated implantation, (3) variation with respect to thickness and number of intervening tissue and matrix layers, (4) occurrence of epitheliochorial, hemochorial, and possibly endotheliochorial placentae, (5) presence or absence of umbilical stalk appendiculae, and (6) hemotrophic and paraplacental modes of nutrient transfer.

b. Morphology and Morphogenesis. The shark yolk-sac placenta, as the name indicates, is derived from the yolk sac. In most placental sharks, prior to implantation, embryos pass through a lecithotrophic phase of about 3 months during which they float freely in the uterus. During this phase, they obtain nutrients from stored yolk reserves.

---

three embryos have been partially dissected from the uterus. The embryos are attached via appendiculae-bearing umbilical cords to the embryonic placental attachment site (arrow) on the uterus. Scale bar is equivalent to 10 cm. (b) A mid- to late-term embryo (140 mm total length) of the placental shark *Mustelus* sp. Portions of the egg envelope (ee) are associated with the rugose (r) portion of the embryonic placenta. A smooth-surfaced umbilical cord extends from the embryo to the smooth portion of the placenta. (Photographs by courtesy of J. I. Castro.)

Subsequently, the yolk sac establishes a definitive, long-term, stable contact with the uterine wall and differentiates into the fetal portion of the placenta. In placental sharks, the yolk sac undergoes two phases of differentiation. In the first phase, changes in the population of yolk-sac cells and the development of the vitelline circulation facilitate yolk utilization. The second phase results in the formation of the functional fetal yolk-sac placenta. In contrast, nonplacental lecithotrophic species only undergo the first set of changes but may later amplify the vitelline circulation for increased efficiency of respiration (Wourms, 1981).

The yolk sac of sharks is a trilaminar extension of body wall and gut, thus consisting of extraembryonic somatopleure, coelom, and splanchnopleure, as well as the circumscribed yolk mass. During early development, phases of differentiation and developmental changes take place that are associated with yolk utilization and respiration. These changes are nearly the same in both the nonplacental lecithotrophic *Squalus acanthias* (Jollie and Jollie, 1967a) as well as in five species of placental sharks, that is, blacktip *Carcharhinus limbatus*, blacknose *C. acronotus*, Atlantic sharpnose *R. terraenovae*, hammerhead *S. mokarran*, and blue *P. glauca* (Hamlett and Wourms, 1984). Typically, the yolk sac of carcharhinid embryos about 10 cm long consists of six regions: (1) somatic ectoderm, (2) somatic mesoderm, (3) extraembryonic coelom, (4) capillaries, (5) endoderm, and (6) yolk syncytium. The ectoderm is a simple, low-cuboidal epithelium that consists of flat cells with conspicuous ridge-like boundaries that contain chains of desmosomes. Interlocking microplicae extend over the cell surface. There is no evidence of endocytosis. Ectodermal cells contain the usual complement of cell organelles, lipid inclusions, a well-developed rough endoplasmic reticulum with dilated cisternae and vesicles, Golgi complexes, and large populations of coated and uncoated vesicles. The organization is similar to that of classical protein-synthesizing and -secreting cells. The ectodermal epithelium rests on a basal lamina, below which lies a collagenous stroma that contains dense bodies of various diameters that may be polyphosphate granules. The second, third, and fourth regions comprise the somatic mesoderm, an intervening narrow extraembryonic coelom, and the splanchnic mesoderm—vitelline capillary region. The morphology of the somatic and splanchnic mesoderm is similar to that of a classic mesothelium. Each region consists of a monolayer of spindle-shaped cells that have flattened nuclei and longitudinally oriented cytoplasmic fibrils. Surfaces of adjacent cells interdigitate but show few desmosomes. Endocytotic activity characterizes the upper and lower surfaces of mesodermal cells, while their cytoplasm contains numerous smooth-walled and coated vesicles as well as dense membrane-bound granules. Yolk-sac capillaries are adjacent to the inner surface of the splanchnic mesodermal layer. They are lined by a continuous layer of endothelium. Smooth-walled endocytotic vesicles occur on both the outer and inner surfaces of the endothelium. The fifth region, the endoderm, is in intimate contact with the basal lamina of the capillary endothelium. Its cells contain many mitochondria and polyribosomes, a heterogenous population of smooth-walled vesicles, and many yolk degradation vesicles. The innermost region, the yolk syncytial layer, contains many morphologically diverse yolk granules in various stages of degradation. The embryo is connected to the yolk mass by a yolk stalk that contains a vitelline artery, vitelline vein, and vitello–intestinal duct. The latter structure is confluent with the embryonic gut. In placental sharks, the internal yolk sac usually is either greatly reduced or absent (Teshima, 1981; Hamlett and Wourms, 1984). Yolk platelets are transported by ciliary movement up the duct and into the gut (Schlernitzauer and Gilbert, 1966; Baranes and Wendling, 1981). The vitelline artery and vein branch repeatedly on the yolk sac. They are connected to a network of smaller vitelline vessels established during early development. Solubilized yolk components are transported into the vitelline circulation and transmitted to the embryo. The pattern of vitelline circulation is retained and amplified in the placenta.

Following the lecithotrophic phase in most placental sharks, the yolk sac enters its second phase of differentiation stage and implants. In most placental sharks, implantation takes place after 2–3 months, but in some (e.g., *Scoliodon laticaudus*), with yolk-deficient eggs, embryos implant very early in gestation. The implanted yolk sac differentiates into a fetal placenta consisting of a relatively smooth-surfaced proximal portion from which the umbilical stalk, the former yolk stalk, extends to the embryo and a rugose, distal portion that establishes contact with the uterine wall. Yolk, formerly contained in the yolk sac, is used up, either by the time of implantation or shortly thereafter. As a result, the former yolk sac, especially the proximal portion, becomes a hollow spherical bag that contains a support system of connective-tissue struts and an extensive vitelline vascular system linked to the vitelline artery and vein. Subsequently, increased ramification of these vessels in the distal rugose placenta combined with surface unfolding and connective tissue hyperplasia gives this region its characteristic spongy texture. The maternal portion of the placenta consists of the uterine wall at the zone of contact. Initially, this region contains all of the cell layers that characterize the uterine wall at nonattachment sites. In the different classes of placentae that occur among sharks, the maternal portion of the placenta may be relatively little changed or undergo considerable alteration.

c. *Placental Diversity.* Five classes of shark placentae were originally recognized on the basis of the morphology of the maternal–fetal junction (Teshima, 1981). An amended version with six classes is presented here. In this system, there is a sequential decrease in the number and thickness of cell and extracellular matrix layers constituting the maternal–fetal placental barrier and an increase in the area of tissue contact by means of interdigitation. The egg envelope that occurs in some placentas should be regarded as a type of extracellular matrix. In oviparous species it contains collagen and probably one or more other structural proteins. Its role as a possible barrier to placental transport is enigmatic since little is known of its permeability properties.

Class 1. The ectodermal epithelium of the fetal placenta is reduced to a thin layer of extremely flat cells. The extraembryonic coelom is obliterated by mesodermal fusion. Many capillaries lie directly beneath the thin ectodermal epithelium. The egg envelope intervenes between fetal and maternal tissues. The maternal placental epithelium is essentially unmodified and consists of a single layer of columnar cells, directly beneath which there are many capillaries. Maternal and fetal tissues do not interdigitate; rather, the fetal placenta "rests" on the maternal placenta. Examples: *Carcharhinus dussumieri* and *C. falciformis* (Teshima and Mizue, 1972; Gilbert and Schlernitzauer, 1966).

Class 2. Greatly reduced epithelia of the fetal and maternal placentas are in contact with the egg envelope. At their junction, maternal and fetal placental tissues interdigitate. The epithelium of the fetal placenta is composed of extremely flat, elongated, squamous cells instead of the tall columnar cells found in the proximal region. The extraembryonic coelom has been obliterated by mesodermal fusion, and there is a marked increase in vascularization. The cells of the epithelial surface of the maternal placenta are greatly reduced in size and number and are extremely flat. A well-developed capillary network lies beneath the epithelium. Example: *Sphyrna tiburo* (Schlernitzauer and Gilbert, 1966).

Class 3. Maternal and fetal tissues are in direct contact and interdigitate with one another. The egg envelope is absent.
The maternal placental epithelium is reduced to a simple columnar epithelium that is underlain by a capillary
network and loose connective tissue. The fetal placenta
consists of four recognizable tissues: epithelium, capillary network, loose connective tissue, and mesothelium
lining the extraembryonic coelom. The epithelium is bi-
layered, consisting of an outer layer of binucleate cells
with microvillar apical borders, underlain by extremely
flattened cells. Example: *Prionace glauca* (Calzoni,
1936; Otake and Mizue, 1985).

Class 4. The much-reduced fetal epithelium degenerates in
some regions, permitting the fetal capillary network to
contact the egg envelope directly. The greatly reduced
but intact, thin, squamous epithelium of the maternal
placenta and its associated underlying capillary bed contact the egg envelope. Maternal and fetal tissues interdigitate. Example: Teshima (1981) established this class
using *Mustelus laevis* as an example. However, the description and illustrations of the placenta of *M. laevis* in
Ranzi (1934) clearly assign *M. laevis* to class 5. Although
an example is not now available, it seems advisable to
retain class 4 until further evidence indicates that such a
category either exists or does not exist.

Class 5. The greater portion of the already much-reduced, thin,
squamous epithelia of the maternal and fetal placental
tissues degenerates so that the capillary networks of
both tissues come in direct contact with the egg envelope at the fetal–maternal placental junction. Fetal and
maternal tissues interdigitate. The fetal capillary network is better developed than the maternal one. Examples: *M. laevis* (Ranzi, 1934) and *M. griseus* (Teshima,
1981).

Class 6. In this highly specialized form of placentation, found
only in a few sharks, the fetal yolk sac fits into and interdigitates with a trophonematous cup, a modified region
of the uterine wall. The egg envelope is absent. The
yolk-sac epithelium is much reduced, consisting of thin,
granular, squamous ectodermal cells that extend over a
deeply fissured, undulating surface. The yolk sac is
filled with a mass of capillaries. The fetal placental epithelium is in direct contact with maternal placental tis-

sues. At the point of contact, the maternal placental epithelium consists of a single layer of columnar cells. According to Setna and Sarangdhar (1948), the basal surfaces of these cells are continuously bathed with free maternal blood derived from the capillaries of the trophonematous cord. According to Mahadevan (1940) and Setna and Sarangdhar (1948), the fetal placental epithelium contacts an intact epithelium of the maternal placenta. In contrast, Teshima *et al.* (1978), stated that the maternal epithelium degenerates and the fetal epithelium contacts the blood-filled maternal connective tissue of the trophonematous cord. If these observations are correct, then the morphology closely approaches that of the classical mammalian hemochorial placenta. Example: *Scoliodon laticaudus.*

It is tempting to extrapolate greater efficiency in placental transfer from the gamut of increasing intimacy of contact between parent and offspring. In the absence of physiological studies on the transfer process, however, such speculation is not justified. Moreover, there are unresolved paradoxes. The European smooth dogfish *M. laevis* has a class 5 placenta that one might presume to be highly efficient. Nevertheless, it also produces one of the richest histotrophes known. Its small brood size and modest increase in embryonic mass, about 1050% (considerably less than that of some trophodermic rays), are difficult to reconcile with the presence of two presumably efficient placental transfer routes, transplacental and paraplacental.

d. Appendiculae. In placental sharks, an umbilical stalk, which may attain a length of more than 25 cm in some carcharhinids, connects the embryo to the proximal portion of the fetal yolk-sac placenta. The umbilical stalk is a modified yolk stalk that contains the vitelline artery, vitelline vein, and vitello–intestinal duct. The latter structure is absent in *S. laticaudus,* since whatever yolk is present in its yolk-deficient egg disappears early in development. In most sharks, the umbilical stalk is an unadorned, smooth-surfaced cylindrical tube. It is bounded by a multilayered squamous epithelium that may possess microvilli and cilia on the outermost surface (Gilbert and Schlernitzauer, 1966; Wourms and Hamlett, 1978). In some sharks, especially within the genera *Rhizoprionodon, Scoliodon, Paragaleus,* and *Sphyrna,* the umbilical stalk is adorned with finger-like processes termed "appendiculae" by Alcock (1890). Within at least one genus, *Sphyrna, S. tiburo* and *S. lewini* possess appendiculae while *S. mokarran* lacks them.

Appendiculae of different species exhibit a marked diversity in form and histological organization: (1) shape and length of processes, (2) simple versus branched structure, (3) presence of a connective tissue core or central blood vessel, (4) pattern of vascularization, and (5) organization of the surface epithelium (Budker, 1958; Mahadevan, 1940; Thillayampalam, 1928). At this time, we are able to recognize at least seven major types [cf. Southwell and Prashad (1919) and Thillayampalam (1928) for earlier classificatory schemes]. *Type 1* is the simplest and presumably most primitive. The umbilical cord is thick, thrown into folds, and bears a few small, flat processes. It was first described in *Scoliodon* sp. (= *Loxodon macrorhinus?*). *Type 2* occurs in *Rhizoprionodon acutus*. The appendiculae are small, flattened, lobulate processes with constricted bases. They lack major blood vessels but have a central core of connective tissue and a multilayered surface epithelium. *Type 3* occurs in *Sphyrna tiburo,* in which simple, unbranched, thick, finger-like processes of moderate length extend from the umbilical stalk. They possess a connective tissue core. Several blood vessels extend into these processes, apparently ramifying to form a dense peripheral capillary bed that lies immediately beneath a simple surface epithelium composed of short columnar cells (Schlernitzauer and Gilbert, 1966). *Type 4* occurs in *S. lewini.* Simple, unbranched, highly flattened processes extend from the umbilical stalk. The smaller ones are lamelliform and of equal width along their axis, while the larger are proximally constricted to form a stalk and distally compressed to form a flat disc. All are heavily vascularized by vessels that pass through the stalk and ramify into capillary beds extending over the surface of the terminal disc (J. P. Wourms, unpublished). *Type 5* occurs in *R. terraenovae.* The umbilical stalk is densely covered with appendiculae. The latter are elongated, flattened, much branched structures. Blood vessels extend into the base of the appendiculae, and ramify into a peripheral capillary plexus that underlies the surface epithelium. There is a central connective tissue core. The outer surface is bounded by a simple, low columnar epithelium. It is only in *R. terraenovae* (Fig. 2) that there is any information about the ultrastructure and possible function of the epithelial cells. Examination of 5- to 7-cm embryos revealed that most epithelial cells are covered with microvilli and that their cytoplasm is separated into distinct cortical and endoplasmic regions. The cortical region is electron-dense and contains small vesicles and membrane-bound elements. There is little or no evidence of endocytosis or an endocytotic complex. The endoplasm is characterized by grossly distended cisternae of the endoplasmic reticulum that contain flocculent material. Moderately large, electron-dense granules of somewhat variable size

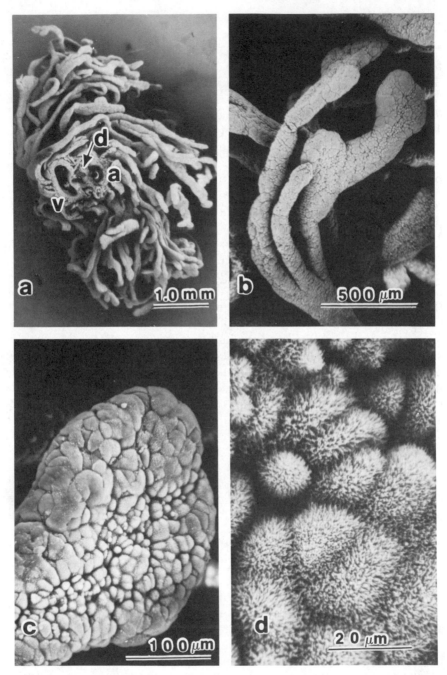

Fig. 2. (a) Scanning electron micrograph of a transverse section through the umbilical cord of the placental sharpnose shark *Rhizoprionodon terraenovae*. The cord contains an umbilical artery (a), umbilical vein (v), and the vitello–intestinal duct (d),

and mostly exhibiting circular profiles are associated with, and apparently derived from, the ER. These granules appear to be lipid. The basal portion of each cell contains a cluster of mitochondria (J. Lombardi and J. P. Wourms, unpublished; J. I. Castro and J. P. Wourms, unpublished). Preliminary experiments with trypan blue and peroxidase indicated that some epithelial cells have a limited capacity for endocytosis (J. Lombardi and J. P. Wourms, unpublished). *Type 6* occurs in *Paragaleus graueli*. The umbilical stalk is densely covered with appendiculae that radiate from it in a stellate fashion. Appendiculae are elongated, simple, unbranched structures that lack a connective tissue core. They appear unique inasmuch as each appendicula receives a branch of the umbilical vein that extends to its distal end. The surface is bounded by a simple epithelium consisting of uni- or binucleate columnar cells. The apical region of these cells is hyaline while the basal region is dense and contains vacuoles (Budker, 1953). *Type 7* occurs in *Scoliodon laticaudus* and *Rhizoprionodon oligolinx* (= *S. palasorrah*). The appendiculae of these species are numerous, elongated (up to 60 mm in *S. laticaudus*), thread-like processes that are either simple and mostly unbranched (*S. laticaudus*) or display dichotomous branching (*R. oligolinx*). The appendiculae of both species are histologically alike. The processes are somewhat flattened. In the latter stages of development, the connective tissue core is much reduced. Its place is occupied by a large vein and artery. There is an extraordinary development of a peripheral capillary bed just beneath the surface epithelium. The surface epithelium is reduced to a simple squamous epithelium (Mahadevan, 1940; Setna and Sarangdhar, 1948; Teshima *et al.*, 1978). Since their discovery, several functions (respiration and nutrient absorption) have been attributed to the appendiculae. The extensive vascularization found in some appendicular types is compatible with either function. Preliminary experiments tentatively have confirmed their absorptive role. If appendiculae are active in the absorption of molecules from uterine fluid, it will be essential to determine the balance between appendicular absorption and transplacental nutrient transfer. Similar conclusions have been reached by Hamlett (1986).

    e. Ultrastructure. There are only two ultrastructural studies of the selachian yolk-sac placenta, one on the sandbar shark *C. plumbeus*

---

which lies between the artery and vein. Numerous, elongate appendiculae extend from cord. (b) A single appendicula exhibits tripartite branching. (c) The flattened spatulate tip of the right-hand branch of the appendicula in (b). Cells appear to vary in size. (d) The apical surface of appendicular cells is amplified by numerous microvilli.

(Hamlett *et al.*, 1985a,b,c) and the other on the blue shark *P. glauca* (Otake and Mizue, 1985, 1986). The placenta of *C. plumbeus* belongs to class 2, that is, an egg envelope intervenes between interdigitating maternal and fetal tissues and there is relatively little reduction of cell layers. The blue shark has a class 3 placenta, that is, one similar to class 2 but without an intervening egg envelope.

In *C. plumbeus*, the distal rugose portion of the placenta forms the fetal attachment site. It abuts against an egg envelope and interdigitates with maternal placental tissues. The rugose placenta consists of (1) surface epithelial cells, (2) a collagenous stroma with vitelline capillaries, and (3) an innermost boundary layer. The surface epithelium consists of a single layer of elongated, peg-shaped columnar cells that adhere to the inner surface of the egg envelope. Wide spaces occur between the lateral margins of these cells. The apical surface is highly irregular, consisting of a continuous system of tubular invaginations of the apical cell membrane. The apical cytoplasm contains a network of anastomosing, smooth-walled, membranous canaliculi. The cytoplasm contains the usual complement of cytoplasmic organelles, including both coated and uncoated vesicles, in the juxtanuclear region. The basal cytoplasm is characterized by large numbers of coated and uncoated vesicles that are closely associated with microtubules. It also contains many "whorl-like configurations" that display a periodic substructure. Since these bear a close resemblance to similar structures believed to be yolk precursors in amphibians and trout, it has been suggested that they represent yolk precursors continuously synthesized and transported to the embryo during gestation. Vitelline capillaries are intimately associated with the basal surface of the epithelial cells. The basal lamina of the capillary endothelium is in contact with the basal surface of the adjacent epithelium. The thin-walled capillary endothelium displays a high degree of endocytosis and vesicular transport. A collagenous stroma separates the surface epithelium and associated capillaries from an endodermal boundary layer. Cells of the latter layer are joined by desmosomes, are almost devoid of cell organelles, and have only a few microvilli on their free surface.

The distal portion of the placenta in *P. glauca* displays differences from, as well as similarities to, that of *C. plumbeus* (Otake and Mizue, 1985). First, the egg envelope is lacking. The fetal placental epithelium is either in direct contact, suggestive of tight adhesion, or else is separated by a narrow space filled with electron-dense, periodic acid–Schiff (PAS) positive material and numerous small particles. The latter configuration is more frequent. The fetal epithelium is bilaminar, composed of an outer layer of low cuboidal giant cells (50–70 × 75–

100 μm) with an underlying layer of extremely flattened cells. The free surface of the giant cells is covered by numerous microvilli. The apical portion of the giant cells contains many surface invaginations, numerous small tubular structures, and spherical coated vesicles that contain electron-dense material similar to that found in the extracellular space. Cells of the maternal epithelium appear to be the source of the electron-dense material. Elements of the rough and smooth endoplasmic reticulum are scattered throughout the giant cells. Mitochondria are most numerous in their basal portions, which display many projections and interdigitate with the underlying cells. The lateral surfaces of the giant cells are smooth and in close contact with adjacent ones. The cells of the underlying squamous epithelium contain an extensive RER and smooth endoplasmic reticulum (SER), but otherwise nothing remarkable. Otake and Mizue have suggested that exogenous macromolecules are synthesized and secreted by the maternal placental epithelium and subsequently endocytosed by the giant cells of the fetal placental epithelium. The capillary network lies immediately beneath the surface epithelium and is invested in a loose collagenous stroma and separated from the extraembryonic coelom by endodermal endothelium. The capillary endothelium is fenestrated and is in direct contact with the basal lamina of the epithelial cell layer. The configuration is suggestive of active transport.

The smooth, proximal portion of the fetal placenta of *C. plumbeus* is a thin, flaccid, almost transparent, hollow sac. Cells on the exterior surface are polygonal. Most of them have a dense covering of microvilli, but some posesss ridge-like microplicae. The surface epithelium is one to three cell layers thick. It contains two cell types at the surface. Low cuboidal cells with ovoid nuclei and a dome-shaped apical surface covered with microvilli predominate. These cells contain many lipid droplets, but the endoplasmic reticulum and Golgi complexes are relatively uncommon. A second type has a flattened apical surface with either microvilli or cilia and a convoluted nucleus. The innermost cells of the surface epithelium rest on a prominent basal lamina. A collagenous stroma lies between the epithelial basal lamina and the basal lamina of the mesothelium (= endodermal endothelium) that forms the lining of the placental portion of the extraembryonic coelom. The cells forming this boundary layer are squamous and contain many pinocytotic pits and vesicles. Ultrastructural tracer studies show that cells of the smooth placenta do not absorb horseradish peroxidase or trypan blue (Hamlett *et al.*, 1985b). In *P. glauca*, the structures of the proximal and distal portions of the placenta are fundamentally the same (Otake and Mizue, 1985). The apical region of the surface cells exhibits an apical endocytotic complex, an indication that macromolecules may be taken up at this point. On the other hand, the presence of open intercellular spaces and other features suggests an exchange of low-molecular-weight constituents between the uterine fluid and the proximal placenta. This interpretation would be consistent with the observations that HRP and trypan blue are not endocytosed by the proximal portion of the placenta of *C. plumbeus.*

Because of the intimate anatomical relations and physiological interactions of the maternal and embryonic components of the yolk-sac placentas, it is more logical to discuss the maternal structures and functions here rather than in the section devoted to maternal specializations. In sandbar and blue sharks, the uterine tissues that form the maternal portion of the placenta are modified similarly (Hamlett *et al.,* 1985c; Otake and Mizue, 1986). In both there are increased vascularization and a reduction in the epithelial and mesodermal components compared with nonplacental regions (Otake and Mizue, 1986; Jollie and Jollie, 1967b). With the exception of the highly specialized placenta of *Scoliodon,* these changes appear to be similar in most placental sharks. In the sandbar shark, the maternal placental attachment sites are highly vascular, rugose elevations of the maternal uterine lining. The attachment site consists of a simple low columnar epithelium underlain by an extensive vascular network. The juxtaluminal epithelium, although reduced in number of cell layers and cell size, is intact. While the thickness of the mesodermal layer has been decreased, the degree of vascularization has been significantly increased. Juxtaluminal epithelial cells possess branched microvilli, saccular invaginations of the apical surface, coated pits, numerous coated vesicles, lipid-like inclusions, a prominent rough endoplasmic reticulum, and many free ribosomes. A basal lamina separates the juxtaluminal epithelium from the underlying profusion of blood vessels that lie in a much reduced connective tissue stroma. Capillaries are closely apposed to the basal surface of the epithelial cells. The capillary endothelium possesses numerous pits and vesicles on both its basal and luminal surfaces, a condition indicative of pinocytotic activity. The morphology of the juxtaluminal epithelial cells is consistent with that of a cell engaged in continuous transport or secretion, as opposed to one in which secretory products are stored prior to exocytosis. The overall cellular organization of the maternal placental attachment site in the blue sharks is virtually identical with that of the sandbar shark. A capillary network underlies the simple columnar epithelium. There are, however, some differences in epithelial cell ultrastructure. In the blue shark, the apical portion of epithelial cells contains numerous PAS-positive and PAS-negative granules and lipid droplets. Each granule is membrane-bound and contains numerous particles that vary in size and electron density. Exocytosis of granules and discharge of particulate contents into the extracellular space was observed, and particles in the extracellular space are similar to those in granules within the cells. The RER occupies a subnuclear position, the Golgi complexes a supranuclear zone. Many vesicles, some bristle-coated and containing highly electron dense material, also were observed in the Golgi zone. The morphology of these cells is consistent with that of a secretory cell in which secretion product is stored prior to exocytosis. Capillaries of the capillary network are in close apposition with the basal lamina of the epithelial cells, a configuration conducive to molecular transport. Information on the organization of nonplacental regions in the uterine wall is lacking for the sandbar shark. In the blue shark, however, Otake and Mizue (1986) have reported that the organization is similar to that of the nonplacental viviparous shark *S. acanthias* (Jollie and Jollie, 1967b)—that is, there is a highly vascularized bilaminar epithelium. Epithelial cells of the outer layer are characterized by wide intercellular spaces between cells and numerous mitochondria distributed in the basal and lateral cytoplasm. There is no evidence of secretory activity. Cell and tissue morphology are consistent with that of a structure engaged in osmoregulation of the uterine fluid and gas exchange. It should be noted that in both *S. acanthias* and *P. glauca*, the uterine fluid is sparse and apparently not rich in organic molecules, unlike that of *Mustelus canis* (Needham, 1942).

    f. Function and Efficiency. Although experimental studies on placental function in chondrichthyans are very limited, it is worth recalling that placental shark embryos undergo a considerable increase in organic mass due to the transfer of maternal nutrients either by transplacental or paraplacental routes. In the sandbar shark *C. plumbeus*, Hamlett *et al.* (1985a) demonstrated that in vitro exposure of full-term placentas to solutions of trypan blue and horseradish peroxidase for short (10–20 min) time intervals resulted in little uptake by the smooth portion of the placenta but rapid absorption by the surface epithelial cells of the distal rugose portion. HRP enters these cells by an extensive system of smooth-walled, anastomosing apical canaliculi and tubules. In a short-term experiment, Hamlett *et al.* (1985a) traced the HRP to the basal region of the cell where small transport vesicles were budded off. Further interpretation proved difficult without long (60 min) exposures. Was HRP hydrolyzed in a lysosomal system or was exposure too short to ascertain the fate of HRP? T. Otake (unpublished) found evidence for differences in HRP uptake in the placentae of different species of sharks. Obviously, further experimental work is required. The ultrastructural observations previously described provide strong evidence either for secretion and endocytosis of maternal materials (Otake and Mizue, 1985) or for their transport and endocytosis (Hamlett *et al.*, 1985a). The suggestion by the latter group that yolk proteins are transported is particularly intriguing. Elsewhere, we have alluded to preliminary observations that the umbilical stalk appenduclae of *R. terraenovae* are able to take up HRP and trypan blue. The appendiculae and smooth proximal placenta most likely are of importance in nonhemotrophic, paraplacental transport (for further details, cf. Wourms, 1981).

Some problems do arise in the consideration of placental transport, one of which is the role of the egg envelope. This structure intervenes between the maternal and fetal tissues in four of the six classes of placentas. The egg envelope consists of collagen and one or more additional structural proteins. In viviparous sharks, the banding pattern characteristic of the egg envelope collagen of oviparous species is absent, and the egg case is much reduced in thickness. Collagen could either have been lost or else assembled in a nonbanded form (Wourms, 1981; Hunt, 1985). Relatively little is known of the permeability properties of the egg envelope even in oviparous species. The egg case of *Scyliorhinus* is permeable to water, sodium ions, urea, and various organic molecules (Hornsey, 1978; Foulley and Mellinger, 1980). It is presumed that reduction in the thickness of the egg envelope and possible loss or macromolecular rearrangement of collagen in viviparous species might enhance permeability. On the other hand, Hamlett *et al.* (1985a) showed that ruthenium red, a relatively small molecule, failed to traverse the egg envelope. Their experiment, however, may provide more information about surface charges on the egg envelope than its porosity. Two other observations should be recalled: Graham (1967) indicated that there were changes in the structure and histochemistry of the egg envelope during gestation in *M. canis*. Do these changes correspond with changes in permeability? Finally, viviparous embryos of the tiger shark *Galeocerdo* are aplacental and develop to term within an egg envelope that often contains a liter or more of periembryonic fluid. Embryos undergo massive growth, attaining a length of 70 cm or more; this increase in size is the same as or greater than that for some placental sharks. It is reasonable to assume that there is an efficient transfer of nutrient molecules from the uterine fluid across the egg envelope and into the periembryonic fluid. Although the egg envelope is a potential permeability barrier, it appears to permit the ready passage of small molecules, but may hinder the passage of macromolecules. The cutoff values in terms of molecular weight or molecular radius are unknown.

No real measure of the efficiency of placental transport has yet been achieved. Thus far, the only available indications have been based on the dogfish shark *M. canis* and the blue shark *P. glauca*, which during gestation show increases in embryonic dry weight of 1050% and 840%, respectively (Wourms, 1981). These increases do not compare favorably with the 1700–5000% increases recorded for the aplacental rays. Such efficiency values, however, are obviously subject to sampling error and a lack of perspective. Differences in reproductive strategy must exert unknown influences on any consideration of placental efficiency. *Mustelus canis* is a small shark that produces a few (six to eight) modestly sized embryos, whereas the larger blue shark produces large numbers (50–75) of embryos of equivalent size. In some of the other placental carcharhinid sharks, a few large embryos are produced (Ballinger, 1978). Although final results have not been obtained, weight increases in the order of 6000–10,000% would not be surprising (J. P. Wourms and J. I. Castro, unpublished), thus indicating that the total amount of maternal nutrients transported to the brood of large sharks may be substantial.

Another point worth recalling is that efficiency estimates are relative and depend a great deal on egg size. In most placental sharks, the size of the egg is about the same as that in oviparous species (sometimes even larger); approximately 2–3 cm in diameter with a dry weight of 2.8–3.4 g. There is no information on gestational weight increases of placental sharks with small eggs. The selachian yolk-sac placenta probably has attained the pinnacle of its evolutionary development in the spadenose shark *Scoliodon laticaudus* (= *S. sorrakawah*). Here it appears to function with the same degree of efficiency as a mammalian placenta. The eggs are very small, less than 1.00 mm in diameter, have a dry weight of 0.06–0.07 mg (J. P. Wourms and T. Otake, unpublished), and are nearly devoid of yolk. The embryo implants very early in gestation. A highly specialized placenta that may be hemochorial develops. Nutrient transport is transplacental and hemotrophic during much of gestation; the extent of paraplacental transport is unknown. Embryos attain a length of 130–150 mm at term (Mahadevan, 1940; Setna and Sarangdhar, 1948; Teshima *et al.*, 1978). A determination of placental efficiency in *Scoliodon* is now underway (J. P. Wourms and T. Otake, unpublished). It would not be surprising to find that gestational weight increases in the order of 1,000,000%, based on an extrapolation from the weight–length relationships found in *Anableps* (Knight *et al.*, 1985) and *Scyliorhinus* (Needham, 1942).

C. Maternal Specializations

1. INTRODUCTION

Specializations of maternal structure that directly participate in viviparity include (1) the ovary and ovarian cycles; (2) the uterine wall, including modifications such as trophonemata, placental attachment site, and compartments, as well as uterine derivatives such as histotrophe; and (3) the shell gland with its product, the egg envelope. Other than what has been stated in Section I,B, endocrine glands will not be considered further (cf. Dodd, 1983). Nor will the liver be discussed, even though it appears to be a source of yolk proteins (Dodd, 1983). This organ also seems to be involved vitally in fetal nutrition, based on correlations between liver size and the temporal state of gravidity (Needham, 1942; Amoroso, 1960). Unfortunately, the physiological role of the liver in gestation has yet to be effectively addressed.

2. OVARY AND OVARIAN CYCLE

The ovary is both the site of egg formation and hormone synthesis. In most sharks and rays, there is a discrete ovarian cycle. This and the duration of gestation are key factors in determining the duration of the female reproductive cycle. The latter is characteristic for each species and may range from 2–3 months in some rays to 3–4 years in the Australian soupfin shark *Galeorhinus* (Wourms, 1977; Dodd, 1983; Dodd and Dodd, 1986). In most viviparous sharks and rays, once a clutch of eggs has been released, ovulation ceases during gestation and oocytes are either temporarily arrested or else pass through a slow process of vitellogenic growth during gestation.

Ova released from the ovary and fertilized contain yolk, which serves as the nutrient substrate for lecithotrophy or the lecithotrophic phase of matrotrophy. In most oviparous and viviparous chondrichthyan fishes, there seems to be a remarkable constancy of egg size, a diameter of about 2–4 cm. Marked deviations from these values, however, occur in some viviparous sharks and rays and in the coelacanth. Egg size may range from extremely small size due to extreme yolk reduction (e.g., the 1.0-mm eggs of *Scoliodon*) or gigantism (e.g., the coelacanth *Latimeria* and the sharks *Chlamydoselachus, Centrophorus*, and *Ginglymostoma*). Two specialized strategies appear to be at work here, namely, elimination of yolk in extreme matrotrophy and massive yolk accumulation for extreme lecithotrophy. The latter strategy is probably primitive and is subject to constraints imposed by the physical limitations of cell size. In this respect, it would be interesting to know more about the size of the eggs and reproductive pattern of the whale shark *Rhincodon* (Wourms, 1977).

What may well be the prime example of adaptive modification of the ovarian cycle for viviparity occurs in the oophagous sharks. In most of them ovulation continues through almost all of gestation, in contrast to other viviparous sharks, in which it typically ceases. The oophagous situation is actually more closely analogous to what occurs in oviparous sharks and skates, in which egg-laying often extends over a period of several months (Wourms, 1977; Dodd, 1983). The best-studied example of oophagy is in the sand tiger shark *Eugomphodus* (=*Odontaspis*). Gilmore *et al.* (1983) have shown that during gestation at least six distinct egg capsule types are produced, apparently correlated with the nutritive phase and developmental state of the oophagous embryos. Egg capsule types differ with respect to (1) size and shape, (2) number, size, and developmental potential of the eggs within the capsule, and (3) presence or absence of ovalbumin or mucous in the capsule. Production of different types of egg capsules is due to variation in the ovulation rate and shell gland activity. Eggs produced during late gestation are not fertilized. This could be due to the absence of sperm. On the other hand, it would not be surprising to discover that a discrete population of nonfertilizable nutritive eggs is being produced (J. P. Wourms, unpublished), a phenomenon that is well-documented in some prosobranch gastropod molluscs.

3. UTERINE WALL: TROPHONEMATA AND HISTOTROPHE

The uterine wall of most viviparous elasmobranchs and the coelacanth both delimits and defines the embryonic environment. The most spectacular maternal specializations for uterine gestation involve the uterine wall and involve (1) amplification of the surface area in the form of folds, villi, or trophonemata, (2) production of histotrophe or uterine milk, (3) compartmentalization of embryos, and (4) development of placental attachment sites. Placental attachment has been considered in conjunction with the yolk-sac placenta (cf. Section II,B,5,e).

The process of compartmentalization is intriguing. Each embryonic shark or coelacanth is afforded its own discrete developmental chamber, which in some instances is replete with an egg-envelope reservoir (Needham, 1942; Budker, 1958; Smith et al., 1975). Unfortunately, virtually no information is available other than that contained in a few anatomical studies. Compartmentalization must be a dynamic process. There are no compartments prior to the entry of fertilized eggs into the uterus. Moreover, compartment orientation changes during gestation (Schlernitzauer and Gilbert, 1966), and compartment orientation is a function of the number of embryos present (Teshima, 1981). Further information on the control of compartmentalization is lacking.

Ranzi (1934) was able to make significant correlations between the degree of embryonic matrotrophy and both the structure of the uterine wall and the composition of the uterine fluid. Subsequently, these were amended or repeated in Needham (1942), Amoroso (1960), Hoar (1969), and Wourms (1981). Ranzi (1934) recognized three sets of uterine environments, which he designated types Ia and b, II, and III. Histotrophe was categorized according to its quantity, total organic content, and whether the predominant molecular species was protein or lipid. The degree of complexity of the uterine wall was analyzed on the basis of the mucosal epithelial organization, the presence or absence of villi, and the number and size of villi. Only type III, will receive a detailed discussion here (for further details, cf. Ranzi, 1934; Needham, 1942; Hoar, 1969; Wourms, 1981).

In species with the type Ia uterine environment (e.g., the spiny dogfish S. acanthias), the embryos usually lose a considerable amount of weight (15–55%) during gestation. The uterine epithelium is either smooth, as in S. acanthias or forms short villi, as in Centrophorus. Secretion or fluid transport is minimal or absent. Glandular structures are reduced or absent. Ultrastructural studies by Jollie and Jollie (1967b) revealed that in S. acanthias there is an emormous increase in the relative surface area of the mucosa during gestation, an extensive vascularization by a system of juxtaepithelial capillaries, and an overall reduction in the number of cell layers and the amount of connective tissue lying between the mucosal epithelium and capillary endothelium. Although no evidence of secretory activity could be found, the presence of extensive apical canaliculi in the epithelial cells suggests endocytosis. Jollie and Jollie (1967b) and Ranzi (1934) attributed both respiratory and osmoregulatory functions to the epithelium. Evans and Oikari (1980) postulated that the composition of the uterine fluid of S. acanthias is controlled by active and passive transport across the uterine lining. Type Ib has been described in electric rays (e.g., *Torpedo ocellata*) and the guitarfish *Rhinobatus*. Torpedos undergo a weight loss of −23% and −34%. The uterus is lined with numerous villi of moderate length. A single-layered, glandular epithelium produces an abundant amount of a dilute (1.2–2.8%) organic material, that is, serous histotrophe.

The type II environment occurs in some aplacental sharks that undergo a moderate weight increase during gestation, for example 11% in *Galeorhinus* (=*Galeus*) *canis*, 110% in *M. antarcticus*, and 369% in *M. mustelus*. The uterine lining is more smooth-surfaced than villous. An abundant quantity of histotrophe is transported through or secreted by so-called "mucoid" cells. Histotrophe has an organic content of 4.9–7.1%, is lipid-free, and is composed of a protein–carbohydrate complex. The type II pattern is also found in placental sharks, like *M. laevis* and *P. glauca*, which undergo weight increases of 1050% and 840%, respectively. While the amount of histotrophe in *P. glauca* is meager and dilute (Otake and Mizue, 1985), it is abundant in *M. laevis* and has a high (9.1%) organic content (Ranzi, 1934; Needham, 1942). A balance sheet between transplacental and paraplacental nutrient transfer has yet to be worked out.

The type III uterine environment has been found in the rays *Dasyatis violacea*, *Myliobatis bovina*, *Gymnura micrura*, *Rhinoptera bonasus*, the manta *Mobula diabola*, and the sawfish *Pristis cuspidatus* (cf. Wourms, 1981, for references). In these species, the uterine epithelium forms tufts of long, glandular villi termed trophonemata by Wood-Mason and Alcock (1891). In some species, such as *G. micrura* (Fig. 3), the trophonemata enter the embryo through the spiracles and pass into the esophagus where they release their secretory products into the gut. This arrangement constitutes a branchial placenta. In all of these batoids, there is considerable increase in embryonic weight during gestation (e.g., 5000% in *G. micrura*). Histotrophe is abundant and rich in organic material (e.g., 13% in *D. violacea* of which 8% is fat).

Wourms and Bodine (1983, 1984, and unpublished) have described the ultrastructure of *G. micrura* trophonemata during early gestation and Hamlett *et al.* (1985d) trophonematal ultrastructure of the cownose ray *R. bonasus* during late gestation. In both species, the trophonemata are spatulate villiform processes, about 20 × 1 mm in *G. micrura* and 20–30 mm long in *R. bonasus*. Their basic organization is similar. Scanning electron microscopy (SEM) reveals that the surfaces of trophonemata comprise a network of anastomosing, cable-like ridges with intervening pits. This massive capillary network extends between two arteries that run along each internal margin of the

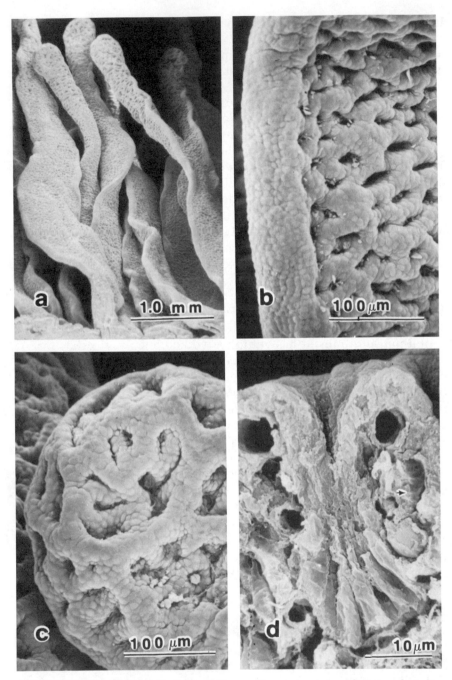

Fig. 3. (a) The uterine wall of a butterfly ray *Gymnura micrura* during early gestation. The lumenal epithelium has differentiated into villiform trophonemata, or
"growth threads." At this stage trophonemata are about 20 mm long by 1 mm wide. The

trophonema and a large, central, axial vein. Within each ridge in G. *micura*, two capillaries run parallel to the surface and to each other but at different levels. Outer and inner capillaries are closely apposed. A simple cuboidal epithelium invests the outer capillaries, while cells of the inner region have complex basal folds. In both species, the surface epithelial cells are smooth-surfaced. Their lateral margins interdigitate and exhibit typical junctions. Their cytoplasm is ribosome-rich and contains ellipsoidal, low-density inclusions. The capillary endothelium contains many membrane-bound vesicles and ellipsoidal, low-density inclusions like those of epithelial cells. During the early gestation of G. *micrura*, functionally differentiated secretory acinar glands occur infrequently. In contrast, there are numerous acinar secretory units in the late-stage trophonemata of R. *bonasus*. In both species, the secretory unit is a discrete acinar gland composed of 8–10 cells joined by extensive junctional complexes. The acinus often contains flocculent material presumed to be a secretory product. In R. *bonasus*, transmission electron microscopy (TEM) reveals that the cuboidal acinar cells closely resemble classic protein-synthesizing and -secreting cells. The rough endoplasmic reticulum is well developed. At the cell periphery, RER cisternae are grossly dilated and filled with low-density, flocculent material. There are many polyribosomes and an extensive and elaborate supranuclear Golgi complex. The bulbous ends of the Golgi saccules contain material of medium electron density. Uncoated vesicles are given off by it. In the apical region, there is a progressive increase in the electron density of material contained within vacuoles and presecretory granules associated with the Golgi complex. Membrane-limited, mature presecretory granules accumulate in the apical portion of the cell (Hamlett *et al.*,

---

basal region is ribbon-like, while the apical region is spatulate. Trophonemata are the source of histotrophe or "uterine milk." The rest of the uterine wall consists of highly vascularized muscle and connective tissue layers. (b) Trophonemata are flat and ribbon-like. Prominent smooth-contoured structures, the outline of arteries, extend along each lateral margin. The surface of the trophonema is a convoluted network of "cables" that consists of a capillary network invested by surface epithelial cells. (c) The apical end of the trophonema is spatulate. While the basic organization of an anastomosing capillary bed predominates, acinar glands (lower right) are occasionally observed. Floculent material and spherical particles are associated with the lumen of the putative acinar glands and may be their secretory products. (d) A freeze-fracture preparation. The surface network of cables is a capillary bed, displayed in transverse section, that is invested by a surface epithelium displayed in both surface and transverse section. Several capillaries run parallel to one another, but at different levels within the cable. Epithelial cells in the basal region of the capillary bed have extremely complex basal folds or foot processes.

1985e). On the basis of light microscopy and histochemistry, late-stage *G. micrura* trophonemata probably have a similar organization (Wood-Mason and Alcock, 1891; Ranzi, 1934).

In both of these rays, the trophonemata pass through two discrete differentiated states, one in early, the other in mid to late gestation. Differences in these states are reflected in both the structure and the function of the trophonemata and in qualitative and quantitative changes in the composition of the histotrophe. Trophonemata originate from the adlumenal uterine epithelium, which in immature or nongravid females is kept in an indifferent state of differentiation. With the advent of sexual maturity or during early gestation, the uterine epithelium undergoes its first phase of differentiation by forming trophonemata that basically are massive capillary beds, surmounted by a simple epithelium, as in *G. micrura*. At this stage, two trophonematal functions may be postulated, (1) that the capillary bed with its investing epithelium is the site of transport of amino acids, lipids, and proteins from the maternal blood to the histotrophe, and (2) that the capillary bed, operating on the countercurrent principle, functions in gas exchange and waste absorption. During mid to late gestation, trophonemata enter their second differentiative phase, at which time some epithelial cell populations develop into acinar cells that synthesize and secrete proteins into the histotrophe. Thus, histotrophe has a dual origin: namely, transport from the maternal blood and synthesis and secretion by trophonematal cells. It is assumed that histotrophe transport and gas exchange and waste absorption continue during mid to late gestation (Wourms and Bodine, 1983, 1984, and unpublished).

Histotrophe is the term applied to the fluid found, during gestation, in the uterus, ovarian lumen, and follicles that participate in intrafollicular gestation. In uterine gestation, histotrophe has been called "uterine milk." Histotrophe may serve several functions, one of which is nutrition (Amoroso, 1952). Our knowledge of the biochemical composition and cellular origin of histotrophe is in a state of flux. The excellent, pioneering studies of Ranzi (1932, 1934) established the modern field of inquiry. His work has been reviewed several times (Needham, 1942; Amoroso, 1960; Hoar, 1969; Wourms, 1981). Unfortunately, for a period of almost 50 years, there was little inclination to carry forward Ranzi's fascinating and pioneering discoveries. Ranzi's main conclusions are worth reiterating, namely, that the degree of matrotrophy is correlated with the qualitative and quantitative composition of the uterine fluid and that this, in turn, is correlated with the structure, and presumably function, of the uterine wall. Trophodermy that involves the uptake of uterine fluid is an efficient form of matrotrophy. Histotrophe is an excellent nutrient; in some instances, such as *M. laevis* and *D. violacea*, its total organic content exceeds that of the well-known mammalian nutrient, cow's milk.

More recently, there has been a revival of interest in histotrophe. Price and Daiber (1967) reported that the uterine fluid of *M. canis* closely resembles the maternal serum. Thorson and Gerst (1972) found that the uterine fluid of *Carcharhinus leucas* was also similar to maternal serum except for its very low protein content. Recently Bodine and Wourms have begun an extensive investigation of the composition of shark and ray histotrophe using contemporary methods. Brief reports of their work have appeared (Wourms and Bodine, 1983, 1984), and we summarize and extend them here. Histotrophe obtained during the early gestation (20–25 mm tailbud embryos) of the butterfly ray *G. micrura* was a dilute (1–2% total organic content), serumlike, white fluid with a pH of 7.4 and a total protein content of 2.38 mg ml^{-1}. Polyacrylamide gel electrophoresis revealed the presence of three or four protein bands (molecular weight 68 kDa to 350–400 kDa) tentatively assigned to serum albumin, immunoglobulin M, and macroglobulin, respectively. Electrophoresis of maternal blood serum revealed bands with similar electrophoretic mobility. Analysis of free amino acids showed significant amounts of phosphoethanolamine (26 mg per 100 ml) and urea. Taurine, gamma-aminobutyric acid (GABA), citrulline, alanine, glutamine, and valine were also present. The composition of free amino acids, especially phosphoethanolamine, is strikingly similar to that found in mammalian amniotic fluid, as well as zoarcid and embiotocid histotrophe (vide infra). Proline and aromatic amino acids were absent. The total lipid content was 0.5–0.6%. In the lipid fraction, gas–liquid chromatography of fatty acids revealed 19% C-8, C-9, C-10 fatty acids; 15% myristic acid; 26% palmitic; 8% palmitoleic; 7% stearic; 17% oleic; 7% linoleic; and 2.3% arachidonic. Such levels of myristic acid and C-8, C-9, C-10 fatty acids are higher than average values for vertebrate body fluids. Trace amounts of five phospholipids were found. Comparative studies of ray and shark histotrophe indicate that the organic content increases during gestation. Inasmuch as the histotrophe of full-term *G. micrura* has been described as a viscous, creamy, yellow fluid (Wood-Mason and Alcock, 1891), it is likely that its organic content will exceed the 13% value reported by Ranzi (1932, 1934) for *D. violacea*.

It is reasonable to conclude that histotrophe serves several functions. It is a nutrient, providing both a source of energy and molecular constituents for the synthetic aspects of cell replication and cell and tissue growth. The presence of maternal blood serum proteins, especially IgM, suggests an immunological role as well as the presence of specialized transport functions. The finding of substantial amounts of estrogens in the histotrophe of placental sharks (A. B. Bodine and J. P. Wourms, unpublished) indicates a possible endocrine function.

4. NIDAMENTAL GLAND AND EGG ENVELOPE

The nidamental or shell gland is responsible for the production of the egg case or capsule, a characteristic feature of oviparous sharks, skates, and chimeras. Thus, the egg capsule is a tertiary egg envelope. The egg case and the shell have been subject to recent reviews (Wourms, 1977; Dodd, 1983; Hunt, 1985; Rusaouen-Innocent, 1985), and only topics that pertain to viviparity will be dealt with here. Even though only a limited number of species have been sampled, the egg capsules of sharks and skates appear to contain a unique form of collagen or collagen-like protein. Other structural proteins also may be present (Hunt, 1985). The collagen-like protein occurs as an imperfectly ordered orthogonal array of structural components, one of which displays a 40-nm periodicity in lateral register over considerable distance (Wourms, 1977; Hunt 1985). In contrast, the egg capsules of chimeras have a different macromolecular organization and probably are composed of different structural proteins (Wourms, 1977). In viviparous sharks and rays, diverse reproductive strategies govern the fate of the egg cases, but they generally are reduced or totally absent. In some instances, such as the stringray *Urolophus* (Babel, 1967) and bat ray *Myliobatis*, no egg case is formed. In many species of sharks, such as *Squalus* and the oophagous *Eugomphodus*, and rays, such as *Gymnura,* the fertilized egg is enclosed within a temporary egg case from which the embryo emerges to complete development in utero. Finally, the egg case may be retained during the entire period of uterine development (e.g., in the tiger shark *Galeocerdo*) or even incorporated into the placenta (e.g., *M. canis*) (Wourms, 1977, 1981). Changes in the size and thickness of the egg capsule are reflected in alterations of shell-gland structure. When the egg case is reduced in thickness or only temporarily present, the characteristic banding pattern of the collagen-like protein is absent. In all instances where the developing embryo is surrounded by an intact egg envelope, the permeability characteristics of the envelope are of paramount importance, especially in matrotrophic species. Unfortunately, the limited body of information on this subject deals only with the egg cases of oviparous species (reviewed by Hunt, 1985). The most informative work is that of Foulley and her co-workers, who investigated the diffusion characteristics of the egg capsule of *Scyliorhinus* using tritiated water, urea, acetate, mannitol, glucose, glycerol, and glyceric acid. In their early report, Foulley and Mellinger (1980) disproved the classic view that the egg case acted as an osmotic protective device. They found that the egg case is extremely permeable to water and various organic molecules and relatively more permeable to sodium ions than urea. Evans (1981), however, found that the egg case of *Raja erinacea* is able to maintain significant osmotic and ionic gradients between the surrounding seawater and the egg case fluids, despite its extremely high permeability to salts, urea, and, presumably, water. Subsequently, in unidirectional osmotic or diffusional flux studies, Foulley *et al.* (1981) discovered an asymmetric permeability in the egg envelope. Passive permeability was highest in the inward direction, that is, toward the embryo. The existence of this phenomenon needs to be tested in viviparous species, especially matrotrophic ones. The obvious implication is that transport of nutrient metabolites to the embryo may be energetically favored, based on an inward, asymmetric transenvelope flux with the embryo functioning as a metabolic sink. If applicable, this model would explain the enigmatic matrotrophy of tiger shark embryos, which undergo massive growth while encased in an egg envelope.

III. TROPHIC SPECIALIZATIONS FOR OVARIAN GESTATION

A. Intralumenal Gestation

1. INTRODUCTION

The ovary of viviparous teleost fishes is unique among vertebrates, since it is both the site of egg production and the site of gestation. Gestation occurs either in the ovarian lumen and is termed "intralumenal gestation," or in the ovarian follicle and is termed "intrafollicular gestation." Intralumenal gestation is the most prevalent mode of development in viviparous teleosts. It occurs in somewhat more than half the viviparous species and in 10 of the 14 or 15 families in which viviparity is known (Wourms, 1981). In most teleosts with intralumenal gestation, fertilization and embryonic development commence in the ovarian follicle and proceed to completion in the lumen (cf.

Section I). Intralumenal gestation is known to occur in several zoarcids, at least some parabrotulids, more than 90 species within the ophidiiform families Bythitidae and Aphyonidae, the approximately 40 species of goodeids, more than 100 species of scorpaenids, the two species of Lake Baikal comephorids, the 23 species or more of embiotocids (surfperches), the three or four species that comprise the anablepid genus *Jenynsia*, and at least one hemiramphid, *Nomorhamphus hageni* (Wourms, 1981).

Intraluminal viviparity has been examined most extensively in the zoarcid *Zoarces viviparus*, the goodeids, the embiotocids, some scorpaenids of the rockfish genus *Sebastes*, a limited number of species within the ophidiiform families Bythitidae and Aphyonidae, and in the anablepid *J. lineata* (cf. Wourms, 1981 for a detailed systematic treatment). Specializations for intralumenal gestation have been described in all of the families except the Comephoridae. Unfortunately, the extent of our current knowledge may be deceptive, since what is known about some groups is based mostly on the study of preserved specimens. This material has proved suitable for a limited amount of scanning electron microscopy (Wourms and Cohen, 1975). Experimental investigations, however, have been largely confined to several species of goodeids, especially *Ameca splendens* (Fig. 4), some scorpaenids in the genus *Sebastes*, some surfperches of the genera *Cymatogaster*, *Embiotoca*, and *Rhacochilus*, and the eelpout *Z. viviparus*.

As one would anticipate from their systematic diversity, patterns of embryonic nutrition vary among the fishes with intralumenal gestation. In almost all species for which embryonic weight data are available, embryonic nutrition appears matrotrophic. This may reflect a sampling bias, however (Table IV). More likely, there may be a range from lecithotrophy through advanced matrotrophy with matrotrophy predominating. In the scorpaenid genus *Sebastes*, *S. marinus* was reported as markedly lecithotrophic (Hsaio, unpublished, cited in Needham, 1942, and Scrimshaw, 1945), and for the next 40 years, lecithotrophy was extrapolated to all other members of the genus. More recently, Boehlert and Yoklavich (1984) and Boehlert *et al.* (1986) have presented evidence for matrotrophy in *S. schlegeli* and probable matrotrophy in *S. caurinus* and *S. melanops* (Cf. Section I). On the basis of gestational weight changes, matrotrophy has been demonstrated in *Z. viviparus* (+1100%) (Kristoffersson *et al.*, 1973), the embiotocid *E. lateralis* (+20,400%) (cf. Wourms, 1981), the anablepid *J. lineata* (+24,000%) (Richter *et al.*, 1983), and the goodeids *Goodea atripinnis* (+1100%), *Chapalichthys encaustus* (+2700%), and *Ameca spendens* (+15,000%) (Lombardi, 1983; Lombardi and

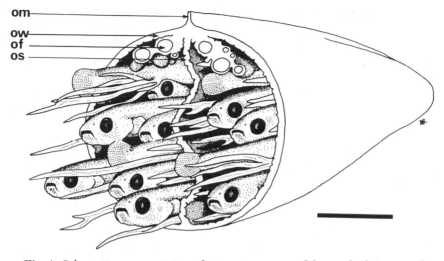

om
ow
of
os

Fig. 4. Schematic representation of a near-term ovary of the goodeid *Ameca splendens* illustrating the positional relationships of embryos to the internal structures of the ovary. The anterior third of the ovary has been deleted to illustrate internal ovarian anatomy; of, oocytes; om, ovarian mesentery; os, ovarian septum; ow, ovarian wall. Scale bar equals 1 cm.

Wourms, 1985a, and unpublished). On the basis of other evidence, such as oophagy, gestational weight stasis, or massive increase in embryonic size, matrotrophy also occurs in some ophidiiform fishes within the families Bythitidae and Aphyonidae and probably occurs in some parabrotulids, hemiramphids, and comephorids (Wourms, 1981, and *vide infra*).

All four modes of embryonic nutrition are found in the species that have intralumenal gestation, and diverse embryonic and maternal specializations have evolved to facilitate nutrient transfer. The four modes comprise (1) lecithotrophy, the first nutritional state in all viviparous teleosts and the dominant mode in some species; (2) oophagy, adelphophagy, and matrophagy; (3) trophodermy via the general body surface, fins, and gut; and (4) placentotrophy, that is, the trophotaeniae and trophotaenial placenta and the buccal and branchial placentae. There are a number of structural and physiological specializations of the ovary and gonoduct for nutrient transfer, such as (1) hypertrophied lumenal epithelium and vascular supply of the ovigerous folds, such as the maternal portion of the branchial placenta of *Jenynsia;* (2) ovigerous folds that enclose embryos in intimately apposed compartments; (3) nutrices calyces, a combination of ovarian follicle and ovarian wall; and (4) villous extensions of the gonoducal lumenal epithelium.

Lecithotrophy will not be considered here, since there is a consensus that yolk utilization in the lecithotrophic phase of development is carried out in the same way, in both viviparous and oviparous teleosts (Terner, 1979; Boulekbache, 1981). For example, Shimizu and Yamada (1980) examined ultrastructural aspects of yolk utilization in the vitelline syncytium of the viviparous rockfish *S. schlegeli* and concluded that the pattern of utilization was very similar to that in oviparous fishes. Among the viviparous teleosts in which extreme matrotrophy has evolved, there has been a tendency for the amount of yolk to be reduced, thus shortening the lecithotrophic phase with an accelerated onset of matrotrophy, as in the poeciliid *H. formosa*, the anablepids *Anableps* and *Jenynsia*, some goodeids (e.g., *Girardinichthys*), and the embiotocids (Table IV and Wourms, 1981). Embiotocids actually lack the classical vertebrate yolk proteins (de Vlaming *et al.*, 1983).

2. EMBRYONIC SPECIALIZATIONS

a. Oophagy, Adelphophagy, and Matrophagy. Oophagy and adelphophagy do not appear to be as prominent in the nutrition of viviparous teleosts as they are in lamniform sharks. Specific instances of oophagy comparable to those in sharks are hard to document. The ingestion of fragmented or partially cytolyzed eggs appears to take place in some but not all goodeids (Turner, 1933, 1937, 1947) and some ophidiiform fishes (Wourms, 1981). On a similar basis, Boehlert and Yoklavich (1984) and Boehlert *et al.* (1986) have suggested that moribund eggs may be a nutrient source in the rockfish *Sebastes*. In *Jenynsia*, the possibility of oophagy or adelphophagy has been suggested since the number of embryos that survive to term is much less than the number of ova fertilized (Turner, 1940b, 1957). Adelphophagy was first described among the viviparous brotulas by Gilchrist (1905), who observed and illustrated an advanced embryo of *Cataetyx memoriabilis* swallowing a somewhat smaller one. Meyer-Rochow (1970) corroborated this observation by showing that ovarian embryos of this species prematurely extruded at a similar developmental stage were able to feed in vitro. Based on a reduction in the number of embryos during gestation, oophagy or adelphophagy has been proposed for another viviparous brotula, *Dinematichthys* (Wourms, 1981; Wourms and Bayne, 1973). The existence of matrophagy, a process in which embryos actively attack and devour maternal tissues, is problematic among these fishes, however. Late-term embryos of the brotulas *Ogilbia cayorum* and *Lucifuga subterraneus* have been observed with ovigerous bulbs in their well-developed mouths (Lane, 1909; Suarez, 1975). These bulbs are fluid-filled vesicles composed of ovarian epithelium and capillaries, and we have called this arrangement a "buccal placenta." Suarez (1975) proposed that the embryos actually suckle at the ovigerous bulbs, but there is an alternative possibility. Late-term embryos with bulbs in their mouths may have been preserved in the act of biting off the bulbs prior to ingesting them. Examination of embryonic gut contents is needed.

b. General Body Surface and Fins. Trophodermy involving the passage of nutrient molecules across the general body surface has been demonstrated in the viviparous teleosts *Clinus superciliosus* and *Heterandria formosa* (cf. Section III,B). Uptake of HRP by the external gill filaments of embryonic sharks and uptake of IgM by the internal gills of juvenile teleosts has also been demonstrated (cf. Section II,B). In the case of *C. superciliosus* and *H. formosa,* the cells of the absorptive epithelium were characterized by their well-developed apical microvilli. In teleosts with intralumenal gestation, evidence to substantiate nutrient uptake via the general body surfaces is circumstantial and based on an association of structure with function with little or no experimental evidence. The interpretation of what information is available can be hampered by systematic differences in physiology, ontogenetic changes in body surface function during gestation, and differences in the biochemical processing of small molecules like amino acids and large ones like proteins. By way of illustration, during development, *H. formosa* suffers a progressive loss of the microvillous surface cells that transport glucose and glycine (Grove, 1985; Grove and Wourms, 1983, and unpublished). Similarly, in late-term embryos of the goodeid *Ameca splendens,* trophotaeniae are the chief site of HRP uptake, and there is no evidence for uptake by the general body surface (Lombardi and Wourms, 1985c).

Finfolds and hypertrophied vertical fins have been postulated to serve in nutrient uptake in three teleost groups. During gestation, the vertical fins of embiotocid embryos hypertrophy, develop spatulate extensions, and become highly vascularized (Turner, 1952; Webb and Brett, 1972a,b; Dobbs, 1974). These specializations of the fins dramatically increase the embryonic surface area, and some investigators have suggested that the fins assist in nutrient uptake and gas exchange (Turner, 1952; Webb and Brett, 1972a,b). However, scanning electron microscopy (SEM) of the vertical fins and spatulate fin extensions of the embiotocid *Micrometrus minimus* revealed a surface epithelium devoid of microvilli but rich in microplicae (Dobbs, 1974). Microplicae characterize the epidermal surfaces of many aquatic vertebrates and are considered a boundary epithelium of limited permeability. Microvilli, on the other hand, are a characteristic feature of absorptive epithelial cells; their absence suggests that embiotocid embryonic fins serve in gas exchange only (Dobbs, 1974). Wourms and Lombardi (1985) found that in embryos of *Rhacochilus vacca* the structure of the cells that cover the spatulate extensions of the fins, which they called "epaulettes," was distinctly different from that of the fins. Epithelial cells of the epaulettes had short, stubby microvilli, whereas epithelial cells on the fins had only microplicae. Epaulettes possess a capillary plexus supplied by a large artery and vein that run along each fin ray. Epaulettes are considered to be a type of dermotrophic placenta (Wourms and Lombardi, 1985) that functions both in respiration (Webb and Brett, 1972a,b) and in the uptake of small nutrient molecules, most likely amino acids and sugars, from the histotrophe. The difference in surface structure of the epaulettes of *M. minimus* and *R. vacca* may reflect a greater energy requirement of the larger *R. vacca* embryos. Since microvilli amplify surface area, their presence would facilitate transport in *R. vacca*. The relationship between uptake via epaulettes and uptake via the gut or trophotaeniae remains a matter of conjecture. It is important to note that epaulettes, like many other embryonic absorptive structures, atrophy just prior to parturition. Atrophy prevents a massive influx of environmental water through the specialized transport structures into the embryo (Webb and Brett, 1972b; Hogarth, 1976).

In embryos of the bythitid *Diplacanthopoma*, Alcock (1895) postulated that the surface epithelium of the hypertrophied vertical finfold, which extends from the head around the tip of the tail to the anus, is involved in nutrient uptake. A similar condition prevails in embryos of some goodeids, such as *Goodea luitpoldii* (Mendoza, 1958). Unfortunately, there is no ultrastructural or experimental work to confirm this, and the trophic role of the embryonic finfolds in these two species remains problematic.

   c. Branchial and Buccal Placentae. Branchial placentae occur in some rays (e.g., *Gymnura*) (cf. Section II,B) and in three families of teleosts. A structural association that we interpret as a buccal placenta occurs in two species of ophidioids. In both instances, there is close association between the maternal ovarian lumenal epithelium and either the embryonic branchial and pharyngeal epithelium or the buccal epithelium.

Turner (1940b, 1952) suggested that the embryonic gill epithelium may be a site of nutrient uptake in the anablepid genus *Jenynsia* and in some embiotocids. In these fishes, there is an association between the gill tissue and the lining of the ovarian lumen. In *Jenynsia*, club-shaped processes of the lining enter the embryonic opercular cleft and fill the mouth and the pharyngeal cavity. Turner (1940b, 1947) referred to this association as a branchial placenta. It occurs in almost every embryo. Turner (1940b, 1947) reported that the processes were highly vascularized and covered by an active secretory epithelium. He postulated that secretory products of the processes are the major source of embryonic nutrition. Recently, Richter *et al.* (1983) demonstrated that embryos of *J. lineata* undergo a 24,000% increase in dry weight during gestation due to maternal nutrient transfer. Information needed to establish a balance sheet between the amount of nutrients taken up by trophodermy and the nutrients derived from ingested eggs or egg fragments is lacking. During development, folds of ovarian tissue extend into an enlarged opercular cleft, but only on one side. Precocious head development in early (3 mm) embryos precedes the accommodation of maternal tissues. Scanning and light microscopy confirmed previous reports of an intimate association between maternal ovarian and embryonic bucco–pharyngeal tissues. The embryonic pharynx is lined with an epithelium whose cells display a cuboidal to low columnar shape. Microvilli occur only on the dorsal surface of the mouth and pharynx. The ovarian tissue processes that extend through the opercular clefts occlude the bucco-pharyngeal cavity. These club-shaped or flap-like processes comprise a simple squamous epithelium that surrounds a highly vascularized, hypertrophied connective-tissue core. The epithelial cells have a smooth, non-amplified apical surface. At the level of enlargement of light microscopy, there is no evidence of secretory activity. Thus, the ovarian epithelium appears to function in molecular transport rather than secretion.

In embiotocids, the embryos are encapsulated in compartments formed by ovigerous folds (Turner, 1938b; McMenamin, 1979). Turner (1952) reported that the epithelium of the ovigerous folds in *Cymatogaster* is closely apposed to the embryonic gill tissue, forming placental-type association in about 50% of the embryos. Wourms and Lombardi (1985) recorded the same situation in *R. vacca*. Gardiner (1978) described the ultrastructure of the internal ovarian epithelium and implied that it may synthesize and secrete proteins. Cells in his micrographs, however, lack many of the characteristics of typical protein-synthesizing and secreting cells (cf. Section III,A,3).

Another instance of a branchial placenta occurs in the goodeid

Hubbsina turneri. Mendoza (1956) reported that, in 50% of the embryos examined, thick, spongy, vascular folds of the median ovarian septa pass under the operculum and enter the branchial chamber of the embryo, where they lie in close apposition to the gill epithelium. He attributed trophic, respiratory, and excretory transport functions to this anatomical association. Unfortunately, except for Gardiner's studies, neither ultrastructural information regarding this maternal–embryonic association nor experimental documentation of the absorption of organic compounds by embryonic gill tissue is available. In the ophidioid species *Ogilbia cayorum* and *Lucifuga subterraneus,* late-stage embryos appear to grasp club-shaped projections of the ovarian wall, the "ovigerous bulbs," with their mouths. We regard this association as a buccal placenta. Ovigerous bulbs are fluid-filled vesicles composed of ovarian epithelium and capillaries and are presumed to supply nutrients to late-stage embryos. Suarez (1975) suggested that the embryos "suckle" on the ovigerous bulbs. Again, nothing is known of the ultrastructure or physiological function of these maternal specializations.

d. Gut. The embryonic hindgut is a major site of nutrient absorption in embiotocids and the eelpout *Zoarces.* It also appears to be important in the rockfish *Sebastes,* some ophidioids, and some goodeids (cf. Section III,B,2 for a discussion of the hypertrophied gut of *Anableps*). An ontogenetic and evolutionary derivative of the gut, the trophotaeniae, are a characteristic feature of embryonic nutrition in goodeids. We shall emphasize the role of the gut in trophodermy and not comment on its obvious (and, presumably, more primitive) role in oophagy (cf. Section II,B).

Hypertrophy of the embryonic hindgut is characteristic of embiotocid development, and several investigators have postulated that it is involved in nutrient absorption (Eigenmann, 1892; Turner, 1952; Igarashi, 1962; Engen, 1968; Dobbs, 1974). Eigenmann (1892) presents a remarkable set of in vitro observations on the functional differentiation of the gut in *Cymatogaster* embryos. The gut develops precociously, making its appearance at the 12- to 15-somite stage of embryos less than 1 mm long. The anterior end becomes ciliated, while the hindgut expands and differentiates into high columnar cells. Although the anus is formed at this stage, the anterior opening to the gut is through the gill slit, not the mouth, which does not appear until the 4-mm stage. During this period, histotrophe, including supernumerary sperm, is drawn in through the gill slit and passed along to the hindgut by means of the ciliary action of the pharyngeal cells. Concurrent with the development of the mouth, the hindgut begins to hypertrophy, and large numbers of long villi develop within it (Eigenmann, 1892). In *Amphistichus,* peristaltic movements of the hindgut commence when embryos attain a length in excess of 4 mm (Triplett, 1960). Histotrophe passes through the gut during most of development and serves as a source of nutrients (Eigenmann, 1892; Triplett, 1960; Igarashi, 1962; Engen, 1968). During the early and middle stages of development, hypertrophy of the gut is so extensive that it distorts the midventral body wall, causing it and the enclosed hindgut to protrude in a sac-like fashion. At later stages of development, extensive growth of the embryo as a whole restores the gut to its more appropriate position. Long, vascularized villi also have been reported in the embryonic hindgut, for example, in *Neoditrema ransonneti* (Igarashi, 1962), *Micrometrus minimus* (Dobbs, 1974), and *R vacca* (Wourms and Lombardi, 1985). Scanning electron microscopy has revealed that microvilli are present on the hindgut epithelial cells in *M. minimus* (Dobbs, 1974) and *R. vacca* (Wourms and Lombardi, 1985). The older literature contained enigmatic drawings and descriptions that suggested that intestinal villi or similar structures extended out from the perianal region in the embryos of some species (Igarashi, 1962). Recently, the existence of externalized intestinal villi and prototypic trophotaeniae has been demonstrated in embryos of the pile perch *R vacca* (Wourms and Lombardi, 1985).

In embryos of the eelpout *Z. viviparus,* a greatly enlarged hindgut with a hypertrophied intestinal epithelium is believed to be the primary site of embryonic nutrient absorption (Stuhlmann, 1887; Kristofferson *et al.,* 1973); the embryonic intestinal epithelium is organized into villi (Wourms, 1981) and the apical surfaces of the hindgut epithelial cells possess microvilli (Kristofferson *et al.,* 1973). Nutrients are obtained from ingested histotrophe (Korsgaard, 1986).

Boehlert and Yoklavich (1984) and Boehlert *et al.* (1986) present evidence that the embryonic hindgut in several species of the scorpaenid genus *Sebastes* is the site of embryonic nutrient absorption late in development. The hindgut of near-term embryos of *S. schlegeli* contained opaque material that moved during peristaltic contraction. The hindgut was well developed and its epithelial cells exhibited microvillous apical cell surfaces. Transmission electron microscopy revealed in the apical cytoplasm the presence of large vacuoles that were filled with an electron-dense homogeneous material. A series of tubular invaginations and small round vesicles were associated with the apical cell surface. Small vesicles, containing moderately electron-dense material, formed a transition zone between the apical cytoplasm and the large vacuoles. This structure is consistent with that of cells known to be engaged in endocytosis.

Although trophotaeniae have been the focus of interest in the study of goodeid embryonic nutrition, the gut has received some attention, in part due to its putative role as the evolutionary and developmental precursor of the trophotaeniae and also as a site of nutrient uptake, especially during late development, for example, in *Goodea luitpoldii* (Turner, 1940c). As might be expected from the variety of goodeid (almost 40 species in 18 genera), there is a diversity in the structure and function of both trophotaenia and gut. Full-term embryos of *Ameca splendens* have well-develped and ultrastructurally differentiated gut cells that possess distinctive microvilli, a prominent endocytotic complex, and many endosomes (Lombardi, 1983). They do not appear to function during gestation, however, since quantitative experiments on HPR uptake indicate that the trophotaeniae are the prime, and possibly the only, site of uptake (Lombardi and Wourms, 1985c). Further research is needed to ascertain whether these cells are functional but at a low level of activity or whether they are primed to function upon parturition. In another goodeid, *Alloophorus robustus*, at least three regions of the gut have become differentiated: a tubular anterior region, a distended middle region with a large lumen, and a straight hind segment with a smaller lumen. The epithelium of the middle segment contains many goblet cells and what appear to be absorptive cells. The latter possess a dense, regular brush border, but lack any apical endocytotic complex. Their subcortical cytoplasm contains electron-lucent vacuoles, which probably contain lipid. There is no evidence of apical lysosomes. Even after 2 h of continuous exposure, HRP was not endocytosed. In the hind segment, goblet cells are rare or even absent. The brush border of the absorptive cells is less densely packed with microvilli than in the midsegment cells. Individual microvilli, however, are regularly arrayed and of equal size. The apical canicular system is prominent. HRP was taken up and the reaction product was localized in apical tubules and small vesicles that are often fused or continuous with endosomes. In the endosomes, HRP reaction product tended to be associated with the limiting membrane, whereas in lysosomes it was peripherally located but otherwise remained free in the lumen as a flocculent aggregate. This pattern of uptake and intracellular transport with the hindgut cells was the same as seen in *Allophorus* trophotaeniae after 30 min of continuous HRP incubation (Hollenberg and Wourms, 1985, 1986).

In the gut of the embryos of viviparous teleosts, nutrient uptake appears to involve both the transport or diffusion of small molecules and the endocytosis of macromolecules especially proteins, and their subsequent intracellular digestion. Watanabe (1982) has demonstrated HRP uptake in the gut of a number of larval teleosts, and Iida and Yamamoto (1985) demonstrated HRP uptake by intestinal absorptive cells in adult goldfish. These and other studies (Govani et al., 1986) indicate that the intestinal absorptive cells of some embryonic and adult teleosts are functionally differentiated in the "open configuration" that characterized embryonic mammalian intestinal absorptive cells and that they do not enter into a "closed," mature state. Several functional advantages may result from this pattern of "intracellular digestion." (1) In many species of viviparous teleosts, embryos lack a histologically differentiated stomach, a condition that undoubtedly imposes constraints on the physiology of digestion. (2) Nutrients are normally available in the form of a liquid concentrate of proteins, lipids, and other molecules that are easily taken up by endocytosis, transport, or diffusion. (3) Patterns of embryonic nutrition, using the gut, that are based on intracellular digestion reduce or eliminate the physiological problem of waste disposal.

e. *Trophotaeniae and the Trophotaenial Placenta.* External, rosette- or ribbon-like structures that extend from the embryonic hindgut into the ovarian lumen characterize embryos of several species of teleosts. These structures were termed "trophotaeniae" by Turner (1937, 1940c, 1947), who postulated that they were embryonic trophic adaptations. Trophotaeniae consist of a simple surface epithelium surrounding a highly vascularized core of loose connective tissue. Trophotaeniae are found in the ophidioid species *Microbrotula randalli* and *Oligopus longhursti* (Wourms and Cohen, 1975; Cohen and Wourms, 1976), the parabrotulid *Parabrotula plagiophthalmus* (Turner, 1936; Wourms and Lombardi, 1979a), and all but one of the species of goodeids (Turner, 1937; Lombardi and Wourms, 1979). Light microscopy and SEM reveal that the major structural details of ribbon-like trophotaeniae of *M. randalli* and *O. longhursti* are virtually identical with those of *Parabrotula* and the ribbon-like trophotaeniae of many goodeids (Lombardi and Wourms, 1979; Wourms and Lombardi, 1979a,b; Lombardi, 1983). Recently, Wourms and Lombardi (1985) discovered prototypic trophotaeniae of the ribbon-type in embryos of the embiotocid *R. vacca*. Epithelial cells of trophotaeniae from *M. randalli*, *O. longhursti*, *P. plagiophthalmus*, and *R. vacca* are densely covered by microvilli, indicating an absorptive function (Lombardi and Wourms, 1979; Wourms and Lombardi, 1979b, 1985;

Lombardi, 1983). The remarkable similarity in trophotaenial structure among these widely divergent taxonomic groups is a clear illustration of convergent evolution of an embryonic trophic adaptation for viviparity (Wourms, 1981).

Trophotaeniae (Fig. 5) are the chief site of nutrient absorption in embryo goodeids and account for their massive (15,000% in *Ameca splendens*) increase in weight (Lombardi and Wourms, 1985c). With the exception of *Ataeniobius toweri*, trophotaeniae are found in the embryos of the approximately 40 species that comprise 18 genera of

Fig. 5. Scanning electron micrograph of a midterm embryo of the goodeid *Chapalichthys encaustus*. Well-developed trophotaeniae extend outward from the anal region. Scale bar equals 5 mm.

goodeids. Goodeid trophotaeniae are either rosette or ribbon-like structures that extend from the perianal region of the embryo into the ovarian lumen where they contact the internal ovarian epithelium. Apposition of the trophotaenial epithelium constitutes a placental association that was termed "the trophotaenial placenta" (Lombardi and Wourms, 1985b). Rosette trophotaeniae consist of a series of short, blunt, lobulated processes united at their bases and attached to the posterior end of the gut. They occur in the genera *Allotoca, Goodea, Neoophorus,* and *Xenoophorus.* As the name implies, ribbon trophotaeniae consist of long, slightly flattened processes that originate from a tube-like mass of tissue that extends outward from the perianal region of developing embryos. They are found in the genera *Allophorus, Ameca, Chapalichthys, Characodon, Girardinichthys, Hubbsina, Ilyodon, Skiffia, Xenotoca,* and *Zoogoneticus.* Considerable diversity in size, shape, and number of appendages exists within the two types of trophotaeniae (Turner, 1933, 1937, 1940c; Mendoza, 1937, 1956; Wourms, 1981). Both types of trophotaeniae are derivatives of the hindgut that begin development in the same way, that is, by hypertrophy of the hindgut epithelium, expansion of the perianal lips, and eversion and lobulation of the externalized hind gut epithelial surfaces. They differ in the degree of axial elongation during their final growth phase (J. Lombardi and J. P. Wourms, unpublished).

Although functions of gas exchange and nutrient absorption were attributed to trophotaeniae by Turner (1933, 1937, 1940c) and Mendoza (1937, 1956), until recently there was no experimental evidence to confirm these hypotheses. Ultrastructural or experimental studies now have been carried out in *Ameca splendens, Alloophorus robustus, Girardinichthys viviparus, Goodea atripinnis,* and *Xenotoca eiseni.* With the exception of *Goodea,* the embryos of these species have ribbon trophotaeniae.

Transmission and scanning electron microscopy have revealed that epithelial-cell morphology of *Ameca* trophotaeniae is nearly identical with that of embryonic mammalian intestinal absorptive cells. Trophotaeniae (Fig. 6) consist of a vascularized core of loose connective tissue surrounded by a simple surface epithelium. Trophotaenial epithelial cells are cuboidal and possess apical microvilli that form a loosely organized brush border characteristic of embryonic gut cells. Adjacent cells are either tightly apposed along their lateral margins or separated by enlarged intercellular spaces. In the latter instance, apical and basal margins are tightly apposed. The apical surface and cytoplasm contain apical canaliculi, smooth-surfaced invaginations of the cell membrane, clathrin-coated pits and vesicles, and smooth-sur-

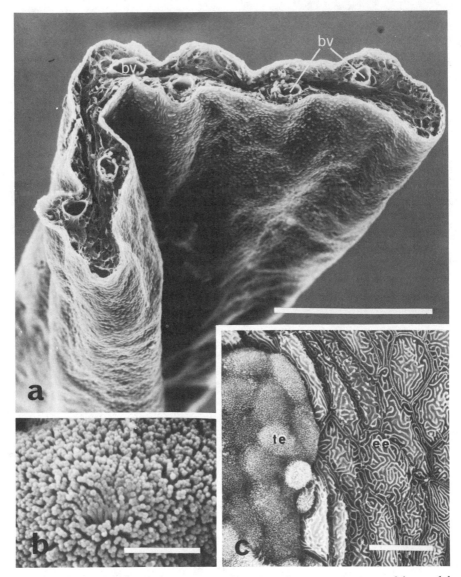

Fig. 6. (a) Scanning electron micrograph of a cryofracture preparation of the caudal trophotaenial ribbon of *Ameca splendens*, revealing its internal anatomy. Numerous blood vessels (bv) are situated within the connective tissue core. Scale bar equals 0.5 mm. (b) Scanning electron micrograph of the apical surface of a typical cell in the trophotaenial epithelium. Cells possess a dense mat of apical microvilli. Scale bar equals 4 μm. (c) Scanning electron micrograph of cell surfaces at the outer base of the trophotaenial peduncle in *A. splendens*. Note the sharp boundary between cells of the embryonic epidermis (ee) possessing microplicae and trophotaenial epithelium (te) possessing microvilli. Scale bar equals 20 μm. [From Lombardi and Wourms (1985b).]

faced vesicles (Lombardi and Wourms, 1985b). There is also an apical tubular membrane system (=transfer tubules) consisting of tubules (Fig. 7) with a uniform diameter, ~90 nm. Tubules contain electron-dense particles 8 × 16 nm arranged with regular 8-nm periodicity along the inner membrane of the lumenal surface of the tubule (Wourms and Staehelin, 1985, and unpublished). Freeze-fracture studies reveal that apical tubules are usually straight but may branch. They are discrete, with round, blunt ends. Continuity between the lumen of tubules and the interior of endosomes occurs. As indicated by localized distortion of the endosomal membrane, tubule fusion and separation are dynamic (Wourms and Staehelin, 1985, and unpublished). The supranuclear cytoplasm contains endosomes, large, membrane-bound, electron-lucent vacuoles, which are acidic prelysosomal compartments, and usually one or more large lysosomal vacuoles containing particulate material. The nucleus occupies the middle region of the cell, while Golgi complexes and endoplasmic reticulum (ER) tend to be localized in the basal portion. High-voltage electron microscopy (EM) of thick sections revealed two populations of mitochondria. One is a large apical, mitochondrial cluster associated with the apical tubules, endosomes, and lysosomal vacuoles. The second population is intimately associated with extensive invaginations of the basolateral surface membrane, in a configuration often described as a mitochondrial pump (J. P. Wourms, unpublished). A basal lamina, 100 mm thick, is apposed to the basal surface of the epithelial cells. Trophotaenial capillaries, lined with a continuous endothelium that contains smooth-surfaced vesicles, abut the basal lamina at frequent intervals. The connective tissue core contains the trophotaenial arteries and veins embedded in a fibrous extracellular matrix and its associated fibroblasts (Lombardi and Wourms, 1985b). The ribbon trophotaeniae of X. eiseni are much like those of Ameca splendens except for two features. First, the former has two populations of epithelial cells on the surface. One population on the ventral surface consists of trophotaenial absorptive cells whose ultrastructure is identical with that of A. splendens. This cell layer is continuous with the hindgut mucosa. The second population, which occurs on the dorsal and lateral surfaces, is a smooth-surfaced, squamous epithelium continuous with the embryonic epidermis. Second, X. eiseni trophotaeniae are of the sheathed type, that is, a wide tissue space intervenes between the connective-tissue core and the surface epithelium, whereas the trophotaeniae of A. splendens are unsheathed (Mendoza, 1972; Lombardi and Wourms, 1985b). The ultrastructure of trophotaenial absorptive cells of Alloophorus is quite similar to that of Ameca. Alloophorus cells are characterized by numerous endocytotic pits and vesicles and

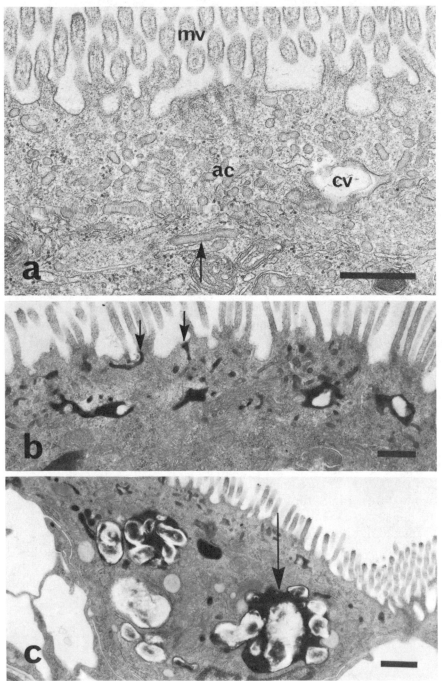

Fig. 7

a prominent apical canalicular system. Lysosomes, endosomes, and the typical cell organelles are present. There are, however, relatively few mitochondria (Hollenberg and Wourms, 1985). In the trophotaeniae of *Girardinichthys viviparus*, the surface epithelium is composed of two cell types. The minor component consists of squamous cells with microplicae on their apical surfaces and extensive Golgi complexes and much smooth ER in their cytoplasm. These cells form occasional stratified patches. They are considered to arise in connection with preparturitional trophotaenial aging. The major cell population consists of absorptive cells that possess a number of irregularly arrayed microvilli on their apical surfaces. Apical canaliculi are absent and endocytotic pits are rare. A distinct terminal web is present. Mitochondria are abundant and are confined to an apical–supranuclear region. There is a smooth-surfaced membranous "tubulolamellar network" associated with the lateral cell surface, and mitochondria are frequently associated with elements of this system. The basal half of the cell is dominated by the nucleus. Golgi complexes and coated and uncoated vesicles are present in the cytoplasm. The basal cell membrane displays considerable micropinocytotic activity (Schindler and de Vries, 1986). The absorptive cells of the rosette trophotaeniac of *Goodea atripinnus* closely resemble those of *Girardinichthys*. The cells are cuboidal, possess a well-developed brush border, and have a wide terminal web that is remarkably uniform in appearance and virtually free of cell organelles. There are few endocytotic pits or vesicles, and the apical canalicular system is poorly developed. In effect, the apical endocytotic complex is absent, so the cells more closely resemble neonatal mammalian intestinal absorptive cells after "closure." The apical cytoplasm is densely populated with mitochondria that are often closely associated with elements of the ER. Large supra-

---

Fig. 7. (a) Transmission electron micrograph of the apical region of a trophotaenial absorptive cell of *Ameca splendens*. Microvilli (mv) extend from prominences on a convoluted apical cell surface. An apical canalicular system (ac) (=endocytotic complex) containing tubular profiles (arrow) is associated with collecting vesicles (cv). Scale bar equals 1 μm. (b) Transmission electron micrograph of the apical region of a trophotaenial absorptive cell fixed after 10 min of continuous exposure to horseradish peroxidase (HRP). Uptake occurs via endocytosis at the apical cell surface (arrows). Reaction product is present within apical canaliculi and collecting vesicles. Scale bar equals 1 μm. (c) Transmission electron micrograph of a trophotaenial absorptive cell fixed after 1 h of continuous exposure to HRP. Reaction product is present within apical canaliculi, collecting vesicles, and greatly enlarged supranuclear lysosomes (arrow). This is also the appearance of cells fixed after 2–6 h of continuous exposure to HRP. Scale bar equals 1 μm. [From Lombardi and Wourms (1985b,c).]

nuclear lysosomes are absent. The subnuclear cytoplasm contains many small vesicles, some of which are coated, a number of Golgi complexes, and a profusion of elements of the ER, mostly SER (Hollenberg and Wourms, 1985, and unpublished).

The use of ultrastructural tracers (Fig. 7), such as horseradish peroxidase (HPR), native ferritin, cationized ferritin, and ruthenium red, as well as acid and alkaline phosphatase (Ac- and Al-Pase) cytochemistry revealed aspects of protein uptake and transport. Microvilli of the brush border were Al-Pase–positive in both *Ameca* and *Girardinichthys* (Lombardi and Wourms, 1985c; Schindler and de Vries, 1986). Lombardi and Wourms (1985c) have shown that trophotaenial absorptive cells of *Ameca* endocytose HRP and degrade it in the lysosomes. Initially (1.5–10 min) HRP was taken up via apical canaliculi, smooth-surfaced invaginations of the cell membrane, and passed into the 90-nm transfer tubules. By 10 min, HRP had passed into the lysosomes. At 15 min, HRP appeared in the moderate sized and Ac-Pase–positive vacuoles, the lysosomes. From 20 min on, HRP appeared in the large, "standing," supranuclear lysosomes. After 3 h in the standing lysosome, nearly all HRP activity had been lost, presumably from enzymatic degradation. The presence of Golgi complexes, residual bodies, and secretory granules in the infranuclear region suggested that products of protein absorption and hydrolysis were discharged from the basal cell surface (Lombardi and Wourms, 1985c). Cationized ferritin was found by Wourms and Staehelin (1985) to enter the cell via coated pits and vesicles and then to pass sequentially into the transfer tubules, endosomes, and lysosomal compartment. Identical results were obtained in *Ameca* by Grosse-Wichtrup and Greven (1985) with the use of native and cationized ferritin. High-voltage EM of aldehyde-fixed cells, subsequently postfixed with ruthenium red, revealed that ruthenium red enters apical canaliculi that deeply penetrate the cell apex but that it does not enter the transfer tubules (Wourms and Staehelin, 1985). Using a different technique, Grosse-Wichtrup and Greven (1986) found that ruthenium red did not enter the apical endocytotic complex but did enter the tubular infolding of the lateral cell membranes, thus establishing their continuity with intercellular spaces. Wourms and Staehelin (1985) concluded that proteins are endocytosed by two routes: some, such as HRP, enter via apical canaculi, while others, such as cationic ferritin, enter via coated pits and vesicles. Both classes of proteins enter the transfer tubules and endosomes. Passage from one compartment to another involves either vesicular transport or fusion of compartmental membranes. In *Alloophorus*, HRP could be demonstrated in the apical canalicular system, endosomes, and lysosomes after a 30-min incubation. Under the same experimental conditions, in which both *Ameca* and *Alloophorus* trophotaeniae absorbed HRP, there was no indication that *Goodea* trophotaeniae took up HRP even after 30 min of continuous exposure. It was concluded that the rosette trophotaeniae of *Goodea* do not endocytose proteins; instead, they function in the transport of small molecules (Hollenberg and Wourms, 1985, and unpublished). Based on their ultrastructural organization, lack of Ac-Pase–positive lysosomes, and failure to endocytose any appreciable amount of cationized ferritin within 10 min (Schindler and de Vries. 1986), it would appear that the epithelial cells of the ribbon trophotaeniae of *Girardinichthys* also function primarily in the transport of small molecules.

Kinetic studies using HRP, glycine, and glucose have been carried out on *Ameca* trophotaeniae. Trophotaeniae of embryos, incubated in vitro in HRP saline, take up HRP at an initial rate of 13.5 ng HRP (mg trophotaeniae protein)$^{-1}$ min^{-1}. The system becomes saturated after 3 h. Trophotaeniae incubated at 4°C showed little or no uptake. In trophotaemiae continuously pulsed with HRP for 1 h and then incubated in HRP saline, levels of absorbed peroxidase declined at a rate of 0.5 ng (mg trophotaenial protein)$^{-1}$ min^{-1}. A comparison of uptake in embryos incubated in HRP with and without trophotaeniae revealed that without trophotaeniae, embryos exhibited a consistently higher level of peroxidase activity than did controls, but that this value did not increase over a 60-min incubation. These data were interpreted as evidence that HRP, and presumably other proteins as well, enter *Ameca* embryos only by the trophotaenial, not by extra-trophotaenial, routes (Lombardi and Wourms, 1985c). Using glycine and glucose radioisotopes, Lombardi (1983) demonstrated that trophotaeniae were the sole site of absorption in *Ameca* embryos. Preliminary studies (F. Hollenberg and J. P. Wourms, unpublished) have suggested that this may not be the case in other goodeids, in which an intestinal route for protein absorption may be significant.

The occurrence of an embryonic adaptation as spectacular as trophotaeniae in four distantly related orders of teleost fishes (Ophidiiformes, Cyprinodontiformes, Perciformes, and Gadiformes) invites inquiry into their evolutionary origin. Wourms and Lombardi (Wourms, 1981) have proposed that trophotaeniae are the culmination of an evolutionary sequence of convergent adaptations of the embryonic gut. Drawing on both developmental and comparative studies, they proposed the following sequences: (1) origin from a simple tubular embryonic gut; (2) precocious enlargement of the hindgut, as seen for instance in embiotocids and *Anableps;* (3) hypertrophy of intestinal villi or lamellae, as seen for instance in embiotocids, zoarcids, and clinids; (4) externalization of the hind gut epithelium via differential growth to form short trophotaenial buds, for example, as seen in goodeid embryos; (5) increase in number and radial growth to form rosette trophotaeniae; and (6) increase in number, curtailed radial growth, and increased axial elongation to form ribbon trophotaeniae. The recent discovery of externalized intestinal villi and short prototypic trophotaeniae in the embiotocid *R. vacca* is of special interest since these structures correspond to steps 4 and 5 in the preceding hypothetical sequence. Trophotaenial evolution appears to be heterochronic, probably involving the accelerated expression of the genes regulating the onset, rate, and extent of intestinal morphogenesis and cell differentiation. A nonallometric hypertrophy of gut tissues results. Since trophotaeniae are efficient in the uptake of maternal nutrients, they presumably confer a selective advantage on those embryos with them. The evolution of trophotaeniae is but one aspect of the evolutionary and developmental plasticity of the teleost gut. Cursory examination of the literature (e.g., Moser, 1981), indicates that the larvae of some species of oviparous deep-sea fishes display spectacular modifications of the gut, including long extensions of the gut and the investing body wall beyond the body's ventral profile. Some of these structures may prove to act like trophotaeniae. Balon (1986) has independently made the same suggestion.

3. MATERNAL SPECIALIZATIONS

The epithelial lining of the ovarian lumen unquestionably plays an important role in supplying nutrients to embryos and has given rise to some remarkable structural specializations. Association of the lumenal epithelium with the branchial region of the embryo to form a branchial placenta (e.g., in the anablepid *Jenynsia*, the goodeid *Hubbsina*, and the embiotocids *Cymatogaster* and *Rhacochilus*) and between the lumenal epithelium and the buccal region of the embryo to form a buccal placenta (e.g., in *Ogilbia* and *Lucifuga*) were discussed in Section III,A,2,c. The structure, possible function, and chemical composition of histotrophe is best known in the embiotocids, goodeids, and the eelpout *Zoarces*. Maternal specializations for maternal–embryonic nutrient transfer are poorly known in other taxonomic groups that exhibit intraovarian gestation. The gross morphology of the ovary has been described for some hemiramphids (Mohr, 1936; Mohsen, 1962) and scorpaenids (Lüling, 1951; Moser, 1967a,b; Magnusson, 1955), but the ovarian structure of the parabrotulids, many of the ophidioids, and the comephorids is for the most part unknown.

During gestation in embiotocids, epithelial cells that line the ovarian cavity and ovigerous folds hypertrophy, and the ovarian wall and ovigerous folds undergo extensive vascularizaiton (Turner, 1938b; Gardiner, 1978). The ovigerous folds form distinct compartments, which enclose individual embryos and may facilitate maternal—embryonic nutrient transfer (McMenamin, 1979). Embryos undergo a significant (20,000% or more) increase in dry weight during gestation due to absorption of maternally derived nutrients (Wourms, 1981). In *Cymatogaster aggregata,* the epithelium lining the ovigerous folds undergoes an annual cycle of morphological changes (Gardiner, 1978). Turner (1938b) suggested that during early gestation, embryos depend on secretions of the ovarian epithelium, while in later phases, embryonic dependency shifts to nutrients that are transferred across the internal ovarian epithelium from the maternal circulatory system. Wiebe (1968) attributed a secretory role to the internal ovarian epithelium of embiotocids. In the first ultrastructural study of the embiotocid ovary, Gardiner (1978) described the structure of ovarian epithelial cells and implied that they may synthesize and secrete proteins. The cells, however, possess few of the features characteristic of secretory cells. Gardiner (1978) also described dilations in the internal ovarian epithelium that are prominent during early stages of gestation but subsequently decrease in size. Dilations occur between the lateral surfaces of adjacent epithelial cells. They contain extracellular material that is presumably synthesized and secreted by the cells.

Recently, de Vlaming *et al.* (1983) analyzed the composition of the ovarian fluid that bathes the developing embryos of *C. aggregata,* *Hysterocarpus traski,* and *Micrometrus minimus.* Their results are somewhat perplexing. They found that ovarian fluid of *C. aggregata* was devoid of maternal serum polypeptides, possessing instead unique polypeptides. They suggested that the internal ovarian epithelium synthesized and secreted these polypeptides. In contrast, the ovarian fluid of *Hysterocarpus traski* and *Micrometrus minimus* contained only serum polypeptides, suggesting that the internal ovarian epithelium of these species transports maternal serum proteins, rather than newly synthesized polypeptides. In all three species, the amino acid composition of the ovarian fluid matched that of the maternal serum.

In *Z. viviparus,* ovigerous processes that contain developing follicles extend into the ovarian lumen during oogenesis. Following ovulation and fertilization, these processes persist and greatly increase the surface area of the lining of the ovarian lumen. Nutrients are believed to pass from the vasculature of the ovigerous processes to the ovarian fluid (Stuhlmann, 1887; Bretschneider and Duyvené de Wit, 1947; Kristofferson *et al.*, 1973). According to Kristofferson *et al.* (1973), the concentration of polypeptides in the ovarian fluid was extremely low. In addition, the amino acid concentration in ovarian fluid was less than that in maternal serum, but it was greater than the polypeptide concentration. On the basis of this, Kristofferson *et al.* (1973) postulated that amino acids were more important for embryonic nutrition. In contrast, lipid levels in ovarian fluid were high, attaining a peak level of 12.2 mg ml^{-1} during late gestation, and this suggested that lipid may be important in embryonic nutrition. Serum lipid levels reached peak levels simultaneously (Korsgaard and Petersen, 1979), and presumably lipid passes across the lining of the ovarian lumen. These authors also reported that during gestation lipids were synthesized in the liver and released into the blood in response to elevated levels of estradiol. Zoarcid histotrophe also contains cellular material, including erythrocytes (Kristoffersen *et al.*, 1973).

Recently, Korsgaard (1986) reported that the marked increase in dry weight of *Zoarces* embryos that takes place after hatching is complemented by a sudden shift either in the production or distribution of ovarian fluid. The period prior to hatching is characterized by the presence of large fluid-filled follicles and very little fluid in the ovarian cavity. After hatching, follicles appear more or less empty and the amount of ovarian fluid in the ovarian cavity has increased considerably. The fluid is the medium by which maternal nutrients are transferred to the embryo. Due to its hypertrophied condition, and in the absence of any other obvious embryonic specializations, the gut is considered the site of nutrient uptake. In 1983, Korsgaard demonstrated that only small molecules were able to pass from the maternal serum to the ovarian fluid. Electrophoretic protein bands of maternal serum and follicular fluid were found to be identical whereas no protein bands could be detected in the ovarian fluid. The osmolarity and concentration of glucose, free fatty acids, and ninhydrin-positive substances (a measure of free amino acids) were similar in both serum and follicular fluid but considerably lower in ovarian fluid, whereas the concentration of chloride ions was found to be almost identical in the three compartments. Furthermore, by intramuscular or intraovarian loading of gravid females with glucose or a mixture of amino acids and subsequently monitoring the rate of clearance from maternal serum, ovarian fluid, or embryonic serum, Korsgaard (1983) was able to demonstrate that the ovarian fluid is not a static pool. Rather, it undergoes rapid turnover with exchange of low-molecular-weight compounds such as amino acids, glucose, or free fatty acids. Subsequently, Korsgaard and Andersen (1985) and Korsgaard (1986) showed that during early gestation embryos took up very small amounts of glucose, glycine, and taurine, in that order. Later in gestation, the embryos accumulated substantial amount of these tracers. In the case of glucose, there is a 10-fold difference, that is, 1.48×10^{-4} μmol g^{-1} h^{-1} versus 2×10^{-3} μmol g^{-1} h^{-1}. Release of both $^{14}CO_2$ and dissolved organic carbon from embryos into the medium was low, indicating that most of the glucose and amino acids were used in synthetic activities. Some (about 14%) of the glucose taken up was, however, oxidized to provide energy. After hatching, total carbon and nitrogen, expressed as milligrams per whole embryo, increased linearly throughout gestation, as did the dry weight. These observations indicate an extensive nutritive relationship between mother and embryos that commences immediately after hatching.

The structural organization of the reproductively active zoarcid ovary is unusual. In nongravid females, oocyte growth and differentiation occur within ovarian follicles that lie at or near the basal surface of the internal ovarian epithelium. Each egg is surrounded by a thin, well-vascularized layer of follicle cells. Following ovulation, empty follicles do not regress but are retained throughout gestation. During this period, follicles undergo a remarkable reorganization and hypertrophy to form extensively vascularized villous projections, termed "calyces nutriciae" (Bretschneider and Duyvené de Wit, 1947). The villi that festoon the inner surface of the ovarian wall are approximately 1 cm in length and greatly amplify the surface area available for metabolic exchange between mother and embryo. A large, centrally situated artery enters the basal portion of each calyx, extends toward the apex, and undergoes arborization to form an apically situated *rete mirabile*. A return network of venules and veins is peripherally situated. The outer surface of the calyx is bounded by a simple, squamous epithelium, and its interior is characterized by numerous vascular elements and by a large extracellular lymph space devoid of connective tissue (Stuhlmann, 1887; Kristoffersen *et al.*, 1973). Little information is available on the ultrastructure of the calyces nutriciae. The transfer of soluble nutrients, chiefly of low molecular weight, into the histotrophe appears to involve the same processes as do transfers between blood plasma and other extracellular body fluids (Kristoffersen *et al.*, 1973). The ovarian calyces are functionally equivalent to the maternal portion of a mammalian epitheliochorial placenta, but morphologically they resemble the chorionic villi of the mammalian chorioallantoic placenta. The villiform calyces of *Zoarces* are also strikingly similar to the trophonemata of the cownose ray *Rhinoptera* (Hamlett *et al.*, 1985e) and butterfly ray *Gymnura* (Wourms and Bodine, 1983).

Embryos of goodeid fishes develop to term within the ovarian lumen, where they undergo considerable increase in weight due to transfer of maternal nutrients across the trophotaenial placenta (Fig. 8). The placenta consists of an embryonic component, the trophotaeniae, and a maternal component, the ovarian lining (Wourms, 1981; Lombardi and Wourms, 1985a,b; cf. Section III,A,2,e). The ultrastructure and micromorphology of both gravid and nongravid ovaries of *Ameca splendens* have been studied recently by Lombardi and Wourms (1985a). The single median ovary of *Ameca* is a hollow structure whose central lumen is divided into lateral chambers by a highly folded longitudinal septum. The heavily vascularized ovarian wall comprises a peritoneum, muscle layer, connective-tissue stroma, and internal ovarian epithelium. Ovigerous tissue is confined to the folds of the ovarian lining that extend into each of the lateral chambers. Matrotrophic development takes place within these ovarian chambers. During gestation, the lining of the ovarian lumen is in direct apposition to the body surfaces and trophotaenial epithelia of developing embryos. Although the structure of the *Ameca* ovary is similar to that of other goodeids, specific differences in the gross morphology of goodeid ovaries have been reported in the literature and are considered to be of taxonomic significance. Hubbs and Turner (1939) distinguished between two ovarian types in which the ovigerous tissue lies either within the connective tissue layer of the ovarian wall or within the connective tissue stroma of the ovarian folds (cf. Lombardi and Wourms, 1985a). Moreover, the overall configuration of the goodeid ovary is not static. In gravid females, it undergoes marked cyclical

---

Fig. 8. (a) Scanning electron micrograph of the internal ovarian epithelium that lines the ovarian lumen in *Ameca splendens*. The epithelial surface is thrown into a series of irregular folds. Marginal clefts delineate individual cells and give the epithelium a "cobblestone" appearance. Scale bar equals 0.2 mm. (b) Scanning electron micrograph of the lumenal surface of phase 1 cells of the internal ovarian epithelium of *A. splendens*. Cell apices are crenulated and separated by deep marginal furrows (arrows). The apical plasma membrane is usually devold of surface membrane specializations, although occasional microvilli are observed. Scale bar equals 3 μm. (c) Scanning electron micrograph of the lumenal surface of a phase II cell of the internal ovarian epithelium of *A. splendens*. Numerous spherical inclusions are evident within the apical cytoplasm of individual cells. Cells are tightly apposed along their apical margins (arrows), and cell apices bulge outward into the ovarian lumen. Scale bar equals 10 μm. [From Lombardi and Wourms (1985a).]

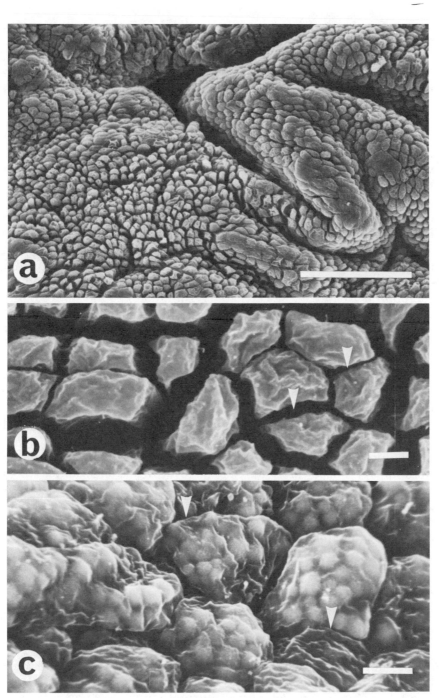

Fig. 8

changes in structure (Mendoza, 1940, 1943). The chief modifications, summarized by Amoroso (1960), are that (1) epithelial cells lining the ovarian lumen increase in height and become more glandular; (2) connective tissue becomes swollen; (3) free cells appear in the ovarian fluid; and (4) the vascularity of the ovary increases and extensive capillary networks form just beneath the epithelium lining the ovarian lumen. In postpartum goodeids, ovarian tissue undergoes involution.

The internal ovarian epithelium in *Ameca* lies above a well-vascularized bed of connective tissue. It is a simple cuboidal epithelium, whose cells have a characteristic convex apical surface. They have a well developed RER and SER, and at times they accumulate numerous large, membrane-bound vesicles in the apical cytoplasm, evidently as an expression of two functional states. Phase I cells contain few, if any, large apically situated vesicles, while phase II cells contain many. Inclusions are of two types: one is a protein or lipoprotein that is presumably derived from the RER and Golgi, and the other appears to be a lipid. Although definitive morphological evidence of secretion is lacking, these cells are considered to be the source of the four or five classes of 80- to 100-kDa proteins that are found in the ovarian fluid. Amino acid transport across the epithelium most likely occurs as well (Lombardi and Wourms, 1985a).

Relatively little information exists on the secretory function of the teleost ovary. Most previous studies have been carried out on oviparous species, and few on viviparous forms, especially those with intralumenal ovarian gestation (Wourms, 1981; Lombardi and Wourms, 1985a). So far, a secretory role has been established for the internal ovarian epithelium of one goodeid and a few embiotocid fishes. Hemotrophic transfer of metabolites across the internal ovarian epithelium appears to be the dominant mode of histotrophe formation in *Zoarces* and *Jenynsia,* as well as the mode of transfer of all amino acids and some proteins in embiotocids. Thus, along with providing a brood chamber for developing embryos, the internal ovarian epithelium also appears to function as a primary source of the nutrients required for embryonic growth.

B. Intrafollicular Gestation

1. Introduction

Intrafollicular gestation is known to occur in the clinids, some labrisomids, the poeciliids, and the anablepid *Anableps* (Fig. 9). Viviparity has been examined most extensively in various poeciliids and

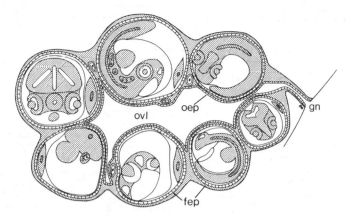

Fig. 9. Diagram of a generalized poeciliid ovary depicted in cross section. Diagram illustrates the relationship between embryonic and maternal tissues in a viviparous teleost displaying intrafollicular gestation. Although not shown in the diagram, the embryonic surface is separated from the follicular epithelium (fep); gn, gonopore; oep, ovarian epithelium; ovl, ovarian lumen; fep, follicular epithelium.

Anableps and, consequently, specializations for intrafollicular gestation are best known in these two groups. Information about reproduction in the clinids and labrisomids, however, is incomplete and mostly scattered through the systematic literature (Breder and Rosen, 1966; Penrith, 1969; Springer, 1970; Rosenblatt and Taylor, 1971; Moser, 1976; George and Springer, 1980). Recent investigations on the maternal–embryonic relationship in *Clinus superciliosus* (Veith, 1979a,b, 1980; Veith and Cornish, 1986) and nutrient transfer in *C. dorsalis* (Cornish, 1985; Cornish and Veith, 1986) constitute the only detailed accounts of viviparity among the clinids.

A variety of embryonic and maternal specializations for intrafollicular gestation has been described. Most remarkable is the postfertilization follicle. In most oviparous and viviparous teleosts, ovulation occurs prior to or soon after fertilization, and the follicle then degenerates. In contrast, the ovarian follicle of species with intrafollicular gestation remains intact after fertilization and undergoes changes to accommodate embryonic development. The follicle wall becomes responsible for metabolite exchange, including gas exchange and nutrient transfer, for maintaining the osmotic environment of the embryo, and possibly for protecting the embryo from immunological rejection (see Section I,B). In species in which intrafollicular gestation has been examined in detail, this functional change is clearly reflected in the structural differences between the pre- and postfertilization follicles. Embryonic specializations include increased vascularization of the embryonic surface and modification of both the embryonic surface and the embryonic gut.

During gestation, metabolite exchange often occurs between the closely apposed follicle wall and the embryonic surface. We refer to this association between the follicle wall and the embryonic surface as the follicular placenta. Turner (1940c) called this maternal–embryonic association the follicular pseudoplacenta, but only in those cases in which nutrient transfer was presumed to occur. We believe that the term placenta is more appropriate for semantic and heuristic reasons. Mossman (1937) has defined a placenta as any combination of closely apposed parental and embryonic tissues that function in maternal–embryonic metabolite exchange. This is a broad definition that includes the cases that Turner (1940c) referred to as pseudoplacentae. Moreover, the prefix "psuedo" in Turner's pseudoplacenta is misleading inasmuch as it implies that something must be false. Mossman's (1937) general definition of the placenta is useful because it shifts the emphasis of the placental concept away from criteria narrowly based on extraembryonic membrane patterns in amniotes and toward broader criteria based on the functional roles of parental and embryonic tissues in physiological exchange. Use of the term follicular placenta emphasizes the functional relationship between maternal and embryonic tissues and also invites comparison with other placental types.

The follicular placental barrier includes (1) maternal capillary endothelium, (2) basal lamina, (3) follicular epithelium, (4) egg envelope, (5) embryonic surface epithelium, (6) basal lamina, and (7) embryonic capillary endothelium. Although Scrimshaw (1944b) and Jollie and Jollie (1964) have reported that the egg envelopes of *Heterandria formosa* and *Poecilia reticulata* disappear during gestation, more recent information indicates that this is not the case, at least for *H. formosa*. In view of Hogarth's (1968) finding that the egg envelope confers protection from immunological rejection in *Xiphophorus*, the egg envelope probably remains intact in most, if not all, species. Paradoxically, it is absent in *Anableps* (Knight *et al.*, 1985)

Embryonic nutrition varies in intrafollicular gestation. Analyses of changes in embryonic dry weight carried out on several poeciliids indicate that embryonic nutrition in this family ranges from lecithotrophy to matrotrophy (Bailey, 1933; Scrimshaw, 1944b, 1945; Thibault and Schultz, 1978; Reznick, 1981). For example, embryonic dry weight decreases approximately 25–45% in *P. reticulata*, *Gambusia affinis*, and *P. monacha* (Thibault and Schultz, 1978; Reznick, 1981). This is similar to the dry weight decreases reported in oviparous species (Paffenhofer and Rosenthal, 1968; Smith, 1957; Terner, 1979), thus indicating that embryonic nutrition in these poeciliids comes almost entirely from yolk stored in the egg. In contrast, embryonic dry weight increases 1800% and 3900% in *P. turneri* and *H. formosa*, respectively, indicating substantial maternal–embryonic transfer (Scrimshaw, 1945; Thibault and Schultz, 1978). Despite this wide range, however, Scrimshaw (1945) reported that embryonic dry weight in many of the poeciliid species he examined remained relatively constant during development, and he interpreted this to indicate the existence of only a moderate maternal–embryonic nutrient transfer. In *Anableps*, embryonic nutrition is highly matrotrophic, with embryonic dry weight increasing as much as 800,000% (Knight *et al.*, 1985). Veith (1979a,b, 1980) demonstrated that embryonic nutrition is matrotrophic in the clinid *Clinus superciliosus*, as it also is in *C. dorsalis* (Cornish, 1985), and a report by Moser (1976) on embryonic size in the clinid *Pavoclinus mus* suggests that embryonic nutrition in this species may be matrotrophic as well. Embryonic nutrition in other viviparous clinids and labrisomids is unknown. Specializations for maternal–embryonic nutrient transfer have been described in several species, however, and are reviewed in the following sections.

2. EMBRYONIC SPECIALIZATIONS

Both the embryonic surface and the embryonic gut have been identified as sites of nutrient absorption in species displaying intrafollicular gestation. The site of nutrient absorption in most embryos has been inferred from their morphological specializations, but some studies have well-documented sites of absorption. Morphological specializations associated with nutrient absorption vary among the species that have been examined, and the site of nutrient absorption may often change during development.

In clinid embryos, body surface and gut both play a role in nutrient absorption. Using autoradiography, Veith (1980) demonstrated that early embryos of *C. superciliosus* absorb histotrophe (follicular fluid) across the epidermis of the dorsal fin, the pericardial sac, and the yolk sac. During the later stages of development, however, embryos absorb histotrophe almost exclusively through their hypertrophied hindgut. This change in the site of absorption is marked by developmental changes in the structure of the embryonic surface. In contrast, Cornish (1985) reported the uptake of [^{3}H]leucine by both gut and epidermis of *C. dorsalis*. The ventral surface of early-stage embryos, particularly in the pericardial and yolk-sac regions, possesses a complex pattern of macroridges. Epidermal cells of the ridges exhibit some microplicae as well as numerous short, stubby microvilli. At later stages of development, macroridges are no longer present on the ventral surface of embryos, and epidermal cells of the pericardial and yolk sacs possess only microplicae, a feature of the juvenile epidermis. In this species, developmental changes in the structure of the hindgut have not been described.

In *Anableps,* the pericardial sac surface and gut also are believed to absorb nutrients during gestation. Early in development, the pericardial sac expands ventrally and posteriorly until it occupies the entire ventral surface of the embryo (Turner, 1938a, 1940b). An extensive plexus of blood vessels, the portal plexus, underlies the surface of the pericardial sac, and numerous vascular expansions associated with this plexus, termed vascular bulbs, give the pericardial sac surface a pebbled appearance (Turner, 1938a, 1940b; Knight *et al.,* 1985). These specializations of the pericardial sac are fully developed by the 5-mm stage (Turner, 1940d). The surface of the pericardial sac is closely apposed to the follicular epithelium, with the vascular bulbs fitting into indentations of the follicular epithelium (Knight *et al.,* 1985). This close association between embryonic and maternal tissues forms the follicular placenta and is believed to be the primary site of maternal–embryonic nutrient transfer (Turner, 1940b; Knight *et al.,* 1985). Knight *et al.* (1985) coined the term pericardial trophoderm to refer to the specialized surface of the pericardial sac.

The embryonic midgut also hypertrophies until it occupies the entire body cavity. According to Turner (1940b), the midgut reaches its maximum size in 21- to 24-mm embryos of *A. anableps* and *A. dowi.* At this stage of development, the mucosa of the midintestine is well differentiated, with numerous elongated villi filling the intestinal lumen. The fact that the expanded portions of the intestine, along with the elongated villi, regress just before birth (Turner, 1940b) suggests that these hindgut modifications are specific embryonic adaptations for nutrient absorption.

At present, the relative importance of the pericardial trophoderm and the embryonic gut in nutrient absorption is unknown, although it is generally believed that maternal–embryonic nutrient transfer occurs primarily across the follicular placenta and that absorption across the hypertrophied gut supplements nutrient absorption (Turner, 1940b; Knight *et al.,* 1985). Embryonic dry weight begins to increase after the appearance of the pericardial trophoderm, but before gut hypertrophy (Knight *et al.,* 1985). Moreover, the close apposition of the pericardial trophoderm to the follicular epithelium suggests that the volume of follicular fluid that enters the embryonic gut would be too small to serve as the major source of nutrients. However, Knight *et al.* (1985) showed with scanning electron microscopy that cells of the pericardial trophoderm of late-stage (60 mm) *A. dowi* embryos lacked microvilli, a characteristic feature of absorptive cells, but possessed microplicae. The presence of microplicae, a feature of differentiated epidermal cells, may indicate that nutrient absorption across the pericardial trophoderm of late-stage embryos is limited. Experimental investigation of the sites of nutrient uptake in *Anableps* embryos will be required in order to clarify the role of the pericardial sac surface during development.

In poeciliid embryos, the surface of the pericardial and yolk sacs is believed to be the primary site of metabolite exchange (Turner, 1940a; Scrimshaw, 1944b, 1945; Fraser and Renton, 1940; Jollie and Jollie, 1964; Dépêche, 1970). The pericardial sac of poeciliid embryos greatly enlarges during development, but, unlike that of embryos of *Anableps,* expands dorsally and anteriorly to invest the developing head completely (Turner, 1940a; Tavolga and Rugh, 1947; Tavolga, 1949; Fraser and Renton, 1940). To emphasize this, Tavolga (1949) and Tavolga and Rugh (1947) referred to the outer surface of the expanded pericardial sac as the pericardial serosa (=chorion) and to the inner surface, adjacent to the embryonic head, as the pericardial amnion. Recent examination of the pericardial sac of poeciliid embryos with electron microscopy, however, has revealed that it differs in several ways from the extraembryonic membranes in amniotes and that the two embryonic structures are only convergently similar (Grove, 1985).

The expanded pericardial sac and the yolk sac are extensively vascularized by a portal plexus that unites the Cuverian ducts (common cardinal veins) with the anteriorly directed sinus venosus of the heart (Turner, 1940a). Close apposition of this vascularized embryonic surface to the follicular epithelium is considered to form the follicular placenta. The fate of the pericardial sac and its relative contribution to the vascularized surface of the embryo vary among species. In lecithotrophic species, the yolk-sac surface forms a large proportion of the vascularized embryonic surface, whereas the yolk sac is greatly reduced in matrotrophic species, and the pericardial sac comprises most of the vascularized embryonic surface (Turner, 1940a; Thibault and Schultz, 1978). In addition, the pericardial sac invests the embryonic head almost to the end of gestation in some species, such as *Heterandria formosa* (Turner, 1940a; Fraser and Renton, 1940; Scrimshaw,

1944b), but in others (e.g., *Poecilia reticulata, Gambusia affinis, Xiphophorus helleri, X. maculatus*), it is ruptured early in development by the enlarging head so that only an ephemeral "neckstrap" remains across the head (Turner, 1940a; Tavolga, 1949; Kunz, 1971).

Ultrastructural studies have revealed differences between the surfaces of lecithotrophic and matrotrophic embryos. Jollie and Jollie (1964) and Dépêche (1970, 1973) reported that an extremely flattened, squamous, bilaminar epithelium covered the pericardial and yolk sacs of the embryos of *P. reticulata*. Both investigators noted numerous vesicles in the surface epithelial cells, and Jollie and Jollie (1964) postulated that these cells were engaged in micropinocytosis. In a study using scanning electron microscopy, however, Dépêche (1973) found that the surfaces of cells covering the pericardial and yolk sacs of these embryos possess a complex network of microridges. Microvilli, a characteristic feature of absorptive cells, were conspicuously absent. Some evidence does suggest that such embryos may absorb exogenous nutrients during the very early stages of development (Dépêche, 1976; Trinkaus and Drake, 1952), but changes in embryonic dry weight indicate that embryonic nutrition is primarily lecithotrophic in this species (Scrimshaw, 1945; Dépêche, 1976; Thibault and Schultz, 1978). Although surface cells may endocytose some follicular fluid components, the absence of microvilli on the embryonic surface reinforces the conclusion that *P. reticulata* embryos derive little or no nutrition from the follicular fluid. Embryos of *G. affinis*, another lecithotrophic species (Chambolle, 1973; Reznick, 1981), display pericardial- and yolk-sac features similar to those found in *P. reticulata* (B. D. Grove, unpublished).

In the matrotrophe *H. formosa*, the embryo's yolk sac is greatly reduced and the pericardial sac makes up a large proportion of the vascularized, embryonic surface (Fig. 10). Recent ultrastructural and experimental studies have revealed that these embryos possess surface specializations for nutrient absorption that are not confined to the pericardial and yolk sacs, however, as was thought by previous investigators. During most of development, the entire surface of the embryos is covered by cells possessing numerous apical microvilli, large apical vesicles, an extensive rough endoplasmic reticulum, and apical coated pits and vesicles (Grove, 1985; Grove and Wourms, 1981, 1982). A series of experimental studies using radioisotopes and macromolecular tracers demonstrated that these cells are absorptive (Grove, 1985; Grove and Wourms, 1982, 1983). Embryos absorbed glycine and glucose at high rates, and it was shown, with the aid of autoradiography, that glycine absorption occurred across the microvillous surface

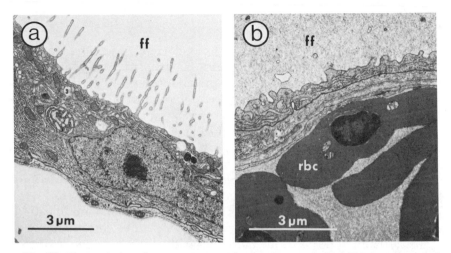

Fig. 10. Transmission electron micrograph of surface epithelia of (a) the pericardial sac of *Heterandria formosa* embryos and (b) the yolk of *Xiphophorus maculatus* embryos. Embryos are at approximately the same stage of development. Note that in *H. formosa* embryos, which are highly matrotrophic, surface cells possess numerous microvilli on their apical surfaces and numerous cell organelles. In contrast, yolk-sac surface cells of *X. maculatus* embryos are more squamous, lack microvilli, and possess fewer organelles. The bulk of embryonic nutrition in this species is derived from yolk; ff, follicular fluid; rbc, red blood cell.

epithelium. Iron dextran and ferritin injected into follicles were also endocytosed by the embryonic surface cells, and within 1 h, ferritin was detected in the embryonic circulation. Although Turner (1940a) reported numerous glandular cells on the surface of the pericardial and yolk sacs and speculated that they modified substances transported from the maternal circulation to the follicular fluid, Grove (1985) found no evidence of secretory activity in embryonic surface cells.

Late in gestation, the pericardial and yolk sacs of *H. formosa* embryos start to regress and surface cells on all regions of the embryo except the pericardial and yolk sacs differentiate into typical epidermal cells. As these cells differentiate, they lose their microvilli and develop microplicae. At the same time, surface cells of the pericardial and yolk sacs retain their microvilli. Presumably, maternal–embryonic nutrient transfer continues across the pericardial- and yolk-sac surfaces. Eventually, these surface cells also lose their microvilli and develop microplicae as they differentiate into typical epidermal cells. Tracer studies have indicated that the fully differentiated epidermal cells are nonabsorptive (Grove, 1985).

3. MATERNAL SPECIALIZATIONS

The follicular epithelium and its associated vasculature are believed to play a major role in the maintenance of the embryonic environment and maternal–embryonic metabolite exchange. In the few species that have been examined, a variety of follicular modifications has been described. These include increased vascularization of the follicle wall and changes in the morphology and physiology of the follicle cells. In some cases, an extensive elaboration of the follicular wall occurs.

Maternal specializations for intrafollicular gestation in the clinids are poorly known. The structure of the follicle wall has not been described; consequently it is not known what structural modifications are associated with intrafollicular gestation in this group. However, Veith (1979b) reported that radioactive thymidine and leucine injected into gravid *C. superciliosus* females crossed the follicular epithelium and were absorbed by embryos. He also analyzed the composition of the follicular fluid in this species and found that both lipid and amino acid concentrations were high, but that protein concentration was significantly lower than in the maternal plasma (Veith, 1979b). Veith postulated that although the follicular epithelium was a barrier to plasma polypeptide movement, it actively secreted amino acids into the follicular fluid to maintain the high intrafollicular amino acid concentration. He also speculated that lipids may be an important source of nutrients for developing embryos.

In *Anableps*, the follicle wall is greatly modified. Turner (1938a) reported that in *A. anableps*, cells of the follicular epithelium associated with unfertilized oocytes degenerate after fertilization and are replaced by free cells from the surrounding stroma. A connective-tissue capsule surrounding the follicle becomes greatly thickened and the vascularization of the follicle wall increases. As gestation proceeds, tissues of the follicle wall are further elaborated to form villi with a vascular, connective-tissue core and covered by a syncitial epithelium. Apparently, these villi are more concentrated in the part of the follicle wall that contacts the trophoderm of the embryo. Knight *et al.* (1985) examined the follicle wall of *A. dowi* with scanning electron microscopy and also reported regional differentiation of the follicle (Fig. 11). The portion of the follicle wall directly apposed to the embryonic trophoderm displayed pits that conform to the vascular bulbs on the trophoderm surface. Other regions of the follicle wall were composed of villi. The apical surfaces of cells lining the inside of the follicle varied and exhibited either short, stubby microvilli, patches of poorly developed microplicae or, in some cases, no surface specializations at all.

Ultrastructural and experimental studies on nutrient transfer across the follicle wall in *Anableps* are lacking. It is clear, however, from embryonic dry weight studies, that nutrients must cross the follicular epithelium in these fishes. Elaboration of the follicle wall into folds and villi undoubtedly enhances maternal—embryonic metabolite exchange by increasing the surface area of the follicular epithelium and ensuring close apposition between the follicular epithelium and the embryonic trophoderm. However, the cellular and the physiological specializations associated with nutrient transfer are unknown.

The extent of follicle wall modification varies among poeciliids. In lecithotrophic species, the wall is not responsible for transporting nutrients from the maternal circulation to the embryo and is considered to represent the least specialized state (Turner, 1940a; Wourms, 1981). In the lecithotrophe *Poecilia reticulata,* the postfertilization follicular epithelium is a relatively unspecialized squamous epithelium (Jollie and Jollie, 1964; Dépêche, 1970). Small vesicles, mitochondria, and a certain amount of rough endoplasmic reticulum have been reported in the follicle cells, but the microvilli and other ultrastructural features typical of transporting epithelial cells are lacking. After fertilization, the follicular circulation increases to form a capillary network that directly underlies the follicular epithelium. The lack of obvious cellular features associated with molecular transport is consistent with lecithotrophy in this species. Nevertheless, the squamous follicular epithelium minimizes the distance between the maternal and embryonic circulation and thereby facilitates gas exchange.

The wall of the postfertilization follicle in matrotrophic poeciliids is much more modified. Turner (1940a) reported that the follicle wall of several species of *Poeciliopsis, Aulophallus retropinna,* and *A. elongatus* becomes heavily vascularized and develops elongated vascular villi covered by secretory cells. He regarded these as the most extreme specializations for maternal—embryonic nutrient transfer. The follicle wall of *H. formosa* is less extensively modified, but the follicular epithelium displays several cellular features typical of metabolically active, transporting epithelia (Grove, 1985; Grove and Wourms, 1982, 1983). These include short apical projections and invaginations, numerous mitochondria, and extensive infolding of the basal cell surface with large numbers of endocytotic vesicles at the base of the cell. Numerous electron-lucent and lysosome-like vesicles are also present, but the rough endoplasmic reticulum is conspicuously sparse, indicating limited protein production by the follicle

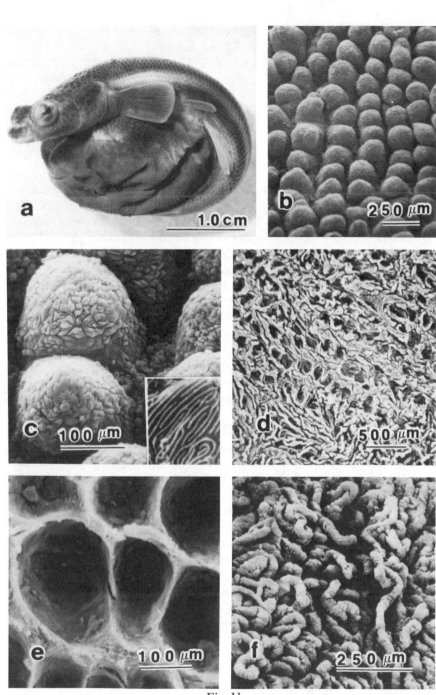

Fig. 11

cells. Capillaries of the follicular circulation are directly adjacent to the basal surface of the follicular epithelium, and their endothelial cells possess numerous transcytotic vesicles.

Experimental studies to elucidate the role of the follicular epithelium in maternal–embryonic nutrient transfer have been limited. Wegmann and Gotting (1971) administered Myofer, an iron dextran tracer, to female *Xiphophorus helleri* by means of intraperitoneal and intramuscular injections. They reported that the tracer was detected with electron microscopy in embryonic tissues both at 20 h and 5 d after injection, indicating macromolecular transport across the follicular epithelium. This is consistent with embryonic dry-weight studies that indicated a moderate level of maternal–embryonic nutrient transfer. Grove (1985) found that ultrastructural tracers injected intraperitoneally into *H. formosa* females were absorbed by follicle cells but were not transported to the intrafollicular compartment. Instead, the tracers accumulated in lysosome-like vesicles. Nevertheless, electrophoretic analysis of follicular fluid and maternal serum revealed that some of the follicular fluid proteins comigrated with the serum proteins. Grove (1985) postulated that the follicular epithelium of *H. formosa* may be a selective barrier to molecular movement that permits only the transport of low-molecular-weight nutrients, such as monosaccharides and amino acids, as well as specific serum macromolecules from the maternal circulation to the follicular fluid. Because the follicle cells absorbed and sequestered macromolecular tracers from the maternal circulation, he also suggested that many maternal serum macromolecules may be absorbed nonselectively by the follicular epi-

---

Fig. 11. (a) *Anableps dowi* 60-mm embryo removed from its follicle. The large prominent pericardial trophoderm bears vascular bulbs on its surface and occupies the same position in which the yolk sac is found in most teleost embryos. (b) Scanning electron micrograph of the vascular bulbs on the surface of the pericardial trophoderm of a 60-mm *A. dowi* embryo. (c) Scanning electron micrograph of several vascular bulbs from the same region of the pericardial trophoderm as in (b). Individual epithelial cells on the surface of the bulbs are delineated by intercellular clefts. *Inset:* Microplicae characterize the surface of the epithelial cells. (d) Scanning electron micrograph of the inner surface of a follicle that contained a 60-mm *A. dowi* embryo. The interior wall of the follicle consists of many villous processes, some of which have been deformed to accommodate vascular bulbs of the pericardial trophoderm. (e) Scanning electron micrograph of the follicle in a region comprised exclusively of pits that interdigitate with the pericardial trophoderm. There is no indication of villar structures. (f) Scanning electron micrograph of the interior lining of a follicle, late in gestation. In this region, the hypertrophied follicular epithelium is comprised entirely of free villi. [From Knight *et al.* (1985).]

thelium, degraded, and the degradation products then transported to the follicular fluid to serve as nutrients.

Although the preceding discussion has concentrated on modification of the follicle wall associated with embryonic nutrition, it is important to note that other physiological requirements have also been associated with intrafollicular gestation. It is clear from work with both lecithotrophic and matrotrophic poeciliids that some of the changes that occur in the follicle wall following fertilization are not directly related to embryonic nutrition. In addition to follicle cells undergoing dramatic changes in cell shape (Jollie and Jollie, 1964), one of the most striking changes in the poeciliid follicular epithelium is the development of junctional complexes between follicle cells immediately after fertilization (Jollie and Jollie, 1964; Grove, 1985). Tight junctions are present in the junctional complexes of the postfertilization follicular epithelium of *Heterandria formosa* (Grove, 1985) and are most likely present in the postfertilization follicle of other poeciliids as well. As a result of the development of these junctional specializations, the follicular epithelium becomes an effective barrier to transepithelial molecular movement, and the contents of the intrafollicular compartment presumably can be regulated independently of the surrounding maternal tissues. This change in the follicular epithelium undoubtedly allows the follicle wall to regulate the osmotic environment for embryonic development, to protect the embryo from immunological rejection, and to transport maternal serum components selectively into the intrafollicular compartment. Little is known about developmental changes in the follicle wall at fertilization in other teleosts with intrafollicular gestation, but because many of these physiological requirements are basic to internal embryonic development, such developmental changes in the follicle wall as the formation of junctional complexes are likely to occur in all of them.

Many questions remain concerning the embryonic and maternal specializations connected with intrafollicular gestion. A more detailed understanding of the embryonic and maternal specializations for maternal–embryonic nutrient transfer in all major taxonomic groups is required. What role does the follicular epithelium play in transporting metabolites to the embryos? What types of molecules are used as nutrients and what is the adaptive significance of utilizing these molecules? What cellular events are involved in the redifferentiation of the follicular epithelium at fertilization, and what controls the onset of follicular epithelium redifferentiation? How has the endocrine control of reproduction been modified in these teleosts? Finally, there are the evolutionary questions concerning the acquisition of embryonic specializations and the phenomenon of follicle redifferentiation.

IV. CONCLUSIONS

This review summarizes the present state of knowledge of maternal–fetal relationships in viviparous fishes. The emphasis has been on the trophic relationship. Several important points emerge. Viviparity made its first appearance in the vertebrate line among fishes. Viviparity has repeatedly and independently evolved from oviparity among species that are widely separated phylogenetically. Nearly all of the adaptations for viviparity that occur in higher vertebrates, including mammals, first appeared in fishes, but fishes also possess unique adaptations. The evolution of viviparity established specialized maternal–fetal relationships, one of the most diverse of which is the trophic relationship. Transfer of maternal nutrients to developing embryos ranges from nil in strictly lecithotrophic species to almost complete nutritional dependence in matrotrophic species. Matrotrophic species have independently evolved many different adaptations to facilitate maternal–embryonic nutrient transfer. These specializations have evolved repeatedly in widely divergent taxonomic groups, giving rise to many examples of convergent adaptation. Trophic specializations often involve remarkable morphological innovations of both maternal and embryonic tissues, but regardless of how unusual these innovations appear, they are ultimately derived from preexisting embryonic and maternal organs and structures. Thus, the acquisition of adaptations for matrotrophy has involved the modification of existing developmental pathways rather than the establishment of radically novel ones. Repeated modification of the same developmental pathway in widely separated taxonomic groups illustrates the conservative nature of the developmental program. There is considerable potential here for the study of the evolution of development. Of particular significance for the comparative study of fish viviparity is the realization that the same constellation of genes that controls the development and differentiation of the uterus, yolk sac, and yolk-sac placenta apparently was able to change in the same or similar ways in several taxa of placental sharks, reptiles, and mammals to form the yolk-sac placenta. Constraints to promote evolutionary/developmental change in one direction, as well as to mitigate against change in others, may well have been put in place with the original set of genes that encoded for the primitive piscine–vertebrate reproductive system. It is clear, though, that the developmental fate of many embryonic structures is remarkably plastic. Consequently, the comparative study of viviparity in fishes has the potential of not only outlining the conservative nature of the developmental program, but also providing insights into what steps in the developmental program are plastic, what constraints are present at different levels of the developmental program, and what evolutionary processes (e.g., natural selection, mutation) are responsible for the changes associated with viviparity.

In 1969, Hoar called attention to the multitude of fascinating problems that await the physiologist, developmental biologist, ecologist, and evolutionary biologist who might become interested in viviparous fishes. During the intervening 17 years, many biologists have responded to his call to action. The study of viviparity in fishes is now undergoing a long-overdue renascence. Research has progressed in two directions, the physiology of development and the ecological and evolutionary considerations of development and life history strategies. The underlying causes for the revival of research interest are both technical and conceptual. First, a variety of new methods, especially in cell biology, have become available and are being used on a widespread basis for experimentation with cells or tissues as well as for the ultrastructural, physiological, and biochemical examination of the small quantities of cells, tissues, and tissue products that characterize so many viviparous embryos and maternal support tissues. Second, there has been the elaboration of new biological concepts and the refinement of old ones. The growth of cell and molecular biology has generated a whole new set of conceptual approaches such as membrane biology and the related areas of physiological transport mechanisms, endocytosis, and intracellular transport of molecules. There has been an increased interest in knowing what sort of molecules are transferred from mother to fetus, such as nutrients, antibodies, and even teratogens. Life history and reproductive strategies have become contemporary fields of interest. Finally, there has been a revival of interest in the evolution of development. Viviparous fishes lend themselves well to studies in these areas. For comparative studies, they supply an almost endless series of adaptations, all derived from a few basic themes. Within some taxa, such as sharks, embiotocids, goodeids, and poeciliids, there occur apparent transitional stages either in the evolution of a specific adaptation or in its subsequent development and differentiation. It is fortunate that some of these same taxa (e.g., the poeciliids, goodeids, and embiotocids) can be readily maintained and will reproduce under laboratory conditions, and hence are suitable for experimental studies.

Future prospects are good. Although much has been done, much more, of course, remains to be done. A successful beginning has been made in formulating problems in the study of fish viviparity in the light of contemporary biology. The functional morphology of trophic adaptations is being worked out at the level of optical microscopic histology, accompanied by an increasing number of ultrastructural studies. Physiological and biochemical studies are becoming more and more numerous. It is important, however, to realize that our basic knowledge of piscine viviparity is quite uneven. While species in some groups such as the goodeids and embiotocids have been subject to experimental study using modern techniques, other groups of viviparous fishes have received only cursory treatment. In some species, it is only possible to speculate on their viviparity! Such uncertainty and unevenness probably will prove exciting to some biologists and annoying to others. Excitement and annoyance, however, have often provided suitable motivation for research.

ACKNOWLEDGMENTS

We are indebted to James W. Atz for critically reading the manuscript and for his many helpful comments during the preparation of this chapter. Research of the authors was supported in part by the following grants or fellowships: Medical Research Council of Canada Postdoctoral Fellowship to B. D. Grove; Sigma Xi Grant-in-aid of Research to J. Lombardi; and Guggenheim Fellowship, National Science Foundation grant PCM-8208525, National Institutes of Health Biomedical Research support grant 2-507-RR07180, and grant NA82AA-D-0057 from the South Carolina Sea Grant Consortium and NOAA National Sea Grant College Program Office, Department of Commerce, to J. P. Wourms.

REFERENCES

Aasen, O. (1966). Brugde, *Cetorhinus maximus* (Gunnerus), 1765. *Saetry. Fiskets Gang* **49**, 909–920.

Alcock, A. (1890). Observations on the gestation of some sharks and rays. *J. Asiat. Soc. Bengal* **59**, 51–56.

Alcock, A. (1895). On a new species of viviparous fish of the family Ophidiidae. *Ann. Mag. Nat. Hist.* [6] **16**, 144–146.

Amend, D. F., and Fender, D. C. (1976). Uptake of bovine serum albumin by rainbow trout from hyperosmotic solutions: A model for vaccinating fish. *Science* **192**, 793–794.

Amoroso, E. C. (1952). Placentation. *In* "Marshall's Physiology of Reproduction" (A. S. Parkes, ed.), Vol. 2, pp. 127–311. Longmans, Green, New York.

Amoroso, E. C. (1960). Viviparity in fishes. *Symp. Zool. Soc. London* **1**, 153–181.

Amoroso, E. C. (1981). Viviparity. *In* "Cellular and Molecular Aspects of Implantation" (S. R. Glasser and D. W. Bullock, eds.), pp. 3–25. Plenum, New York.

Amoroso, E. C., Heap, R. B., and Renfree, M. B. (1979). Hormones and the evolution of viviparity. *In* "Hormones and Evolution" (E. J. W. Barrington, ed.), Vol. 2, pp. 925–989. Academic Press, New York.

Anthony, J., and Millot, J. (1972). Première capture d'une femelle de coelacanthe en état de maturité sexuelle. *C. R. Hebd. Seances Acad. Sci., Ser. D* **224**, 1925–1926.

Anthony, J., and Robineau, D. (1976). Sur quelques caractères juveniles de *Latimeria chalumnae* Smith (Pisces, Crossopterygii, Coelacanthidae). *C. R. Hebd. Seances Acad. Sci., Ser. D* **283**, 1739–1742.

Atz, J. W. (1973). Comments on "Reproductive Endocrinology of Fishes" by Edward M. Donaldson. *Am. Zool.* **13**, 929–931.

Babel, T. S. (1967). Reproduction, life history, and ecology of the stingray, *Urolophus halleri* Cooper. *Calif. Fish Game Bull.* **137**, 1–104.

Bailey, R. J. (1933). The ovarian cycle in the viviparous teleost, *Xiphophorus helleri*. *Biol. Bull. (Woods Hole, Mass.)* **64**, 206–225.

Ball, J. N. (1962). Brood production after hypophysectomy in the viviparous teleost *Mollinesia latipinna* Le Sueur. *Nature (London)* **194**, 787.

Ballinger, R. E. (1978). Variation in evolution of clutch and litter size. *In* "The Vertebrate Ovary" (R. E. Jones, ed.), pp. 789–825. Plenum, New York.

Balon, E. K. (1975). Reproductive guilds of fishes: A proposal and definition. *J. Fish. Res. Board Can.* **32**, 821–864.

Balon, E. K. (1981). Additions and amendments to the classification of reproductive styles in fishes. *Environ. Biol. Fishes* **6**, 377–389.

Balon, E. K. (1984). Patterns in the evolution of reproductive styles in fishes. *In* "Fish Reproduction: Strategies and Tactics" (G. W. Potts and R. J. Wooton, eds.), pp. 35–53. Academic Press, London.

Balon, E. K. (1986). Types of feeding in the ontogeny of fishes and the life-history model. *Environ. Biol. Fishes* **16**, 11–24.

Baranes, A., and Wendling, J. (1981). The early stages of development of *Carcharhinus plumbeus*. *J. Fish Biol.* **18**, 159–175.

Barlow, G. W. (1981). Patterns of parental investment, dispersal and size among coral-reef fishes. *Environ. Biol. Fishes* **6**, 65–85.

Bass, A. J., D'Aubrey, J. D., and Kistnasamy, N. (1975). Sharks of the east coast of southern Africa. II. The families Scyliorhinidae and Pseudotriakidae. *Invest. Rep. Oceanogr. Res. Inst.* **37**, 1–63.

Baughman, J. L. (1955). The oviparity of the whale shark *Rhineodon typus* with records of this and other fishes in Texas waters. *Copeia*, pp. 54–55.

Baylis, J. R. (1981). The evolution of parental care in fishes, with reference to Darwin's rule of male sexual selection. *Environ. Biol. Fishes* **6**, 223–251.

Bertin, L. (1958). Viviparité des Téléostéens. *In* "Traité de Zoologie" (P.-P. Grassé, ed.), Vol. 13, Part 2, pp. 1791–1812. Masson, Paris.

Billingham, R. E., and Beer, A. E. (1985). Reproductive immunology: Past, present, and future. *Perspect. Biol. Med.* **27**, 259–275.

Blackburn, D. G. (1981). An evolutionary analysis of vertebrate viviparity. *Am. Zool.* **21**, 936.

Blackburn, D. G. (1982a). Evolutionary origins of viviparity in reptiles. I. Sauria. *Amphibia-Reptilia* **3**, 185–205.

Blackburn, D. G. (1982b). Classification of vertebrate reproductive modes. *Am. Zool.* **22**, 910.

Blackburn, D. G., Evans, H. E., and Vitt, L. J. (1985). The evolution of fetal nutritional adaptations. *Fortschr. Zool.* **30**, 437–439.

Boehlert, G. W., and Yoklavich, M. M. (1984). Reproduction, embryonic energetics, and the maternal–fetal relationship in the viviparous genus *Sebastes*. *Biol. Bull. (Woods Hole, Mass.)* **167**, 354–370.

Boehlert, G. W., Kusakari, M., Shimizu, M., and Yamada, M. (1986). Energetics during embryonic development in kurosoi, *Sebastes schlegeli*. *J. Exp. Mar. Biol. Ecol.* **101**, 239–256.

Boulekbache, H. (1981). Energy metabolism in fish development. *Am. Zool.* **21**, 377–389.

Brambell, F. W. R. (1970). "The Transmission of Passive Immunity from Mother to Young." North-Holland Publ., Amsterdam.

Breder, C. M., and Rosen, D. E. (1966). "Modes of Reproduction in Fishes." Natural History Press, Garden City, New York.

Bretschneider, L. H., and Duyvené de Wit, J. J. (1947). "Sexual Endocrinology of Nonmammalian Vertebrates." Elsevier, Amsterdam.

Browning, H. C. (1973). The evolutionary history of the corpus luteum. *Biol. Reprod.* **8**, 128–157.

Budker, P. (1953). Sur le cordon ombilical des squales vivipares. *Bull. Mus. Hist. Nat.* [2] **25**, 541–545.

Budker, P. (1958). La viviparité chez les sélaciens. *In* "Traité de Zoologie" (P.-P. Grassé, ed.), Vol. 13, Part 2, pp. 1755–1790. Masson, Paris.

Cadenat, J. (1959). Notes d'ichtyologie ouest-africaine. 20. *Galeus polli espèce* nouvelle ovovivipare de Scylliorhinidae. *Bull. Inst. Fondam. Afr. Noire, Ser. A.* **21**, 395–409.

Calzoni, M. (1936). Ricerche sulla placenta del *Carcharias glaucus. Pubbl. Stan. Zool. Napoli* **15**, 169–174.

Chambolle, P. (1964). Influence de l'hypophysectomie sur la gestation de *Gambusia* sp. (Poisson Téléostéen). *C. R. Hebd. Seances Acad. Sci., Ser. D* **259**, 3855–3857.

Chambolle, P. (1966). Recherches sur l'allongement de la durée de survie apres hypophysectomie chez *Gambusia* sp. *C. R. Hebd. Seances Acad. Sci., Ser. D* **262**, 1750–1753.

Chambolle, P. (1967). Influence de l'injection d' A.C.T.H. sur la survive de *Gambusia* sp. (Poisson Téléostéen) privé d'hypophyse. *C. R. Hebd. Seances Acad. Sci., Ser. D* **264**, 1464–1466.

Chambolle, P. (1973). Recherches sur les facteurs physiologiques de la reproduction chez les poissons "ovovivipares" analyses expérimentale sur *Gambusia* sp. *Bull. Biol. Fr. Belg.* **107**, 27–101.

Cheong, R. T., Henrich, S., Farr, J. A., and Travis, J. (1984). Variation in fecundity and its relationship to body size in a population of the least killifish, *Heterandria formosa* (Pisces: Poeciliidae). *Copeia*, pp. 720–726.

Chernyayev, Zh. A. (1974). Morphological and ecological features of the "Big Golomyanka" or Baikal oil-fish (*Comephorus baicalensis*). *J. Ichthyol. (Engl. Transl.)* **14**, 856–868.

Chieffi, G., and Bern, H. A. (1966). Report on international roundtable on steroid hormones in fishes at the Stazione Zoologica of Naples. *Gen. Comp. Endocrinol.* **7**, 203–204.

Cohen, D. M., and Wourms, J. P. (1976). *Microbrotula randalli*, a new viviparous ophidioid fish from Samoa and New Hebrides, whose embryos bear trophotaeniae. *Proc. Biol. Soc. Wash.* **89**, 81–98.

Compagno, L. J. V. (1973). Interrelationships of living elasmobranchs. *In* "Interrelationships of Fishes" (P. H. Greenwood, R. S. Miles, and C. Patterson, eds.), *Zool. J. Linnean Soc.* **53**, 15–61, Suppl. 1. Academic Press, New York.

Compagno, L. J. V. (1984). "Sharks of the World," FAO Fish. Synop. No. 125, Vol. 4, Parts 1 and 2. FAO, Rome.

Constantz, G. D. (1980). Energetics of viviparity in the gila topminnow (Pisces: Poeciliidae). *Copeia* **4**, 876–878.

Cornish, D. A. (1985). An autoradiographic study of the transfer of nutrients to the embryos of the teleost *Clinus dorsalis. S. Afr. J. Sci.* **81**, 393–394.

Cornish, D. A., and Veith, W. J. (1986). Embryonic adaptations and nutrition in the viviparous teleost *Clinus dorsalis* (Perciformes: Clinidae). *S. Afr. J. Zool.* **21**, 79–84.

Dépêche, J. (1970). L'infrastructure du sac vitelline du capuchon pericardique de l'embryon de *Poecilia reticulata* (Téléostéen vivipare). *C. R. Hebd. Seances Acad. Sci.* **270**, 1352–1355.

Dépêche, J. (1973). Infrastructure superficelle de la vesicule vitellaire et du sac pericardique de l'embryon de *Poecilia reticulata* (Poisson Téléostéen). *Z. Zellforsch. Mikrosk. Anat.* **141**, 235–253.

Dépêche, J. (1976). Acquisition et limites de l'autonomie trophique embryonnaire au cours du développement de poisson téléostéen vivipare *Poecilia reticulata. Bull. Biol. Fr. Belg.* **110**, 45–97.

de Vlaming, V., Baltz, D., Anderson, S., Fitzgerald, R., Delahunty, G., and Barkley, M. (1983). Aspects of embryo nutrition and excretion among viviparous embiotocid teleosts: Potential endocrine involvements. *Comp. Biochem. Physiol. A* **76A**, 189–198.

Devys, M. M., Thierry, A., Barbier, M., and Janot, M. M. (1972). Premières observations sur les lipides de l'ovocyte du Coelacanthe (*Latimeria chalumnae*). *C. R. Hebd. Seances Acad. Sci., Ser. D* **275**, 2085–2087.

Dobbs, G. H. (1974). Scanning electron microscopy of intraovarian embryos of the viviparous teleost *Micrometrus minimus* (Gibbons). (Perciformes: Embiotocidae). *J. Fish Biol.* **7**, 209–214.

Dodd, J. M. (1960). Gonadal and gonadotrophic hormones in lower vertebrates. *In* "Marshall's Physiology of Reproduction" (A. S. Parkes, ed.), 3rd ed., Vol. 1, Part 2, pp. 417–582. Longmans, Green, New York.

Dodd, J. M. (1975). The hormones of sex and reproduction and their effects in fish and lower chordates: Twenty years on. *Am. Zool.* **15**, Suppl. 1, 137–171.

Dodd, J. M. (1977). The structure of the ovary in non-mammalian vertebrates. *In* "The Ovary" (S. Zuckerman and B. J. Weir, eds.), 2nd ed., Vol. 1, pp. 219–263. Academic Press, New York.

Dodd, J. M. (1983). Reproduction in cartilaginous fishes. *In* "Fish Physiology" (W. S. Hoar, D. J. Randall, and E. M. Donaldson, eds.), Vol. 9A, pp. 31–95. Academic Press, New York.

Dodd, J. M., and Dodd, M. H. I. (1986). Evolutionary aspects of reproduction in cyclostomes and cartilaginous fishes. *In* "Evolutionary Biology of Primitive Fishes" (R. E. Foreman, A. Gorbman, J. M. Dodd, and R. Olsson, eds.), pp. 295–319. Plenum, New York.

Dodd, J. M., Dodd, M. H. I., and Dugan, R. T. (1983). Control of reproduction in elasmobranch fishes. *In* "Control Processes in Fish Physiology" (J. C. Rankin, T. J. Pitcher, and R. T. Dugan, eds.), pp. 221–287. Croom Helm, London.

Donaldson, E. M. (1973). Reproductive endocrinology of fishes. *Am. Zool.* **13**, 909–927.

Downhower, J. F., and Brown, L. (1975). Superfoetation in fishes and the cost of reproduction. *Nature (London)* **256**, 345.

Edidin, M. (1976). Cell surface antigens in mammalian development. *In* "The Cell Surface in Animal Embryogenesis and Development" (G. Poste and G. L. Nicolson, eds.), Vol. 1, pp. 127–143. Elsevier/North-Holland, Amsterdam.

Eigenmann, C. H. (1892). *Cymatogaster aggregatus* Gibbons. A contribution to the ontogeny of viviparous fishes. *Fish. Bull.* **12**, 401–478.

Engen, P. C. (1968). Organogenesis in the wallege surfperch, *Hyperprosopon argentium* (Gibbons). *Calif. Fish Game* **54**, 156–169.

Evans, D. H. (1981). The egg case of the oviparous elasmobranch *Raja erinacea* does osmoregulate. *J. Exp. Biol.* **92**, 337–340.

Evans, D. H., and Mansberger, L. (1979). Further studies on the osmoregulation of premature "pups" of *Squalus acanthias. Bull. Mt. Desert Isl. Mar. Lab.* **19**, 101–103.

Evans, D. H., and Oikari, A. (1980). Osmotic and ionic relationships between embryonic and uterine fluids during gestation of *Squalus acanthias. Bull. Mt. Desert Isl. Biol. Lab.* **20**, 53–55.

Fawcett, D. W. (1986). "Bloom and Fawcett: Textbook of Histology," 11th ed. Saunders, Philadelphia, Pennsylvania.

Forster, G. R., Badcock, J. R., Longbottom, M. R., Merrett, N. R., and Thomson, K. S. (1970). Results of the Royal Society Indian Ocean Deep Slope Fishing Expedition, 1969. *Proc. Soc. London, Ser. B* **175**, 367–404.

Foulley, M. M., and Mellinger, J. (1980). La diffusion de l'eau tritiée de l'urée-^{14}C et d'autres substances à travers la coque de l'oeuf de roussette, *Scyliorhinus canicula. C. R. Hebd. Seances Acad. Sci., Ser. D* **290**, 427–430.

Foulley, M. M., Wrisez, F., and Mellinger, J. (1981). Observations sur la permeabilité asymmetrique de la coque de l'oeuf, roussette (*Scyliorhinus canicula*). *C. R. Seances Acad. Sci., Ser. 3* **293**, 389–394.

Fraser, E. A., and Renton, R. M. (1940). Observations on the breeding and development of the viviparous fish, *Heterandria formosa. Q. J. Microsc. Sci.* [N.S.] **81**, 479–520.

Fujita, K. (1981). Oviphagous embryos of the pseudocarchariid shark, *Pseudocarcharias kamoharai* from the central Pacific. *Jpn. J. Ichthyol.* **28**, 37–44.

Gardiner, D. M. (1978). Cyclic changes in fine structure of the epithelium lining the ovary of the viviparous teleost, *Cymatogaster aggregata* (Perciformes: Embiotocidae). *J. Morphol.* **156**, 367–380.

George, A., and Springer, V. G. (1980). Revision of the clinid tribe Ophiclinini including five new species, and a definition of the family Clinidae. *Smithson. Contrib. Zool.* **252**, 1–31.

Gilbert, P. W. (1981). Patterns of shark reproduction. *Oceanus* **24**(4), 30–39.

Gilbert, P. W., and Schlernitzauer, D. A. (1966). The placenta and gravid uterus of *Carcharhinus falciformis. Copeia*, pp. 451–457.

Gilchrist, J. D. F. (1905). The development of South African fishes. Part 2. *Mar. Invest. S. Afr.* **3**, 131–167.

Gilmore, R. G. (1983). Observations on the embryos of the longfin mako, *Isurus paucus* and the bigeye thresher, *Alopias superciliosus Copeia* **2**, 375–382.

Gilmore, R. G., Dodrill, J. W., and Linley, P. A. (1983). Reproduction and embryonic development of the sand tiger shark, *Odontaspis taurus* (Rafinesque). *Fish. Bull.* **81**, 201–225.

Goodrich, E. S. (1930). "Studies on the Structure and Development of Vertebrates." Macmillan, London.

Govani, J. J., Boehlert, G. W., and Watanabe, Y. (1986). The physiology of digestion in fish larvae. *Environ. Biol. Fishes* **16**, 59–77.

Gowan, L. K., Reinig, J. W., Schwabe, C., Bedarkar, S., and Blundell, T. L. (1981). On the primary and tertiary structure of relaxin from the sand tiger shark (*Odontaspis taurus*). *FEBS Lett.* **129**, 80–82.

Graham, C. R. (1967). Nutrient transfer from mother to fetus and placental formation. Ph.D. Dissertation, University of Delaware, Newark.

Gray, J. (1928). The growth rate of fish. II. The growth rate of the embryo of *Salmo fario. J. Exp. Biol.* **6**, 110–124.

Gross, M. R., and Sargent, R. C. (1985). The evolution of male and female parental care in fishes. *Am. Zool.* **25**, 807–822.

Gross, M. R., and Shine, R. (1981). Parental care and mode of fertilization in ectothermic vertebrates. *Evolution (Lawrence, Kans.)* **35**, 775–793.

Grosse-Wichtrup, L., and Greven, H. (1985). Uptake of ferritin by the trophotaeniae of the goodeid fish, *Ameca splendens* Miller and Fitzsimons 1971 (Teleoste, Cyprino-dontiformes). *Cytobios*, **42**, 33–40.

Grosse-Wichtrup, L., and Greven, H. (1986). Ultrastructural aspects of the trophotaenial epithelium in *Ameca splendens* Miller and Fitzsimons 1971 (Teleostei, Cyprino-dontiformes) revealed by electron dense stains. *Cytobios* **45**, 175–183.

Grove, B. D. (1985). "The structure, function, and development of the follicular pla-centa in the viviparous fish, *Heterandria formosa*." Ph.D. Dissertation, Clemson University, Clemson, South Carolina.

Grove, B. D., and Wourms, J. P. (1981). Maternal-foetal relationship in the poeciliid *Heterandria formosa*. *Am. Zool.* **21**, 965.

Grove, B. D., and Wourms, J. P. (1982). Embryonic nutrient absorption in the poeciliid *Heterandria formosa*. *Am. Zool.* **22**, 881.

Grove, B. D., and Wourms, J. P. (1983). Endocytosis of molecular tracers by embryos of the viviparous fish, *Heterandria formosa*. *J. Cell Biol.* **97**, 100a.

Gruber, S. H., and Compagno, L. J. V. (1981). Taxonomic status and biology of the bigeye thresher *Alopias superciliosus*. *Fish. Bull.* **79**, 617–640.

Gubanov, Y. P. (1972). On the biology of the thresher shark *Alopias vulpinus* (Bona-terre) in the Northwest Indian Ocean. *J. Ichthyol. (Engl. Transl.)* **12**, 591–600.

Gudger, E. W. (1912). Natural history notes on some Beaufort, N.C. fishes. 1910–1911. No. 1. Elasmobranchii—with special reference to utero-gestation. *Proc. Biol. Soc. Wash.* **25**, 141–156.

Gudger, E. W. (1940). The breeding habits, reproductive organs, and external embry-onic development of *Chlamydoselachus* based on notes and drawings left by Bashford Dean. *In* "Bashford Dean Memorial Volume—Archiac Fishes" (E. W. Gudger, ed.), Artic. 7, pp. 521–646. Am. Mus. Nat. Hist., New York.

Gulidov, M. V. (1963). Respiratory organs of the embryos of viviparous fishes. *Vopr. Ikhtiol.* **3**, 21–31 (in Russian).

Hamlett, W. C. (1986). Prenatal nutrient absorptive structures in selachians. *In* "Indo-Pacific Fish Biology: Proceedings of the Second International Conference on Indo-Pacific Fishes" (T. Uyeno, R. Arai, T. Taniuchi, and K. Matsuura, eds.), pp. 333–344. Ichthyol. Soc. Jpn., Tokyo.

Hamlett, W. C., and Wourms, J. P. (1984). Ultrastructure of the preimplantation yolk sac of placental sharks. *Tissue Cell* **16**, 613–625.

Hamlett, W. C., Wourms, J. P., and Hudson, J. (1985a). Ultrastructure of the full term shark yolk sac placenta. I. Morphology and cellular transport at the fetal attachment site. *J. Ultrastruct. Res.* **91**, 192–206.

Hamlett, W. C., Wourms, J. P., and Hudson, J. (1985b). Ultrastructure of the full term shark yolk sac placenta. II. The smooth proximal segment. *J. Ultrastruct. Res.* **91**, 207–220.

Hamlett, W. C., Wourms, J. P., and Hudson, J. (1985c). Ultrastructure of the full term shark yolk sac placenta. III. The maternal attachment site. *J. Ultrastruct. Res.* **91**, 221–231.

Hamlett, W. C., Allen, D. J., Stribling, M. D., Schwartz, F. J., and Didio, L, J. A. (1985d). Permeability of external gill filaments in the embryonic shark. Electron micro-scopic observations using horseradish peroxidase as a macromolecular tracer. *J. Submicrosc. Cytol.* **17**, 31–40.

Hamlett, W. C., Wourms, J. P., and Smith, J. W. (1985e). Sting-ray placental analogue: Structure of the trophonemata in *Rhinoptera bonasus*. *J. Submicrosc. Cytol.* **17**, 541–550.

Heller, H. (1972). The effect of neurohypophyseal hormones on the female reproductive tract of lower vertebrates. *Gen. Comp. Endocrinol., Suppl.* **3**, 703–714.

Hisaw, R. L., and Abramowitz, A. A. (1938). The physiology of reproduction in the dogfish, *Mustelus canis*. *Rep. Woods Hole Oceanogr. Inst., 1937*, pp. 21–22.

Hisaw, R. L., and Abramowitz, A. A. (1939). Physiology of reproduction in the dogfishes, *Mustelus canis* and *Squalus acanthias*. *Rep. Woods Hole Oceanogr. Inst., 1938*, p. 22.

Hjorth, J. P. (1974). Genetics of *Zoarces* populations. VII. Fetal and adult hemoglobins and a polymorphism common to both. *Hereditas* **78**, 69–72.

Hoar, W. S. (1969). Reproduction. *In* "Fish Physiology" (W. S. Hoar and J. J. Randall, eds.), Vol. 3, pp. 1–72. Academic Press, New York.

Hogarth, P. J. (1968). Immunological aspects of foetal-maternal relations in lower vertebrates. *J. Reprod. Fertil., Suppl.* **3**, 15–27.

Hogarth, P. J. (1972a). Immune relations between mother and foetus in the viviparous fish, *Xiphophorus helleri* Haeckel. I. Antigenicity of the foetus. *J. Fish Biol.* **4**, 265–269.

Hogarth, P. J. (1972b). Immune relation between mother and foetus in the viviparous fish, *Xiphophorus helleri* Haeckel. II. Lack of status of the ovary as a favorable site for allograft survival. *J. Fish Biol.* **4**, 271–275.

Hogarth, P. J. (1972c). Antigenicity of *Poecilia* sperm. *Experientia* **28**, 463–464.

Hogarth, P. J. (1973). Immune relations between mother and foetus in the viviparous fish, *Xiphophorus helleri* Haeckel. III. Survival of embryos after ectopic transplantation. *J. Fish Biol.* **5**, 109–113.

Hogarth, P. J. (1976). "Viviparity." Arnold, London.

Hollenberg, F., and Wourms, J. P. (1985). Comparative ultrastructure and protein uptake of trophotaenial cells of two goodeid fishes. *Am. Zool.* **25**, 95A.

Hollenberg, F., and Wourms, J. P. (1986). Relationship of embryonic hindgut epithelium to trophotaenial absorptive cells in two goodeid fishes. *Am. Zool.* **26**, 12A.

Hornsey, D. J. (1978). Permeability coefficients of the egg-case membrane of *Scyliorhinus canicula* L. *Experientia* **34**, 1596.

Hubbs, C. L., and Turner, C. L. (1939). Studies of the fishes of the order Cyprinodontes. XVI. A revision of the Goodeidae. *Misc. Publ. Mus. Zool., Univ. Mich.* **42**, 1–80.

Hughes, G. M. (1984). General anatomy of the gills. *In* "Fish Physiology" (W. S. Hoar and D. J. Randall, eds.), Vol. 10A, pp. 1–72. Academic Press, Orlando, Florida.

Hunt, S. (1985). The selachian egg case collagen. *In* "Biology of Invertebrate and Lower Vertebrate Collagens" (A. Bairati and R. Garrone, eds.), pp. 409–450. Plenum, New York.

Igarashi, T. (1962). Morphological changes of the embryo of a viviparous teleost, *Neoditrema ransonneti* Steindachner during gestation. *Bull. Fac. Fish., Hokkaido Univ.* **13**, 47–52.

Iida, H., and Yamamoto, T. (1985). Intracellular transport of horseradish peroxidase, in the absorptive cells of goldfish hindgut in vitro, with special reference to the cytoplasmic tubules. *Cell Tissue Res.* **240**, 553–560.

Ingermann, R. L., and Terwilliger, R. C. (1981a). Oxygen affinities of maternal and fetal hemoglobins of the viviparous seaperch, *Embiotoca lateralis*. *J. Comp. Physiol.* **142**, 523–531.

Ingermann, R. L., and Terwilliger, R. C. (1981b). Intraerythrocyte organic phosphates of fetal and adult seaperch (*Embiotoca lateralis*): Their role in maternal-fetal oxygen transport. *J. Comp. Physiol.* **144**, 253–259.

Ingermann, R. L., and Terwilliger, R. C. (1982). Blood parameters and facilitation of maternal-fetal oxygen transfer in a viviparous fish (*Embiotoca lateralis*). *Comp. Biochem. Physiol. A* **73A**, 497–501.

Ingermann, R. L., Terwilliger, R. C., and Roberts, M. S. (1984). Foetal and adult blood oxygen affinities of the viviparous seaperch, *Embiotoca lateralis. J. Exp. Biol.* **108**, 453–457.

Ishii, S. (1960). Effects of estrone and progesterone on the ovary of the viviparous teleost, *Neoditrema ransonneti*, during gestation. *J. Fac. Sci., Univ. Tokyo, Sect. 4* **9**, 101–109.

Ishii, S. (1961). Effects of some hormones on the gestation of the top minnow. *J. Fac. Sci., Univ. Tokyo, Sect. 4* **9**, 279–290.

Ishii, S. (1963). Some factors involved in the delivery of the young in the top-minnow, *Gambusia affinis. J. Fac. Sci., Univ. Tokyo, Sect. 4* **10**, 181–187.

Jacob, F. (1977). Evolution and tinkering. *Science* **196**, 1161–1166.

Jacobs, J. (1971). "Livebearing Aquarium Fishes" (G. Vevers, trans.). Macmillan, New York.

Janincki, A. (1966). Zyworodnosc u ryb kostnoszkieletowych (Viviparity in teleostean fishes). *Przegl. Zool.* **10**, 272–285.

Jollie, W. P., and Jollie, L. G. (1964). The fine structure of the ovarian follicle of the ovoviviparous poeciliid fish, *Lebistes reticulatus*. II. Formation of follicular pseudoplacenta. *J. Morphol.* **114**, 503–526.

Jollie, W. P., and Jollie, L. G. (1967a). Electron microscopic observations on the yolk sac of the spiny dogfish, *Squalus acanthias. J. Ultrastruct. Res.* **18**, 102–126.

Jollie, W. P., and Jollie, L. G. (1967b). Electron microscopic observations on accommodations to pregnancy in the uterus of the spiny dogfish, *Squalus acanthias. J. Ultrastruct. Res.* **20**, 161–178.

Knight, F. M., Lombardi, J., Wourms, J. P., and Burns, J. R. (1985). Follicular placenta and embryonic growth of the viviparous four-eyed fish (*Anableps*). *J. Morphol.* **185**, 131–142.

Koob, T. J., Laffan, J. J., and Callard, I. P. (1984). Effects of relaxin and insulin on reproductive tract size and early fetal loss in *Squalus acanthias. Biol. Reprod.* **31**, 231–238.

Korsgaard, B. (1983). The chemical composition of follicular and ovarian fluids of the pregnant blenny *Zoarces viviparus* (L.). *Can. J. Zool.* **61**, 1101–1108.

Korsgaard, B. (1986). Trophic adaptations during early intraovarian development of embryos of *Zoarces viviparus* (L.). *J. Exp. Mar. Biol. Ecol.* **98**, 141–152.

Korsgaard, B., and Andersen, F. O. (1985). Embryonic nutrition, growth and energetics in *Zoarces viviparus* (L.) as indication of a maternal–fetal trophic relationship. *J. Comp. Physiol. B* **155B**, 437–444.

Korsgaard, B., and Petersen, I. (1979). Vitellogenin, lipid, and carbohydrate metabolism during vitellogenesis and pregnancy, and after hormonal induction in the blenny *Zoarces viviparus* (L.). *Comp. Biochem. Physiol. B* **63B**, 245–251.

Kristofferson, R., Broberg, S., and Pekkarinen, M. (1973). Histology and physiology of embryotrophe formation, embryonic nutrition and growth in the eel pout, *Zoarces viviparus* (L.). *Ann. Zool. Fenn.* **10**, 467–477.

Kryvi, H. (1976). The structure of the embryonic external gill filaments of the velvet belly shark, *Etmopterus spinax. J. Zool.* **180**, 252–261.

Kudo, S. (1959). Studies on the sexual maturation of female and on the embryos of the Japanese dogfish *Halaelurus buergeri* (Miller et Henle). *Rep. Nankai Reg. Fish. Res. Lab.* **11**, 41–46 (in Japanese).

Kujala, G. A. (1978). Corticosteroid and neurohypophysial hormone control of parturition in the guppy, *Poecilia reticulata. Gen. Comp. Endocrinol.* **36**, 286–296.

Kunz, Y. W. (1971). Histological study of the greatly enlarged pericardial sac in the embryo of the viviparous teleost, *Lebistes reticulatus*. *Rev. Suisse Zool.* **78**, 187–207.

Lane, H. H. (1909). On the ovary and ova in *Lucifuga* and *Stygicola*. *Carnegie Inst. Washington. Publ.* **104**, 226–231.

Laurent, P. (1984). Gill internal morphology. *In* "Fish Physiology" (W. S. Hoar and D. J. Randall, eds.), Vol. 10A, pp. 73–183. Academic Press, Orlando, Florida.

Liley, N. R. (1969). Hormones and reproductive behavior in fishes. *In* "Fish Physiology" (W. S. Hoar and D. J. Randall, eds.), Vol. 3, pp. 73–116. Academic Press, New York.

Lombardi, J. (1983). Structure, function, and evolution of trophotaeniae: Placental ana logues of viviparous fish embryos. Ph.D. Dissertation, Clemson University, Clemson, South Carolina.

Lombardi, J., and Wourms, J. P. (1979). Structure, function, and evolution of trophotaeniae: Placental analogues of viviparous fishes. *Am. Zool.* **19**, 976.

Lombardi, J., and Wourms, J. P. (1985a). The trophotaenial placenta of a viviparous goodeid fish. I. Ultrastructure of the internal ovarian epithelium, the maternal component. *J. Morphol.* **184**, 271–292.

Lombardi, J., and Wourms, J. P. (1985b). The trophotaenial placenta of a viviparous goodeid fish. II. Ultrastructure of trophotaeniae, the embryonic component. *J. Morphol.* **184**, 293–309.

Lombardi, J., and Wourms, J. P. (1985c). The trophotaenial placenta of a viviparous goodeid fish. III. Protein uptake by trophotaeniae, the embryonic component. *J. Exp. Zool.* **236**, 165–179.

Lüling, K. H. (1951). Zur intraovarialen Entwicklung und Embryologie des Rotbarsches (*Sebastes marinus* L.). *Zool. Jahrb., Abt. Anat. Ontog. Tiere* **71**, 145–288.

Lund, R. (1980). Viviparity and intrauterine feeding in a new holocephalan fish from the Lower Carboniferous of Montana. *Science* **209**, 697–699.

McMenamin, J. W. (1979). Functional morphology of sheathed blood vessels in the ovarian curtains of embiotocid fishes. *Am. Zool.* **19**, 947.

Magnusson, J. (1955). Mikroskopisch-anatomische Untersuchugen zur Fortpflanzungs-biologie des Rotbarsches (*Sebastes marinus* Linné). *Z. Zellforsch. Mikrosk. Anat.* **43**, 121–167.

Mahadevan, G. (1940). Preliminary observations on the structure of the uterus and the placenta of a few Indian elasmobranchs. *Proc.—Indian Acad. Sci., Sect. B* **11**, 2–38.

Manwell, C. (1958). A "fetal–maternal shift" in the ovoviviparous spiny dogfish *Squalus suckleyi* (Girard). *Physiol. Zool.* **31**, 93–100.

Manwell, C. (1963). Fetal and adult hemoglobins of the spiny dogfish *Squalus suckleyi*. *Arch. Biochem. Biophys.* **101**, 504–511.

Matthews, L. H. (1950). Reproduction of the basking shark, *Cetorhinus maximus* (Gunner). *Philos. Trans. R. Soc. London, Ser. B* **234**, 247–316.

Matthews, L. H. (1955). The evolution of viviparity in vertebrates. *Mem. Soc. Endocrinol.* **4**, 129–148.

Medawar, P. B. (1953). Some immunological and endocrinological problems raised by the evolution of viviparity in vertebrates. *Symp. Soc. Exp. Biol.* **7**, 320–338.

Mendoza, G. (1937). Structural and vascular changes accompanying the resorption of the proctodaeal processes after birth in the embryos of the Goodeidae. *J. Morphol.* **61**, 95–125.

Mendoza, G. (1940). The reproductive cycle of the viviparous teleost, *Neotoca bilineata*, a member of the family Goodeidae. II. The cyclic changes in the ovarian soma during gestation. *Biol. Bull. (Woods Hole, Mass.)* **78**, 349–365.

Mendoza, G. (1943). The reproductive cycle of the viviparous teleost, *Neotoca bilineata*, a member of the family Goodeidae. IV. The germinal tissue. *Biol. Bull* (*Woods Hole, Mass.*) **84**, 87–97.

Mendoza, G. (1956). Adaptations during gestation in the viviparous cyprinodont teleost, *Hubbsina turneri. J. Morphol.* **99**, 73–96.

Mendoza, G. (1958). The fin fold of *Goodea luitpoldii*, a viviparous cyprinodont teleost. *J. Morphol.* **103**, 539–560.

Mendoza, G. (1962). The reproductive cycles of three viviparous teleosts, *Alloophorus robustus, Goodea luitpoldii*, and *Neoophorus diazi. Biol. Bull.* (*Woods Hole, Mass.*) **123**, 351–365.

Mendoza, G. (1972). The fine structure of an absorptive epithelium in a viviparous teleost. *J. Morphol.* **136**, 109–115.

Meyer-Rochow, V. B. (1970). *Cataetyx memoriabilis* n. sp., ein neuer Tiefsee-Ophidiidae aus dem sudostlichen Atlantik. *Abh. Verh. Naturwiss. Ver. Hamburg* [N.S.] **14**, 37–53.

Miller, R. K., Koszala, T. R., and Brent, R. L. (1976). The transport of molecules across placental membranes. *In* "The Cell Surface in Animal Embryogenesis and Development" (G. Poste and G. L. Nicolson, eds.), Vol. 1, pp. 145–224. Elsevier/North-Holland, Amsterdam.

Millot, T., Anthony, J., and Robineau, D. (1978). "Anatomie de *Latimeria chalumnae*," Vol 3. CNRS, Paris.

Mock, D. W. (1984). Siblicidal aggression and resource monopolization in birds. *Science* **225**, 731–733.

Mohr, E. (1936). Hemirhamphiden-Studien. IV. Die Gattung *Dermogenys* van Hasselt, V. Die Gattung *Nomorhamphus* Weber and de Beaufort. VI. Die Gattung *Hemirhamphodon* Bleeker. *Mitt. Zool. Mus. Berl.* **21**, 34–64.

Mohsen, T. (1962). Un nouveau genre d' hemirhamphides: *Grecarchopterus* nov. gen., basé sur des caractères particuliers du système urogenital. *Bull. Aquat. Biol.* **3**, 109–120.

Moser, H. G. (1967a). Reproduction and development of *Sebastodes paucispinis* and comparison with other rockfishes off California. *Copeia*, pp. 773–779.

Moser, H. G. (1967b). Seasonal histological changes in the gonads of *Sebastodes paucispinis* Ayers, an ovoviviparous teleost (Family Scorpaenidae). *J. Morphol.* **123**, 329–353.

Moser, H. G. (1976). Viviparous reproduction in the South African kelpfish, *Pavoclinus* (*Fucominus*) *mus* (Gilchrist and Thompson, 1908) and comparative comments on reproduction in other members of the Clinidae. *Proc. Annu. Meet. Am. Soc. Ichthyol. Herpetol.*, 1976, p. 56.

Moser, H. G. (1981). Morphological and functional aspects of marine fish larvae. *In* "Marine Fish Larvae: Morphology, Ecology, and Relation to Fisheries" (R. Lasker, ed.), pp. 89–131. Univ. of Washington Press, Seattle.

Moskal'kova, K. I. (1985). "Coprophagy" in embryos of the bony fish *Neogobius melanostomus* (Pallus) (Pisces, Gobiidae). *Dokl. Biol. Sci.* (*Engl. Transl.*) **282**, 333–336.

Mossman, H. W. (1937). Comparative morphogenesis of the fetal membranes and accessory uterine structures. *Contrib. Embryol. Carnegie Inst.* **26**, 129–246.

Myagkov, N. A., and Kondyurin, V. V. (1978). Reproduction of the catsharks, *Apristurus saldancha. Sov. J. Mar. Biol.* (*Engl. Transl.*) **4**, 627–628.

Nagahama, Y. (1983). The functional morphology of teleost gonads. *In* "Fish Physiology" (W. S. Hoar, D. J. Randall, and E. M. Donaldson, eds.), Vol 9A, pp. 223–275. Academic Press, New York.

Nakaya, K. (1975). Taxonomy, comparative anatomy and phylogeny of Japanese catsharks, Scyliorhinidae. *Mem. Fac. Fish., Hokkaido Univ.* **23**, 1–94.

Needham, J. (1942). "Biochemistry and Morphogenesis." Cambridge Univ. Press, London and New York.

Nelsen, O. E. (1953). "Comparative Embryology of the Vertebrates." McGraw-Hill (Blakiston), New York.

Nelson, J. S. (1984). "Fishes of the World," 2nd ed. Wiley, New York.

Ng, T. B., and Idler, D. R. (1983). Yolk formation and differentiation in teleost fishes. *In* "Fish Physiology" (W. S. Hoar, D. J. Randall, and E. M. Donaldson, eds.) Vol. 9A, pp. 373–404. Academic Press, New York.

Okano, S., Otake, T., Teshima, K., and Mizup, K. (1981). Studies on sharks—XX. Epithelial cells of the intestine in *Mustelus manazo* and *M. griseus* embryos. *Bull. Fac. Fish Nagasaki Univ.* **51**, 23–28.

Otake, T., and Mizue, K. (1981). Direct evidence for oophagy in thresher shark, *Alopias pelagicus. Jpn. J. Ichthyol.* **28**, 171–172.

Otake, T., and Mizue, K. (1985). The fine structure of the placenta of the blue shark, *Prionace glauca. Jpn. J. Ichthyol.* **32**, 52–59.

Otake, T., and Mizue, K. (1986). The fine structure of the intra-uterine epithelium during integestation in the blue shark, *Prionace glauca. Jpn. J. Ichthyol.* **33**, 151–161.

Packard, G. C., Tracy, C. R., and Roth, J. J. (1977). The physiological ecology of reptilian eggs and embryos, and the evolution of viviparity within the Class Reptilia. *Biol. Rev. Cambridge Philos. Soc.* **52**, 71–105.

Paffenhofer, G. A., and Rosenthal, H. (1968). Trockengewicht and Kalorien-gehalt sich entwickelnder Heringseier. *Helgol. Wiss. Meeresunters.* **18**, 45–52.

Pang, P. K. T., Griffiths, R. W., and Atz, J. W. (1977). Osmoregulation in elasmobranchs. *Am. Zool.* **17**, 365–377.

Penrith, M. L. (1969). The systematics of the fishes of the family Clinidae in Southern Africa. *Ann. S. Afr. Mus.* **55**, 1–121.

Pickford, G. E., and Atz, J. W. (1957). "The Physiology of the Pituitary Gland of Fishes." N. Y. Zool. Soc., New York.

Price, K. S., and Daiber, F. C. (1967). Osmotic environments during fetal development of dogfish, *Mustelus canis* (Mitchill) and *Squalus acanthias* Linnaeus, and some comparison with skates and rays. *Physiol. Zool.* **40**, 248–260.

Ranzi, S. (1932). Le basi fisio-morfologische dello sviluppo embrionale dei Selaci. Parti I. *Pubbl. Stn. Zool. Napoli* **13**, 209–290.

Ranzi, S. (1934). Le basi fisio-morfologische dello sviluppo embrionale dei Selaci. Parti II and III. *Pubbl. Stn. Zool. Napoli* **13**, 331–437.

Read, L. J. (1968). Orthithine-urea cycle enzymes in early embryos of the dogfish *Squalus suckleyi and* and skate *Raja binoculata. Comp. Biochem. Physiol.* **24**, 669–674.

Reinig, J. W., Daniel, L. N., Schwabe, C., Gowan, L. K., Steinetz, B. G., and O'Byrne, E. M. (1981). Isolation and characterization of relaxin from the sand tiger shark (*Odontaspis taurus*). *Endocrinology (Baltimore)* **109**, 537–543.

Reznick, D. (1981). "Grandfather effects": The genetics of interpopulation differences in offspring size in the mosquito fish. *Evolution (Lawrence, Kans.)* **35**, 921–930.

Richter, J., Lombardi, J., and Wourms, J. P. (1983). Branchial placenta and embryonic growth in the viviparous fish, *Jenynsia. Am. Zool.* **23**, 1017.

Rosen, D. E. (1962). Egg retention pattern in evolution. *Nat. Hist., N.Y.* **71**(10), 46–53.

Rosen, D. E., Forey, P. L., Gardiner, B. G., and Patterson, C. (1981). Lungfishes, tetrapods, paleontology, and plesiomorphy. *Bull. Am. Mus. Nat. Hist.* **167**, 159–276.

Rosenblatt, R. H., and Taylor, L. R. (1971). The Pacific species of the clinid fish tribe Starksiini. *Pac. Sci.* **25**, 436–463.

Rusaouen-Innocent, M. (1985). Nidamental gland secreting the dogfish eggshell. *In* "Biology of Invertebrate and Lower Vertebrate Collagens" (A. Bairati and R. Garrone, eds.), pp. 471–476. Plenum. New York.

Schindler, J. F., and de Vries, U. (1986). Ultrastructure of embryonic anal processes in *Girardinichthys viviparus* (Cyprinodontiformes, Osteichythyes). *J. Morphol.* **188**, 203–224.

Schlernitzauer, D. A., and Gilbert, P. W. (1966). Placentation and associated aspects of gestation in the bonnethead shark, *Sphyra tiburo. J. Morphol.* **120**, 219–231.

Schwabe, C., Steinetz, B., Weiss, G., Segaloff, A., McDonald, J. K., O'Byrne, E. M., Hochman, J., Carriere, B., and Goldsmith, L. (1978). Relaxin. *Recent Prog. Horm. Res.* **34**, 123–199.

Scrimshaw, N. S. (1944a). Superfoetation in poeciliid fishes. *Copeia*, pp. 180–183.

Scrimshaw, N. S. (1944b). Embryonic growth in the viviparous poeciliid, *Heterandria formosa. Biol. Bull. (Woods Hole, Mass.)* **87**, 37–51.

Scrimshaw, N. S. (1945). Embryonic development in poeciliid fishes. *Biol. Bull. (Woods Hole, Mass.)* **88**, 233–246.

Setna, S. B., and Sarangdhar, P. N. (1948). Description, bionomics, and development of *Scoliodon sorrakowah* (Cuvier). *Rec. Indian Mus.* **46**, 25–53.

Shann, E. W. (1923). The embryonic development of the porbeagle shark, *Lamma cornubica. Proc. Zool. Soc. London* **11**, 161–171.

Sherwood, O. D., and Downing, S. J. (1983). The chemistry and physiology of relaxin. *In* "Factors Regulating Ovarian Function" (G. S. Greenwald and P. F. Terranova, eds.), pp. 381–410. Raven Press, New York.

Shimizu, M., and Yamada, J. (1980). Ultrastructural aspects of yolk absorption in the vitelline syncytium of the embryonic rockfish, *Sebastes schlegeli. Jpn. J. Ichthyol.* **27**, 56–63.

Shine, R., and Bull, J. (1979). The evolution of live-bearing in lizards and snakes. *Am. Nat.* **103**, 905–923.

Simpson, G. G. (1950). Evolutionary determination and the fossil record. *Sci. Mon.* **71**, 262–267.

Smedley, N. (1927). On the development of the dogfish, *Scyllium marmaratum* Benn., and allied species. *J. Malay. Branch Asiat. Soc.* **5**, 355–359.

Smith, C. L., Rand, C. S., Schaeffer, B., and Atz, J. W. (1975). *Latimeria*, the living coelacanth, is ovoviviparous. *Science* **190**, 1105–1106.

Smith, S. (1957). Early development and hatching. *In* "The Physiology of Fishes" (M. E. Brown, ed.), Vol. 1, pp. 323–359. Academic Press, New York.

Soin, S. G. (1971). "Adaptational Features in Fish Ontogeny." Israel Program for Scientific Translation, Jerusalem.

Southwell, T., and Prashad, B. (1919). Notes from the Bengal Fisheries Laboratory, No. 6. Embryological and developmental studies of Indian fishes. *Rec. Indian Mus.* **16**, 215–240.

Springer, S. (1948). Oviphagous embryos of the sand shark, *Carcharias taurus. Copeia*, pp. 153–157.

Springer, S. (1979). A revision of the catsharks, family Scyliorhinidae. *NOAA Tech. Rep. NMFS CARC* **422**, 1–97.

Springer, V. G. (1970). The western South Atlantic clinid fish *Ribeiroclinus eigenmanni*, with discussion of the intrarelationships and zoogeography of the clinidae. *Copeia*, pp. 420–436.

Stearns, S. C. (1976). Life-history tactics: A review of the ideas. *Q. Rev. Biol.* **51**, 3–47.

Stevens, J. D. (1983). Observations on reproduction in the shortfin mako, *Isurus oxyrinchus*. *Copeia*, 126–130.

Stuhlmann, F. (1887). Zur Kenntnis des ovariums der Aalmutter. (*Zoarces viviparus* Cuv.). *Abh. Naturwiss. Ver. Hamburg* 10, 3–48.

Suarez, S. S. (1975). The reproductive biology of *Ogilbia cayorum*, a viviparous brotulid fish. *Bull. Mar. Sci.* 25, 143–173.

Sund, O. (1943). Et Brugdebarsel. *Naturen* 67, 285–286.

Tamura, E., Honma, Y., and Kitamura, Y. (1981). Seasonal changes in the thymus of the viviparous surfperch *Ditrema temmincki*, and special reference to its maturity and gestation. *Jpn. J. Ichthyol.* 28, 295–303.

Taniuchi, T., Kobayashi, H., and Otake, T. (1984). Occurrence and reproductive mode of the false cat shark, *Pseudotriakis microdon*, in Japan. *Jpn. J. Ichthyol.* 31, 88–92.

Tavolga, W. N. (1949). Embryonic development of the platyfish (*Platypoecilus*), the swordtail (*Xiphophorus*), and their hybrids. *Bull. Am. Mus. Nat. Hist.* 94, 161–229.

Tavolga, W. N., and Rugh, R. (1947). Development of the platyfish, *Platypoecilus maculatus*. *Zoologica* (N.Y.) 32, 1–15.

Terner, C. (1979). Metabolism and energy conversion during early development. *In* "Fish Physiology" (W. S. Hoar, D. J. Randall, and J. R. Brett, eds.), Vol. 8, pp. 261–278. Academic Press, New York.

Teshima, K. (1973). Studies on sharks. III. The stage of placentation and the umbilical stalk in *Carcharhinus dussmieri*. *J. Shimonoseki Univ. Fish.* 21, 295–309.

Teshima, K. (1975). Studies on sharks. VIII. Placentation in *Mustelus griseus*. *Jpn. J. Ichthyol.* 22, 7–12.

Teshima, K. (1981). Studies on the reproduction of Japanese smooth dogfishes, *Mustelus manazo* and *M. griseus*. *J. Shimonoseki Univ. Fish.* 29, 113–199.

Teshima, K., and Mizue, K. (1972). Studies on sharks. I. Reproduction in the female sumitsuki shark. *Carcharhinus dussumieri*. *Mar. Biol.* (Berlin) 14, 222–231.

Teshima, K., Ahmad, M., and Mizue, K. (1978). Studies on sharks. XIV. Reproduction in the Telok Anson shark collected from Perak River, Malaysia. *Jpn. J. Ichthyol.* 25, 181–189.

Te Winkel, L. E. (1943). Observations on later phases of embryonic nutrition in *Squalus acanthias*. *J. Morphol.* 73, 177–205.

Te Winkel, L. E. (1963). Notes on the smooth dogfish, *Mustelus canis*, during the first three months of gestation. II. Structural modifications of yolk-sacs and yolk-stalks correlated with increasing absorptive function. *J. Exp. Zool.* 152, 123–137.

Thibault, R. E. (1974). Genetics of cannibalism in a viviparous fish and its relationship to population density. *Nature* (London) 251, 138–140.

Thibault, R. E. (1975). Reply to Downhower and Brown. *Nature* (London) 256, 345–346.

Thibault, R. E., and Schultz, R. J. (1978). Reproductive adaptations among viviparous fishes (Cyprinodontiformes: Poeciliidae). *Evolution* (Laurence, Kans.) 32, 320–333.

Thillayampalam, E. M. (1928). "*Scoliodon* (The common shark of the Indian Seas)," Indian Zool. Mem. Indian Anim. Types No. 2. Methodist Publ. House, Lucknow.

Thorson, T. B., and Gerst, J. W. (1972). Comparison of some parameters of serum and uterine fluid of pregnant, viviparous sharks (*Carcharhinus leucas*) and serum of their near-term young. *Comp. Biochem. Physiol. A* 42A, 33–40.

Tortonese, E. (1950). Studi sui Plagiostomi. III. La viviparita: Un fondamentale carattere biologico degli Squali. *Arch. Zool. Ital.* 35, 101–155.

Trinkaus, J. P., and Drake, J. W. (1952). Role of exogenous nutrients in the development of the viviparous teleost, *Lebistes reticulatus*. *Anat. Rec.* 112, 435.

Triplett, E. L. (1960). Notes on the life history of the barred surfperch, *Amphistichus argenteus* Agassiz, and a technique for culturing embiotocid embryos. *Calif. Fish Game* **46**, 433–439.

Triplett, E. L., and Barrymore, S. (1960a). Tissue specificity in embryonic and adult *Cymatogaster aggregata* studied by scale transplantation. *Biol. Bull. (Woods Hole, Mass.)* **118**, 463–471.

Triplett, E. L., and Barrymore, S. (1960b). Some aspects of osmoregulation in embryonic and adult *Cymatogaster aggregata* and other embiotocid fishes. *Biol. Bull. (Woods Hole, Mass.)* **118**, 472–478.

Tsang, P., and Callard, I. P. (1983). *In vitro* steroid production by ovarian granulosa cells of *Squalus acanthias. Bull. Mt. Desert Isl. Biol. Lab.* **23**, 78–79.

Tsang, P., and Callard, I. P. (1984). Ovarian steriod synthesis and secretion during the reproductive cycle of the spiny dogfish, *Squalus acanthias. Biol. Reprod.* **30**, Suppl. 1, 67 (abstr.).

Turner, C. L. (1933) Viviparity superimposed upon ovo-viviparity in the Goodeidae, a family of cyprinodont teleost fishes of the Mexican plateau. *J. Morphol.* **55**, 207–251.

Turner, C. L. (1936). The absorptive processes in the embryos of *Parabrotula dentiens,* a viviparous, deep-sea brotulid fish. *J. Morphol.* **59**, 313–325.

Turner, C. L. (1937). The trophotaeniae of the Goodeidae, a family of viviparous cyprinodont fishes. *J. Morphol.* **61**, 495–523.

Turner, C. L. (1938a). Adaptations for viviparity in embryos and ovary of *Anableps anableps. J. Morphol.* **62**, 323–349.

Turner, C. L. (1938b). Histological and cytological changes in the ovary of *Cymatogaster aggregatus* during gestation. *J. Morphol.* **62**, 351–373.

Turner, C. L. (1940a). Pseudoamnion, pseudochorion, and follicular pseudoplacenta in poeciliid fishes. *J. Morphol.* **67**, 59–89.

Turner, C. L. (1940b). Follicular pseudoplacenta and gut modifications in anablepid fishes. *J. Morphol.* **67**, 91–105.

Turner, C. L. (1940c). Pericardial sac, trophotaeniae, and alimentary tract in embryos of goodeid fishes. *J. Morphol.* **67**, 271–289.

Turner, C. L. (1940d). Adaptations for viviparity in jenynsiid fishes. *J. Morphol.* **67**, 291–297.

Turner, C. L. (1942). Diversity of endocrine function in the reproduction of viviparous fishes. *Am. Nat.* **76**, 179–190.

Turner, C. L. (1947). Viviparity in teleost fishes. *Sci. Mon.* **65**, 508–518.

Turner, C. L. (1952). An accessory respiratory device in embryos of the embiotocid fish, *Cymatogaster aggregata,* during gestation. *Copeia,* pp. 146–147.

Turner, C. L. (1957). The breeding cycle of the South American fish, *Jenynsia lineata,* in the Northern Hemisphere. *Copeia,* pp. 195–203.

Veith, W. J. (1979a). The chemical composition of the follicular fluid of the viviparous teleost *Clinus superciliosus. Comp. Biochem. Physiol. A* **63A**, 37–40.

Veith, W. J. (1979b). Reproduction in the live-bearing teleost *Clinus superciliosus. S. Afr. J. Zool.* **14**, 208–211.

Veith, W. J. (1980). Viviparity and embryonic adaptations in the teleost *Clinus superciliosus. Can. J. Zool.* **58**, 1–12.

Veith, W. J., and Cornish, D. A. (1986). Ovarian adaptations in the viviparous teleosts *Clinus superciliosus* and *Clinus dorsalis* (Perciformes: Clinidae). *S. Afr. J. Zool.* **21**, 343–347.

Wake, M. H. (1977). Fetal maintenance and its evolutionary significance in the amphibia: Gymnophiona. *J. Herpetol.* **11**, 379–386.

Wake, M. H. (1982). Diversity within a framework of constraints. Amphibian reproductive modes. *In* "Environmental Adaptations and Evolution" (D. Mossakowski and G. Roth, eds.), pp. 87–106. Fischer, Stuttgart.

Wake, M. H. (1985). Oviduct structure and function in non-mammalian vertebrates. *Fortschr. Zool.* **30**, 427–435.

Wallace, R. A. (1985). Vitellogenesis and oocyte growth in non-mammalian vertebrates. *In* "Developmental Biology" (L. W. Browder, ed.), Vol. 1, pp. 127–178. Plenum, New York.

Watanabe, Y. (1982). Intracellular digestion of horseradish peroxidase by the intestinal cells of teleost larvae and juveniles. *Bull. Jpn. Soc. Sci. Fish.* **48**, 37–42.

Webb, P. W., and Brett, J. R. (1972a). Respiratory adaptations of prenatal young in the ovary of two species of viviparous seaperch, *Rhacochilus vacca* and *Embiotoca lateralis. J. Fish. Res. Board Can.* **29**, 1525–1542.

Webb, P. W., and Brett, J. R. (1972b). Oxygen consumption of embryos and parents, and oxygen transfer characteristics within the ovary of two species of viviparous seaperch, *Rhacochilus vacca* and *Embiotoca lateralis. J. Fish. Res. Board Can.* **29**, 1543–1553.

Wegmann, I., and Gotting, K. J. (1971). Untersuchungen zur Dotterbildung in den Oocyten von *Xiphophorus helleri* (Heckel, 1848), Teleostei, Poeciliidae). *Z. Zellforsch. Mikrosk. Anat.* **119**, 405–433.

Wiebe, J. P. (1968). The reproductive cycle of the viviparous seaperch, *Cymatogaster aggregata* Gibbons. *Can. J. Zool.* **46**, 1221–1234.

Williams, G. C. (1966). "Adaptation and Natural Selection." Princeton Univ. Press, Princeton, New Jersey.

Wolfson, F. H. (1983). Records of seven juveniles of the whale shark, *Rhiniodon typus. J. Fish Biol.* **22**, 647–655.

Wood-Mason, J., and Alcock A. (1891). On the uterine villiform papillae of *Pteroplatea micrura*, and their relation to the embryo. *Proc. R. Soc. London* **49**, 359–367.

Wourms, J. P. (1977). Reproduction and development in chondrichthyan fishes. *Am. Zool.* **17**, 379–410.

Wourms, J. P. (1981). Viviparity: The maternal-fetal relationship in fishes. *Am. Zool.* **21**, 473–515.

Wourms, J. P. (1987). Mammal-like adaptations in the development of a placental shark. *Am. Zool.* **27** (in press).

Wourms, J. P., and Bayne, O. (1973). Development of the viviparous brotulid fish, *Dinematichthys iluocoeteoides. Copeia*, pp. 32–40.

Wourms, J. P., and Bodine, A. B. (1983). Biochemical analysis and cellular origin of uterine histotrophe during early gestation of the viviparous butterfly ray. *Am. Zool.* **23**, 1018.

Wourms, J. P., and Bodine, A. B. (1984). Structure and function of trophonemata, a placental analogue, during early gestation of the butterfly ray. *In* "International Cell Biology 1984" (S. Seno and Y. Orada, eds.), p. 407. Academic Press, Orlando, Florida.

Wourms, J. P., and Cohen, D. (1975). Trophotaeniae, embryonic adaptations in the viviparous ophidioid fish, *Oligopus longhursti:* A study of museum specimens. *J. Morphol.* **147I**, 385–401.

Wourms, J. P., and Hamlett, W. C. (1978). Cell surfaces during development of shark yolk sac placentae. *Am. Zool.* **18**, 642.

Wourms, J. P., and Lombardi, J. (1979a). Convergent evolution of trophotaeniae and other gut derived embryonic adaptations in viviparous teleosts. *Proc. Annu. Meet. Am. Soc. Ichthyol. Herpetol.* **59**, 94.

Wourms, J. P., and Lombardi, J. (1979b). Cell ultrastructure and protein absorption in the trophotaenial epithelium, a placental analogue of viviparous fish embryos. *J. Cell Biol.* **83**(2, Pt. 2), 399a.

Wourms, J. P., and Lombardi, J. (1985). Prototypic trophotaeniae and other placental structures in embryos of the pile perch, *Rhacochilus vacca* (Embiotocidae). *Am. Zool.* **25**, 95A.

Wourms, J. P., and Staehelin, L. A. (1985). Endocytotic complex in trophotaenial placental cells. *J. Cell Biol.* **101** (5, Pt. 2), 291a.

Wourms, J. P., Stribling, M. D., and Atz, J. W. (1980). Maternal–fetal nutrient relationships in the coelacanth, *Latimeria. Am. Zool.* **20**, 962.

Wourms, J. P., Hamlett, W. C., and Stribling, M. D. (1981). Embryonic oophagy and adelphophagy in sharks. *Am. Zool.* **21**, 1019.

Young, G. (1980). Ultrastructural studies on the pituitary gland of the teleost *Poecilia latipinna* with special reference to reproduction. Ph.D. Thesis, Faculty of Pure Science, University of Sheffield (cited from Young and Ball, 1983b).

Young, G., and Ball, J. N. (1982). Ultrastructural changes in the adenohypophysis during the ovarian cycle of the viviparous teleost *Poecilia latipinna.* I. The gonadotrophic cells. *Gen. Comp. Endocrinol.* **48**, 39–59.

Young, G., and Ball, J. N. (1983a). Ultrastructural changes in the adenohypophysis during the ovarian cycle of the viviparous teleost *Poecilia latipinna.* II. The thyrotrophic cells and the thyroid gland. *Gen. Comp. Endocrinol.* **51**, 24–38.

Young, G., and Ball, J. N. (1983b). Ultrastructural changes in the adenohypophysis during the ovarian cycle of the viviparous teleost *Poecilia latipinna.* III. The growth hormone, adrenocorticotrophic, and prolactin cells and the pars intermedia. *Gen. Comp. Endocrinol.* **52**, 86–101.

FIRST METAMORPHOSIS

JOHN H. YOUSON

Departments of Zoology and Anatomy and Scarborough Campus
University of Toronto,
Toronto, Ontario, Canada M5S 1A8

I. INTRODUCTION

The term metamorphosis, when considered in the broadest sense in the animal kingdom, refers to any abrupt change in the form or structure of an organism during postembryonic development. The

FISH PHYSIOLOGY, VOL. XIB

modification (transformation) of the organism, or of its tissues and organs, is usually in preparation for a change in environment, behavior, or mode of feeding. When one applies this term to vertebrates, it is apparent that only some fishes and amphibians possess a metamorphosis during their postembryonic development. The fact that these two vertebrate groups are both mainly aquatic and share some common evolutionary history is probably highly relevant to their possession of metamorphosis (Szarski, 1957). Every student of biology is aware of the dramatic metamorphosis that occurs between the tadpole and adult stages in anuran amphibians and of the immense value that is laid on the event for studies in developmental biology (Etkin and Gilbert, 1968, Gilbert and Frieden, 1981). However, the same cannot be said for fish metamorphosis. There has certainly been increased research activity in the ontogeny of fishes in recent years (Alderdice, 1985), but data on metamorphosis are sparse, despite the stress that was laid on the need for research by an earlier report in this series (Blaxter, 1969).

The lack of information on metamorphosis has proven to be an obstacle in many research areas of fish biology, particularly in systematics (Cohen, 1984). In some cases fish formerly considered as distinct species are now known to be metamorphosing larvae of another species (de Sylva and Eschmeyer, 1977; Johnson, 1984). Part of this deficiency in our knowledge can be explained by the absence of clearly defined parameters for assessing and comparing metamorphosis among the vast number of fish species. For example, it may seem practical to define metamorphosis in a flounder as the period of eye migration (Richardson and Joseph, 1973), but there is a need for a list of more general characteristics of fish metamorphosis so that this phase can be identified in species with less dramatic events. Furthermore, the existing literature is riddled with copious terms and synonyms for what might be considered as a fish metamorphosis, such as transformer, postlarval stage, transitional stage, transformation stage, and kasidoron stage. These terms reflect the fact that it is recognized that all fish species have an ontogenetic interval where some postembryonic change prepares them for their new role as juveniles or adults and that the degree of change is variable between species. However, it is not clear what postembryonic changes in fishes are true metamorphic events.

The objective of this chapter is to emphasize metamorphosis as a substantial and significant developmental strategy among fishes. This will be accomplished by providing criteria for identifying this ontogenetic interval which follow those used for other vertebrates, by placing metamorphosis in context with the entire ontogenetic process, and then dealing with metamorphic events and their morphological and physiological significance. No attempt is made to provide a compendium of all fish species that undergo true metamorphosis, for this will be left to those who may be stimulated to do so by this essay. Instead, an attempt will be made to provide some terms of reference for such an endeavor, which hopefully will involve a multidisciplinary approach (Alderdice, 1985).

II. METAMORPHOSIS AND FISH ONTOGENY

A. Definition

There have been some earlier attempts to place metamorphosis in fishes in the broad sense of all animal metamorphosis (Szarski, 1957). Metamorphosis is usually defined for all vertebrates (Just et al., 1981) or all protochordates and chordates (Barrington, 1968), and then fishes have been considered within these confines. For example, Barrington (1968) took exception to an earlier definition by Ahlström and Counts (1958) that fish metamorphosis or transitional stage (Ahlström, 1968) is an interval during which marked changes occur in body proportions and structures without any marked increase in length. He felt that physiological change and change of habitat were ignored by this definition. Although these aspects and consequences of fish metamorphosis had not been overlooked (for example, see Szarski, 1957), Barrington (1961, 1968) was particularly interested in including the physiological and behavioral changes of parr–smolt transformation (smoltification) which some call a *second metamorphosis* (Wald, 1958, 1981), a secondary metamorphosis (Balon, 1985b), or a second type of metamorphosis (Norris, 1983). Just et al. (1981) did not feel that parr–smolt transformation met their criteria of chordate metamorphosis, while Wald (1981) suggested that first metamorphosis and second metamorphosis are necessary consequences of the expression of two separate genotypes, which often develop in parallel in the same organism. Many, but not necessarily all, larval genes are repressed at first metamorphosis when the adult genes are expressed progressively. Second metamorphosis occurs when the adult genes controlling sexual maturation and/or migratory behavior and physiology are fully expressed. If one accepts this viewpoint, then first and second metamorphosis should not be considered as similar morphogenetic intervals. Norris (1983) called smoltification a second type of meta-morphosis, which is found in species with a complex life cycle involv-ing a migration between freshwater and marine habitats. Examples include postmetamorphic juveniles of the Salmoniformes, the Anguil-liformes, and the Petromyzontiformes. It is clear that not all species possessing a second metamorphosis undergo a first metamorphosis, so that it should be understood that the term second metamorphosis im-plies only that it is a different type of metamorphosis from the first. Postlarval change or second metamorphosis is not a consideration of this present review, which will be concerned with postembryonic (larval) change or *first metamorphosis*. As outlined below, first meta-morphosis in fishes is an example of *true vertebrate metamorphosis*.

A reiteration of the three criteria for general chordate first meta-morphosis (Just *et al.*, 1981) will be a useful beginning for an attempt at providing guidelines for assessing first metamorphosis in fishes. In abbreviated form these are:

1. A change in nonreproductive structures between embryonic life and sexual maturation but not embryonic development, sexual maturation, or aging.
2. The larva occupies an ecological niche different from the em-bryo and adult because of its form. This assures that late em-bryo and early adult (juvenile) periods are not considered.
3. Morphological change at the end of larval life (climax) is trig-gered by an external (environmental) and/or internal (e.g., hor-monal) cue.

The above criteria when considered together are designed to ex-clude parr–smolt transformation and include all significant morpho-logical change occurring during larval life and not just at climax. Al-though Just *et al.* (1981) state that more than one of their criteria should be met before using the term metamorphosis, all three could be applied to metamorphosis in many larval fishes. However, it is questionable whether progressive morphological change that began in the embryo, such as ossification of the skeleton (Ahlström and Counts, 1958; Richardson and Joseph, 1973; Griswold and McKen-ney, 1984) and continued development of the nervous system (Kawa-mura and Ishida, 1985), are metamorphic events in fishes. In some cases these are found in young of fishes that follow a direct develop-ment from the embryo to a juvenile/adult with no intervening larval interval. The larval interval must be present in the life cycle to have a first metamorphosis. Furthermore, it is not absolutely clear that all cases of parr–smolt transformation would be excluded by these three criteria. Therefore, it seems that three additional standards or characters should be added to the above to more clearly define fish first metamorphosis. These are:

1. Generally characterized by a marked change in form that is not necessarily a rapid process (permits inclusion of transitional larval features).
2. The immediately postembryonic larva and the adult (or juvenile) do not look alike (eliminates all direct postembryonic development).
3. The process does not involve growth (e.g., increased snout–vent length) and may in fact feature a decrease in length (eliminates increase in length during larval life and direct development as a metamorphic event).

Note that transitional larval characters, as in criterion 1, are a difficult issue and will be given further consideration below. It is questionable whether their acquisition is an event of metamorphosis.

B. Place within the Ontogenetic Sequence

1. HISTORICAL CONCEPTS

The task of placing first metamorphosis of fishes within the general context of true chordate metamorphosis has not been difficult. A more onerous task is to position first metamorphosis within the existing ontogenetic sequence of steps (intervals) of fishes. A problem of terminology and sequencing of life-cycle intervals has been with us for some time (Hubbs, 1943), but particularly with regard to those intervals that are postembryonic. Past descriptions indicate at least four developmental pathways (strategies) that various fish species take from the embryo to the adult (Fig. 1).

Type 1. The young at hatching are a replica of the adult except in size and sexual maturity and proceed to adulthood by attaining these characters over variable (and often prolonged) periods of time.

Type 2. The young of type 1 may also proceed to a juvenile for a limited time before gaining through second metamorphosis the definitive adult characters.

Type 3. Exogenously feeding young or larvae arise from the embryo and acquire prejuvenile characters, which gradu-

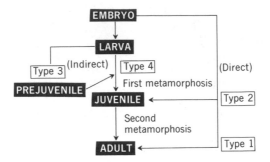

Fig. 1. Four (types 1–4) ontogenetic pathways during the development of a fish embryo into an adult. Types 1 and 2 are direct development, while types 3 and 4 are indirect and involve a larval interval which terminates at this first metamorphosis. A second metamorphosis may occur between juvenile and adult intervals.

                ally transform or are lost during first metamorphosis before those of juvenile and/or adult stages are attained.

Type 4. After a postembryonic period of variable length in which larval form does not change to any appreciable extent, larvae undergo a marked, often instantaneous, first metamorphosis into a juvenile.

The juvenile interval in all larval pathways (types 3 and 4) is sometimes followed by a second phase of change less dramatic than the first called the second metamorphosis (Wald, 1958; Barrington, 1968). Types 1 and 2 are direct development, type 3 is transitory or intermediate, and type 4 is indirect.

    Type 4 (indirect) development in fishes is an example of a vertebrate first metamorphosis. There are only a few groups of fishes that have a first metamorphosis in their life cycle (Balon, 1975). Although well represented among Class Osteichthyes and all Petromyzontiformes, first metamorphosis is not present in either Myxiniformes or Class Chondrichthyes. Unquestionable examples of first metamorphosis among bony fishes are those illustrating dramatic transformation of the body form, such as seen in Anguilliformes (true eels), Notacanthiformes (spiny eels), Elopiformes (bonefish, ten-pounders, and tarpons), and Pleuronectiformes (flatfishes). In the past, the fishes of the type 3 (transitory) pathway have been difficult to categorize in a metamorphic grouping. In Dipneusti (lungfishes), Polypteriformes (bichir), Acipenseriformes (sturgeon), Lepisosteiformes (garpike), and Amiiformes (bowfin), larval organs for respiration, feeding, or locomotion develop to aid a planktonic existence and are lost before a metamorphic climax. These fishes all acquire and lose external gills some-time in early larval life and never undergo an abrupt change into a juvenile (Norman and Greenwood, 1963). These two features are suffi-cient evidence to permit the exclusion of these fishes from examples of Class Osteichthyes with a first metamorphosis. Furthermore, Ophi-diiformes (Gordon *et al.*, 1984) and Lophiiformes (Pietsch, 1984) are examples of Euteleostei that also possess transitory larval structures that persist for some time but are not present in the definitive pheno-type. Their transient features are cenogenetic adaptations (Moser, 1981). Some flatfish larvae also have an elongated second dorsal spine that disappears during metamorphosis (Amaoka, 1972, 1973). There is no question that the above species undergo morphological change during the larva period, but in many cases, such as in the kasidoron stage of gibberichthyids where a long pelvic appendage is acquired for larval life (de Sylva and Eschmeyer, 1977), the changes are not in preparation for adulthood. It is not known whether development of temporary larval structures is initiated during embryonic life and they grow in the larva to a position of prominence. If such is the case, it may be necessary in the future to exclude these from the examples of first metamorphosis. However, any of the above that involve abrupt change from the larval phenotype will continue to meet the criteria.

2. RECENT CONCEPTS

The most recent attempt at describing intervals of ontogeny in fishes is that of Balon (1979, 1981, 1984, 1985a). His saltatory model describes embryo, larva, juvenile, adult, and senescent periods (Fig. 2), separated by major thresholds (i.e., a switch or rapid transition into a new stabilized state). Each period may be divided into phases. Steps are the shortest intervals of ontogeny, are separated by less dramatic thresholds, and are found within a phase. Metamorphosis is a major threshold separating the larva period from juvenile or adult periods. This model fits well with the criteria of metamorphosis as they are presented in this chapter. The postembryonic pathways of fishes pre-sented earlier (Fig. 1) are reconsidered (Fig. 2) using the intervals of ontogeny of Balon (1985a). The early proposal of Balon (1975) did not adequately deal with metamorphosis (Richards, 1976); although the new scheme (Balon, 1985a) places this event in a better perspective with the rest of fish ontogeny, metamorphosis may occupy a consider-able period of time and it is questionable as to whether metamorpho-

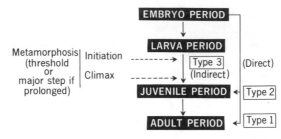

Fig. 2. The pathways of ontogeny in fishes; the terminology of life-cycle intervals of Balon (1985a) is used.

sis should be relegated to a category such as a threshold rather than a phase of larval life.

Other recent schemes of intervals of the early life history of fishes call the larval growth period a premetamorphic interval (Hardy, 1978) and the change from larva to adult a transitional stage of larval life termed the "transformation stage" (Kendall *et al.*, 1984). Metamorphosis is designated as an event of larval life. Therefore, to the present day, there is no term that has been universally accepted for the metamorphic interval of ontogeny, and this interval has not been always considered in the context of larval life. It is the proposal of the present writer that first metamorphosis in fishes be considered in the same context as in amphibian metamorphosis. That is, fish first metamorphosis is a *phase* in the larva period during which time postembryonic changes occur before the juvenile or adult periods are attained (Fig. 3). First metamorphosis is obligatory during indirect development when a larva period is part of the life cycle—that is, the use of the term larva implies that there is a first metamorphosis in the life cycle. In some cases metamorphosis may be initiated early and extend over the entire larva period (for qualification see above comments on transitory larval structures). However, in a more classical sense, first metamorphosis is a second larval phase, which follows a first phase of larval growth (premetamorphic phase) and is marked by an abrupt transformation from the larval phenotype. Stages should be described during the metamorphic phase and should begin with the *initiation event* and end with the completion of the *climax event*, at which time the juvenile form and behavior are present. The time at which these two events occur during metamorphosis should be a major consideration of any life-history study, even if they are, in some species, almost simultaneous events.

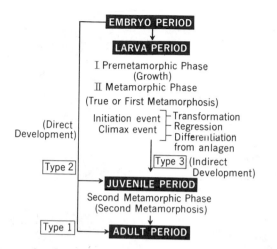

Fig. 3. The intervals of ontogeny in fishes and the three pathways between the embryo and the adult. Direct development from the embryo period leads to a juvenile period (type 2) or an adult period (type 1), while indirect development (type 3) involves a larva period that has a metamorphic phase leading to the juvenile period. The metamorphic phase has initiation and climax events that involve the morphogenetic processes of transformation, regression, and differentiation from anlagen. A second metamorphosis may occur between juvenile and adult periods.

III. STAGING

There should be clearly delineated criteria or parameters that denote the intervals of the metamorphic phase of the larva period. In the past, external metamorphic characters have been used to describe the intervals of metamorphosis, and these intervals have been called *stages*. This author cannot provide any reason for discontinuing the use of this term. However, it should always be understood that ontogenetic processes are continuous and "stage" does not imply an instantaneous state. The Petromyzontiformes represent the only group of fishes where there are standard criteria for staging metamorphosis (Potter *et al.*, 1982). The diversity of Osteichthyes and differences in their ontogeny will probably never permit a standardization of stages across the various orders. However, a series of universal stages at the level of the genus is not beyond comprehension. This will not be difficult if the species within the genus are few and the events of metamorphosis are relatively similar. Cooperation and communication between lamprey biologists in the two hemispheres (Potter *et al.*, 1978a, 1980; Bird and Potter, 1979a,b; Youson and Potter, 1979;

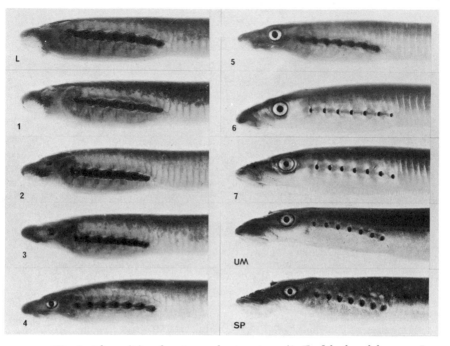

Fig. 4. A larva (L) and metamorphosing stages (1–7) of the brook lamprey *Lampetra planeri* as seen in lateral view. Upstream migrant females are seen in mature (UM) and spent (SP) condition. [From Bird and Potter (1979a).]

Beamish and Thomas, 1984; Tsuneki and Ouji, 1984a; Beamish and Austin, 1985) has resulted in the designation of seven clearly defined stages of metamorphosis (Fig. 4), primarily based on changes in the mouth, eye, branchiopores, dorsal fins, and body pigmentation (Potter *et al.*, 1982). As ultimately happened with amphibians once their metamorphosis was divided into a universally acceptable series of stages, lampreys during their metamorphosis have been introduced as a universal and important experimental tool for studies in developmental biology (Youson, 1985).

Staging of metamorphosis in osteichthians is only rarely provided with descriptions of the early life history. With few exceptions, these have been provided for those species undergoing little larval change prior to a dramatic remodeling of body form at climax. The best-known examples of first metamorphosis in bony fishes are those of Anguilliformes (Fig. 5), Elopiformes (Fig. 6), Notacanthiformes, and Pleuronectiformes (Fig. 7), but even in these cases, data on staging are

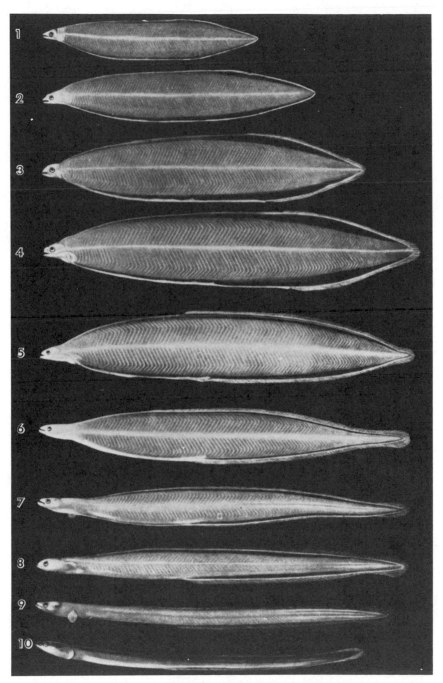

Fig. 5. Larvae or leptocephali (1–5), metamorphosing stages (6–8), glass eel (9), and elver (10) of European eel *Anguilla anguilla*. [From Sterba (1963).]

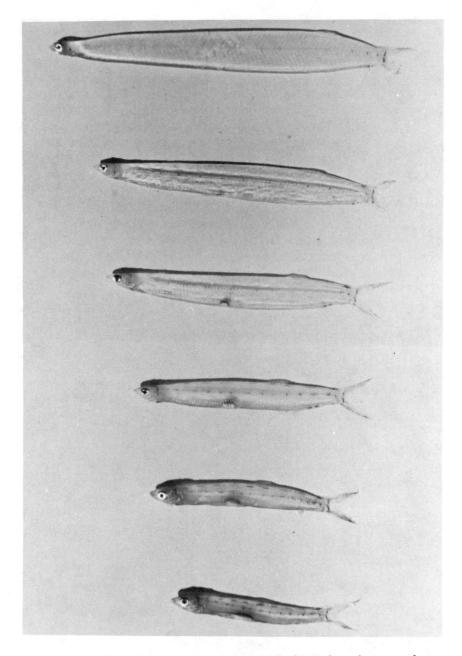

Fig. 6. Morphological changes, covering a period of 8-10 days, that occur during metamorphosis of *Albula* leptocephali. Standard lengths (from top to bottom) are 57, 47, 43, 36, 31, and 27 mm. [From Pfeiler and Luna (1984).]

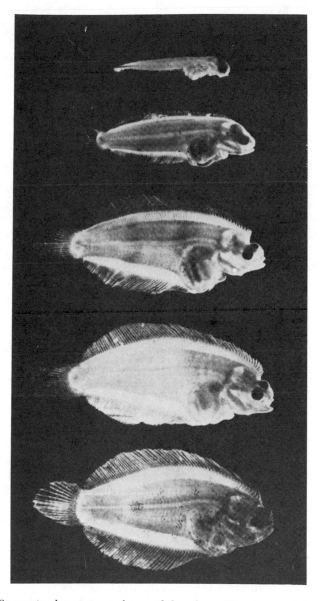

Fig. 7. Stages in the metamorphosis of the plaice, *Pleuronectes platessa*. [From Norman and Greenwood (1963).]

sparse. In individual species of these orders, first metamorphosis is often classified within a continuum of postembryonic developmental stages that extend to the sexually mature adult. Thus, for example, early metamorphosis and late metamorphosis are stages 4a and 4b, respectively, in the postembryonic development of plaice (Ryland, 1966), and prometamorphic, midmetamorphic, and postmetamorphic are stages 2 to 4, respectively, in the series of six postembryonic stages of a gonostomatid (Ahlström and Counts, 1958). Although Fukuhara (1986) and Seikai et al. (1986) use stages F, G, and H to describe early, middle, and late stages, respectively, in metamorphosis of Japanese flounder, Miwa and Inui (1987) refer to these three stages as prometa-morphic, climax, and post climax. Recently, first metamorphosis in bonefish Albula was given as the second phase of a two-phase larval developmental period (Pfeiler, 1986); however, there was no attempt at subdivision into stages. Six stages are described for Anguilliformes (Schmidt, 1906; Kubota, 1961), but the specific criteria are not well defined. One of the most detailed descriptions of first metamorphosis in a bony fish is that available for Megalops atlanticus (Mercado and Ciardelli, 1972). Although metamorphosis is stage 2 of a three-stage life cycle (stage 1, larva; stage 3, adult), this interval is subdivided into four phases, each characterized by a specific body length and both internal and external features. In other orders, definitive staging criteria have not been provided because of the paucity of specimens available during metamorphosis. In the case of Pleuronectiformes, the small numbers of metamorphosing individuals in field samples are explained by the fact that the phase is transitory, avoidance is increased as the animals become larger, and metamorphosing individuals may change habitat (Ahlström et al., 1984). However, there is now excellent potential for rearing flatfish larvae in the laboratory (Policansky and Sieswerda, 1979; Smigielski, 1979; Inui and Miwa, 1985; Fukuhara, 1986; Seikai et al., 1986; Crawford, 1987; Miwa and Inui, 1987b). As the metamorphosing individuals are becoming increasingly available for analysis, it is now important that suitable staging criteria be utilized.

The use of external criteria for staging metamorphosis in most bony fishes has largely been ignored, despite the early encouragement of the value of such features as myotome numbers and fin position in assessing advancing biological age (Matsubara, 1942). Recent evidence indicates that there may be progressive external changes that can be used in some species to permit staging of metamorphosis. For example, in Congridae, the forward advancement of the subterminal anus and its accompanying anal fin orgins and the developing pterygiophores and actinotrichia are key characters (Castle, 1984). Similar morphogenetic events characterize metamorphosis in other eels (Castle, 1970). In *Siganus lineatus* (Bryan and Madraisau, 1977), it would seem that progressive changes in pigmentation could be followed more closely to provide stages. In this species a brown-head omnivorous stage is the only feature that presently separates metamorphosis from a dark-head carnivorous larval stage and a herbivorous juvenile stage. Development of nostrils and olfactory lamellae can be used for staging in some flounders (Amaoka, 1973). To ensure that the earliest stages are determined, there should be a correlation of the time of initiation of both internal and external changes (Mercado and Ciardelli, 1972; Youson *et al.*, 1977). Murr and Sklower (1928) took a similar approach in an early attempt at staging metamorphosis in eel leptocephali.

IV. TIMING

One of the least understood aspects of metamorphosis in fishes is the factors that determine the time at which the phase begins. If the entire postembryonic interval in species with larvae is considered as a first metamorphosis (e.g., species with transitory larval structures), then a determination of the time of the onset of this phase is not a problem. However, if the ontogeny of a fish species is characterized by a climax during which a major remodeling takes place after a long, uneventful larva period, the timing of the initiation of the phase is undoubtedly of great significance to the ultimate survival of the postmetamorphic individual. Thus, in most cases, climax is a highly synchronized event that is directly correlated to the state of physiological preparation of the organism, to environmental conditions, and to availability of food for the recently metamorphosed juvenile.

A. Body Length, Growth Rate, and Age

The ontogenetic pathway (i.e., direct or indirect) followed by fish species is related to the amount of yolk in the egg (Norman and Greenwood, 1963; Balon, 1986). The availability of nutrients to the embryo and larva probably has a bearing also on the length of larval life and the time of the onset of metamorphosis. There is also evidence that some marine fishes may delay metamorphosis in order to maximize dispersal (Victor, 1986). Most fish species must reach "metamorphosing size" before they can undergo a climax change into a juvenile. The body length at which metamorphosis occurs is species-specific and is related to the duration of the larva period, that is, age of the individuals. Data from lampreys serve as a good illustration of this point (Potter, 1980b). The duration of larval life of lampreys can be ascertained by examining length-frequency data in larval (ammocoete) populations in a site which is stable and restricted. This method of age determination indicates that species of lampreys reach metamorphosing size within a range of $2\frac{1}{2}$ years (*Mordacia mordax*) to $6\frac{1}{4}$ years (*Lampetra planeri*). However, it is quite likely that ammocoetes of some species may remain within the final year class for at least another 1–2 additional years, and maybe up to a total of 18 years (Manion and Smith, 1978), without a substantial increase in length. During this "arrested growth phase" (Hardisty and Potter, 1971a,b), ammocoetes prepare themselves for the nontrophic phase of metamorphosis by building up their reserves of lipid (Lowe *et al.*, 1973; O'Boyle and Beamish, 1977; Youson *et al.*, 1979). Length-frequency data for aging of ammocoetes of the sea lamprey (*Petromyzon marinus*) have been supported by measurements of the yearly growth pattern of the larval opisthonephric kidney (Ooi and Youson, 1976). Growth patterns in calcareous otic elements could also prove to be an important determinant of the duration of larval life in lampreys (Volk, 1986).

In lampreys there appears to be no correlation between the size at metamorphosis and the length of sexually mature adults. Thus, ammocoetes of the anadromous form of *P. marinus* (whose adults may eventually reach 83 cm) enter metamorphosis at a smaller size than their landlocked counterparts, whose spawning adults rarely exceed 50 cm in length (Smith, 1971; Beamish, 1980b; Hendrich *et al.*, 1980). Furthermore, there is evidence indicating that in paired species (i.e., a parasitic species and a closely related nonparasitic species), ammocoetes of the nonparasitic species enter metamorphosis at a mean length greater than that of parasitic species (Potter, 1980a). There is no increase in length in nonparasitic species following metamorphosis, for the entire adult period is nontrophic and is characterized by maturation of the gonads and depletion of energy stores.

For any given population, the time (age) at which an ammocoete beings initial metamorphosis is likely to be dependent on larval growth rate (Purvis, 1980). Because growth rates are highly variable, there is a wide range in the time at which metamorphosis can occur within a species. Environmental factors such as population density and temperature are an important influence in this regard (Manion and Smith, 1978; Purvis, 1979, 1980; Malmqvist, 1983; Morman,

1987). Thus, 119–130 mm is the metamorphosing size of lampreys in the small Dennis Stream in New Brunswick (Potter et al., 1978b; Youson and Potter, 1979), whereas in other larger watersheds of that province, where animal density is less, the metamorphosing individuals are generally larger and show a narrower size range (J. H. Youson and G. M. Wright, unpublished data). Experimental evidence also indicates higher growth rates and earlier metamorphosis in low-density populations (Morman, 1987).

The length at which osteichthians enter metamorphosis has also been an important consideration (Ahlström and Counts, 1958) and will likely continue as an important criterion for determining the potential timing of the climax event in individual species. However, as in lampreys, there is considerable individual variation in the age and length at which bony fishes enter metamorphosis. For example, laboratory-reared Japanese flounder *Paralichthys olivaceus* complete metamorphosis between 11.4 and 17.5 mm standard length at age 25–50 days (Fukuhara, 1986). There is a wider range of age and length with each subsequent stage of metamorphosis (Fig. 8a and 8b), but 15 mm is the size at which most larvae metamorphose into juveniles (Fig. 9a). Policansky (1982) demonstrated that size (length), and not age, is the single most important determinant of the timing of metamorphosis in starry flounder *Platichthys stellatus* (Fig. 10). He viewed this as a character of first metamorphosis that is expressed throughout the animal kingdom (Policansky, 1983).

In fishes, length at metamorphosis has been directly correlated with the initiation of internal morphological change (Murr and Sklower, 1928; Kubota, 1961; Mercado and Ciardelli, 1972), with age at metamorphosis as determined with otoliths (Van Utrecht, 1982; Campana, 1984), and with changes in behavior (Breder, 1949; Fukuhara, 1986) and physiology (Forstner et al., 1983). The wide range of sizes at which eel (Smith, 1984; Pfeiler, 1986) and flatfish (Ahlström et al., 1984) species enter metamorphosis has some significance for discussions of evolution, systematics, and adaptation and will be considered later. The basic difference between the dramatic metamorphoses that occur in eels and flatfishes is length of time to climax after hatching. Whereas at least the European-bound 2½-year-old *Anguilla* leptocephali enter climax at about 75 mm length (Schmidt, 1906), larval flatfishes of 15–25 mm climax only several weeks after hatching (Policansky, 1982; Campana, 1984; Tanaka, 1985; Fukuhara, 1986; Seikai et al., 1986; Crawford, 1987). However, in both Anguilliformes and Pleuronectiformes there are exceptions. For example, the moringuid and muraenescoid eels metamorphose at 20 cm and 2–5 months after

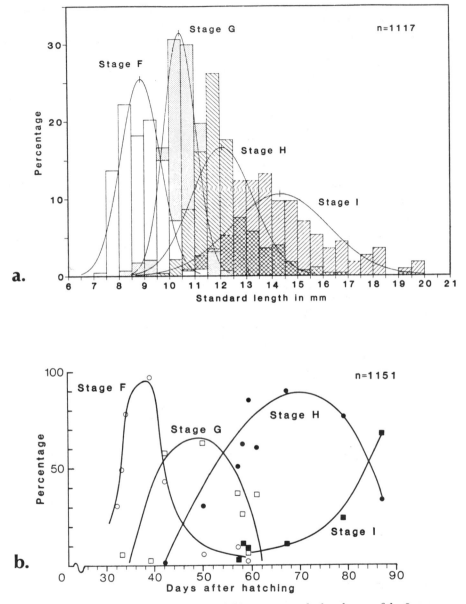

Fig. 8. A broader range of (a) sizes and (b) ages is noted when larvae of the Japanese flounder *Paralichthys olivaceus* advance through early (F), middle (G), and late (H) stages of metamorphosis to the juvenile period (I). [From Fukuhara (1986).]

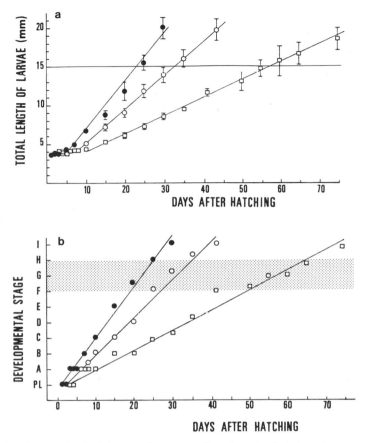

Fig. 9. The age at which larvae of Japanese flounder (*Paralichthys olivaceus*) reach (a) metamorphosing size and (b) the three stages F, G, and H of metamorphosis varies with the rearing temperatures of 13°C (open squares), 16°C (open circles), and 19°C (closed circles) [Modified from Seikai *et al.* (1986).]

hatching (Castle, 1977, 1979) and flatfishes metamorphosing at 12 cm have been reported (Ahlström *et al.*, 1984). There seems to be a wide variation in age at metamorphosis of flatfish larvae in laboratory crosses (Policansky, 1982).

The relationship between age and length of larvae at metamorphosis is an important parameter for consideration in both lampreys and bony fishes. However, the studies of Policansky (1982) on the laboratory-reared starry flounder indicate that, although age and length influence the timing of metamorphoric climax, the influence of length

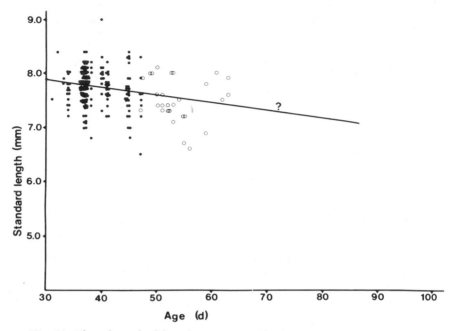

Fig. 10. Plot of standard length versus age (days) at metamorphosis of the starry flounder *Platichthys stellatus*. Closed circles represent animals reared at 12.06 ± 1.35°C and open circles those reared at 9.79 ± 0.71°C. [From Policansky (1982).]

may be strong in bony fishes (Fig. 10). This study appears to contradict an earlier investigation on plaice, which concluded that age and temperature are more important than length in controlling the onset of metamorphosis (Riley, 1966). A more recent report on Japanese flounder indicates that higher temperature results in reduced time (age) to metamorphosis and smaller metamorphosing individuals (Seikai *et al.*, 1986). These data imply that factors other than age, body length, and growth rate influence metamorphosis in flounder. In contrast, carefully controlled field experiments on sea lamprey *Petromyzon marinus* in the Big Garlic River, Michigan (Manion and Smith, 1978), showed that slow growth can extend larval life to up to 18 years from the expected 5–8 years. Also, low population density results in higher mean length at metamorphosis and shorter larval life in this lamprey species (Morman, 1987). These studies may provide us with examples of the controlling effect of growth rate (i.e., the time it takes to reach the critical length) on the time of onset of fish metamorphosis.

B. Physiological Preparation

Although larval age cannot be overlooked, present data indicate that length (size or growth rate) is one of the most important factors in determining the onset of the metamorphic phase. In lampreys, length has been examined in relation to changes in body weight during the larva period. This relationship has been analyzed as a condition factor $\{K = [\text{weight (g)/length (mm)}^3] \times 10^6\}$. A higher condition factor (Fig. 11) is generally found in animals in early stages of metamorphosis compared to later stages (Potter *et al.*, 1978a, 1980, 1983; Beamish and Thomas, 1984). The change in condition factor reflects both the need of immediately premetamorphic ammocoetes to build up lipid reserves and the utilization of this lipid during the nontrophic phase of metamorphosis (Lowe *et al.*, 1973; O'Boyle and Beamish, 1977; Youson *et al.*, 1979). A pronounced transformation of metabolic pathways accompanies the switch to the exclusive use of stored lipid (O'Boyle and Beamish, 1977). Because the weight/length relationship changes quite markedly in some species of lamprey, there have been some

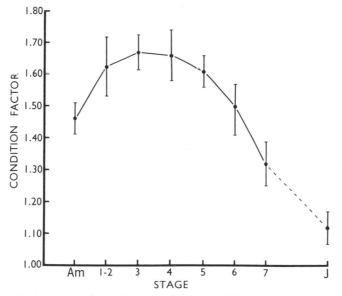

Fig. 11. The mean condition factor (and 95% confidence limits) from anadromous *Petromyzon marinus*. Am, ammocoete of metamorphosing length; stages 1–7 of metamorphosis; J, juvenile 6 months after stage 7. [From Potter *et al.* (1978b).]

recent successful attempts to identify immediately premetamorphic ammocoetes using the condition factor (Cole and Youson, 1981; Joss, 1985). In the past, identification of these individuals has been complicated by the variable time at which ammocoetes in any population enter metamorphosis.

Metamorphic climax is a nontrophic phase in many bony fishes (Kubota, 1961; Tesch, 1977; Moriarty, 1978; Pfeiler, 1986), and major reorganization of metabolism occurs at or near climax (Forstner et al., 1983). It is becoming a well-documented fact that development of specific metabolic pathways and the storage of energy are important features of the larva period in bony fishes (Laurence, 1975; Buckley, 1982; Cetta and Capuzzo, 1982). It would be of interest to investigate whether a parameter such as condition factor can be assessed in order to denote immediately premetamorphic climax in other fishes. It is well established that the bonefish *Albula* sp. utilizes its stored energy during metamorphosis (Rasquin, 1955), and the same may be true for other members of the Superorder Elopomorpha (Pfeiler, 1986). Furthermore, condition factor is presently being used to assess larval growth rates in fishes (Tandler and Helps, 1985) and therefore could be applied through metamorphosis. In addition, RNA–DNA ratios (Buckley, 1980) and both nucleic acid and protein changes (Fukuda et al., 1986) in metamorphosing flatfishes have been useful indicators of alterations in nutritional status and morphogenetic events. Buckley (1980) indicated that RNA–DNA ratios are particularly valuable since they are not affected by either age or size during larval life but increase during metamorphosis. Fukuda et al. (1986) used the protein–DNA ratio to follow increased cell size at metamorphosis following a rapid decline during larval growth.

C. Temperature

Tesch (1977) suggested that the time at which late-metamorphosing leptocephali of *Anguilla* sp. arrive at the mouths of rivers in northern Europe and North America from the Mediterrean can be correlated with temperature. Tucker (1959) reinterpreted the original data of Schmidt (1906) and claimed that temperature conditions encountered by leptocephali resulted in "eco-phenotypes" of *Anguilla anguilla* (for review, see Harden Jones, 1968). However, a recent alternative suggestion is that the varying time of metamorphosis in *Anguilla* sp. may reflect the influence either of stimulating factors in the waters of the continental shelf or river mouths and/or of inhibitory factors of the marine environment (McKeown, 1984). There are no reliable field data on the effects of temperature on the time of metamorphosis in Osteichthyes, but laboratory studies on the starry flounder *Platichthys stellatus* (Policansky, 1982), the winter flounder *Pseudopleuronectes americanus* (Laurence, 1975), and the Japanese flounder *Paralichthys olivaceus* (Seikai *et al.*, 1986) show a marked reduction in age (i.e., time to metamorphosis) and slight increase in length of progeny raised at a higher temperature (Figs. 9a, 9b, and 10). In the winter flounder, an increase in temperature from 5 to 8°C results in a reduction of time to metamorphosis from 80 to 49 days (Laurence, 1975), whereas a similar temperature increase in the Japanese flounder reduces the time to metamorphosis by 31 days (Figs. 9a and 9b). Furthermore, growth rates from hatching to juvenile are 0.22, 0.43, and 0.59 mm/day at 13, 16, and 19°C, respectively (Seikai *et al.*, 1986). The time at which plaice *Pleuronectes platessa* metamorphose may also be controlled by temperature (Riley, 1966). Moreover, a more rapid rate of metamorphosis at higher temperature was described for *Albula* sp. (Rasquin, 1955).

Rising water temperature has been correlated with the initiation event of metamorphosis in lampreys of metamorphosing size (Potter, 1970; Purvis, 1980). Laboratory and field experiments indicate that most lampreys metamorphose at 20–25°C, and a drop in temperature between 7 and 10°C can delay the onset by 4–5 weeks (Potter, 1970).

D. Behavior

Fish behavior during metamorphosis has received little attention, but with increased observations of organisms in laboratory culture, this will likely be a subject of future study. Information that can be gained from such observations may be similar to the following in *Siganus lineatus* (Bryan and Madraisau, 1977). The larvae of this species are carnivores, but when metamorphosis commences they spend less time swimming in search of live food and stop to gaze at algae. Later in metamorphosis they also begin to ingest the algae. This behavior is linked to a lengthening of the intestine and a temporary omnivorous diet during metamorphosis. Adults of *S. lineatus* are herbivores and have developed a long intestine during metamorphosis to aid in assimilation of their new source of nutrition. Increased swimming speed has been noted as an indicator of the onset of metamorphosis in the Japanese flounder and is correlated with morphological change in the finfold (Fukuhara, 1986). Other behavior related to metamorphosis has been noted in lampreys, where metamorphosing individuals seek shelter in a coarser substrate and faster flowing water than premetamorphic animals (Potter, 1970; Potter and Huggins, 1973). Also, changes in swimming behaviour accompany the initiation of metamorphosis in those species of bony fish which switch from a pelagic larva to a demersal juvenile (Caldwell, 1962; Tanaka, 1985).

V. CONTROL

There have been numerous studies (Just et al., 1981; Hoar, this volume) on the factors involved in initiation of the second metamorphosis in fishes, such as parr–smolt or yellow–silver eel transformation, but data on the control of first (true) metamorphosis (larval metamorphosis) are sparse (Norris, 1983). There is little doubt that environmental cues such as water temperature and, perhaps, photoperiod are important but, as in amphibians (White and Nicoll, 1981; Fox, 1984), hormones probably play a significant role. Early studies on both lampreys (for review, see Youson, 1980) and bony fish (for review, see Barrington, 1968; Just et al., 1981) were ambivalent in their assessment of the role of hormones. Most recent studies suggest that both environmental and hormonal cues interact to trigger and maintain fish species during the climax of metamorphosis.

A. Environmental

Environmental temperature not only affects the duration of larval life (Section IV,C) but is also a trigger for metamorphosis. Purvis (1980) subjected ammocoetes of Petromyzon marinus of similar metamorphosing size and at the same time of the year to either 20–21°C (aquarium), to 14–16°C (Big Garlic River) or 7–11°C. (Lake Superior) and found the incidence of metamorphosis to be 75–100%, 46–76%, and 5–10%, respectively. Increased incidence of metamorphosis at higher temperature had been previously shown in the Southern Hemisphere lamprey M. mordax (Potter, 1970). It is now common practice to hold ammocoetes in the laboratory at 20–25°C during the time that they would normally initiate metamorphosis in their natural environment. This dependence on temperature to trigger or to maintain metamorphic climax in lampreys explains why this phase of larval life is synchronized among individual populations of the same species (Purvis, 1980). There is a reasonable amount of evidence from the studies of Policansky (1982) and Sekai *et al.* 1986) that larval life in starry flounders can be shortened by triggering of metamorphic climax with higher temperature (Fig. 10). In the Japanese flounder, 24, 32, and 55 days are the times to metamorphosis at temperatures of 19, 16, and 13°C, respectively (Figs. 9a and 9b).

Eddy (1969) and Cole and Youson (1981) have shown that pinealectomy (removal of both the pineal and parapineal in this case) will prevent metamorphosis in lampreys. These two organs possess cells that have all the potential for photoreception (Cole and Youson, 1981). Although photoperiod may have some involvement in regulating the onset of metamorphosis, carefully controlled experiments suggest that the pineal complex (pineal and parapineal) may monitor other environmental cues, such as temperature, which in turn stimulate internal mechanisms to initiate the event (Cole and Youson, 1981). Survival of larvae of the gilthead sea bream *Sparus aurata* past their metamorphosis is directly correlated with the duration of the photoperiod (Tandler and Helps, 1985). This may be related to growth rates, for there is an increased chance of encountering prey within a longer photoperiod and, under such conditions, metamorphosing size can be attained within a shorter time period.

B. Hormonal

In any experiments concerning the stimulation, retardation, or prevention of fish metamorphosis, it is important to be certain that immediately premetamorphic animals are used. This problem is particularly significant in studies of lamprey metamorphosis, where the duration of larval life varies among individuals in a given population. As noted earlier, condition factor is valuable for identifying immediately premetamorphic ammocoetes (Potter *et al.*, 1978b). We now know that the pineal complex and pituitary have some involvement in metamorphosis. Pinealectomy (Cole and Youson, 1981) and hypophysectomy (Joss, 1985) prevent metamorphosis in ammocoetes which have reached a certain condition factor. Removal of the rostral pars distalis inhibits metamorphosis, while extirpation of only the caudal pars distalis will retard the process but permit its completion. The rostral pars distalis is important in initiating metamorphosis, but the caudal portion must be present to complete the phase. According to Joss (1985), these experimental results imply that more than one hormone is involved in lamprey metamorphosis and that they are likely thyrotrophin, somatotrophin, and gonadotrophin. However, cells of the caudal pars distalis, but not the rostral portion, show increase in activity (cell division and granulation) during the early stages of metamorphosis in *P. marinus* (G. M. Wright, personal communication). Furthermore, present assumptions of Joss (1985) of hypophyseal hormone activity during lamprey metamorphosis are complicated by the fact that growth of the caudal pars distalis during metamorphosis is much greater in nonparasitic *Lampetra planeri* than in parasitic *L. fluviatilis* (M. W. Hardisty, personal communication). This would suggest implication of this region of the pituitary in regulating gonadal growth and sexual maturation, both of which only occur in the nonparasitic species at this time. No doubt future studies on lampreys will be directed toward a definitive identification of the hypophyseal hormones and the cells involved in this stimulus to development. In the meantime, we are only left with an impression that both rostral and caudal pars distalis are evidently involved in lamprey metamorphosis.

Assuming that there is some hormonal control from the pituitary, the mode of action of these hormones may be to act directly on the tissues undergoing the developmental change or, as in amphibian metamorphosis (White and Nicoll, 1981; Fox, 1984), to stimulate other endocrine glands to release hormones that act upon the tissues. In light of the importance of thyroid hormones to amphibian metamorphosis, the only hormones to be examined during fish first metamorphosis are thyroxine (T4) and triiodothyronine (T3). In lampreys, transformation of the larval endostyle to a thyroid gland with follicles is initiated very early in metamorphosis (Youson *et al.*, 1977; Wright and Youson, 1976, 1980) and is accompanied by a dramatic decline (Fig. 12) in serum levels of both T4 (Wright and Youson, 1977) and T3 (Lintlop and Youson, 1983a). The drop in T3 levels is not correlated with increase in binding capacity of this hormone to receptors in hepatocyte nuclei of the liver (Lintlop and Youson, 1983b), but other organs have not been examined. The significance of the marked decline in circulating levels of thyroid hormones to the metamorphic phase is not understood. However, these data, and those from attempts at initiating lamprey metamorphosis by external T4 stimulation (Leach, 1946), do not suggest an involvement of thyroid hormones, at least in the manner that is seen in amphibians (White and Nicoll, 1981).

T4 enhances yolk resorption, growth, development, and survival of larvae of many fish species (Lam, 1980, 1985; Lam and Sharma, 1985; Lam *et al.*, 1985). Feeding of the larvae of the mudskipper with thyroid extracts accelerates their metamorphosis, while leptocephali (Vilter, 1946) may be stimulated into metamorphosis by treatment with

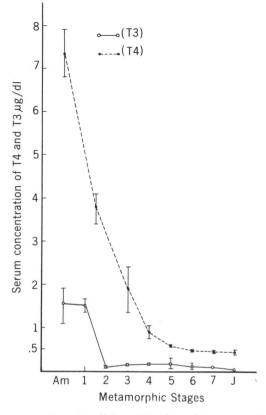

Fig. 12. Sera concentrations (μg dl^{-1} ± 2SE) of triiodothyronine (T3) and thyroxine (T4) in ammocoete (Am), metamorphosing stages (1–7), and juveniles (J) of *Petromyzon marinus*. [Data from Wright and Youson (1977) and Lintlop and Youson (1983a).]

T4. Recent studies on flounder larvae (Inui and Miwa, 1985; Miwa and Inui, 1987a,b) have provided definitive evidence that the thyroid plays a major role in initiating first metamorphosis in bony fish. T4 induces metamorphosis, while antithyroid agents arrest the process (Fig. 13). The effect of T4 is dose dependent with 100 ppb and 10 ppb (but not 1 ppb) in sea water resulting in metamorphosis (Miwa and Inui, 1987b). T3 is several times more potent than T4 in inducing metamorphosis. These experimental studies confirm earlier histological investigations that showed an increase in activity of the thyroid gland during eel and plaice metamorphoses (Murr and Sklower, 1928; Sklower, 1930). Histological and immunohistochemical studies with antisera to thyroxine (T4) has revealed that the thyroid follicular cells

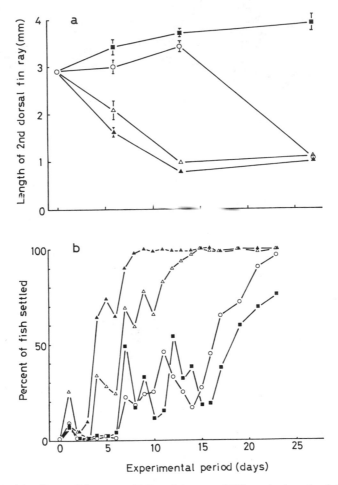

Fig. 13. (a) Effects of thyroxine (T4) and thiourea (TU) on the length of the second elongated dorsal fin ray (○, control; ■, 30 ppm TU; △, 0.05 ppm T4; ▲, 9.1 ppm T4. (b) Effects of T4 and TU on settling behavior of flounder larvae (○, control; ■, 30 ppm TU; △, 0.05 ppm T4; ▲, 0.1 ppm T4. T4 treatment terminated after 16 days. [From Inui and Miwa (1985).]

show high activity during metamorphic climax in Japanese flounder and appear inactive at postclimax (Miwa and Inui, 1987a). These authors also observed that cells in the proximal pars distalis that are immunoreactive to antithyrotropin showed an activity during metamorphosis that reflects the control of the anterior pituitary on thyroid activity. Furthermore, the administration of T4 and thiourea reveals that negative feedback regulation exists in flounder. These studies are a clear indication that a thyroid–pituitary axis is involved in metamorphosis in flounder.

Sterba (1955) was able to produce partial metamorphosis in ammocoetes by corticortrophin injections and there seems to be an increase in activity of the essential enzyme Δ^5-3β-hydroxysteriod dehydrogenase in the presumptive adrenocortical (interrenal) tissue at this time in the life cycle (Seiler et al., 1981). Whether adrenocorticosteriods have any significance in fish first metamorphosis, as they have recently been shown to have in amphibian metamorphosis (Krug et al., 1983), should be a future consideration.

VI. DURATION

The length of time between the initiation of metamorphosis and the appearance of a definitive phenotype (juvenile, smolt) is generally species-specific. In those species that initiate metamorphosis soon after hatching of the embryo in order to remodel transitory embryonic larval structures, this phase takes up most of larval life. In contrast, in species in which metamorphosis is initiated and completed as a single dramatic climax, the event may last only a few days to a few weeks, such as in Pleuronectiformes (Ahlström et al., 1984) and Albula sp. (Rasquin, 1955), or be as long as to 3–4 months, as in most lampreys (Potter, 1980b), and 9–10 months, as in Auguilla sp. (Schmidt, 1906) and a few lampreys (Beamish and Youson, 1987). As metamorphosis is usually an interval of ontogeny in many fishes when there is an alteration in feeding or digestive mechanisms, the termination of the metamorphic phase may be determined by the time at which the animal is able to resume, begin, or alter feeding. Thus, metamorphosis is completed when Siganus lineatus becomes a carnivore (Bryan and Madraisau, 1977), when P. marinus starts feeding parasitically (Potter and Beamish, 1977; Youson and Horbert, 1982; Langille and Youson, 1984a,b), or when Pleuronectiformes begin to inhabit the bottom (Fukuhara, 1986). That the acquisition of the external characters of the juvenile is not sufficient evidence to assume the end of metamorphosis is also illustrated in the ability of the animals to osmoregulate in their new niche. The ability of Anguilla sp. to osmoregulate in fresh water and the concomitant change in behavior are important criteria for assessing the termination of the metamorphosis of glass eels from leptocephali (Tesch, 1977). Metamorphosis in anadromous lampreys involves a transformation of organs of osmoregulation (Beamish,

1980a; Youson, 1980, 1981a,b), and the completion of this development, such as in the esophagus (Richards and Beamish, 1981; Beamish and Youson, 1987) and kidneys (Youson, 1982a,b), dictates the time at which migration into saltwater occurs and the end of the larva period. The duration of the metamorphic phase of larval life in all lamprey species, whether parasitic or nonparasitic, is long and probably is a reflection of the complexity of change that takes place, particularly in those tissues and organs that are (were) involved in osmoregulation in present and ancestral individuals. Thus, all extant nonparasitic species of lamprey undergo the same metamorphic change over a protracted phase common to freshwater and anadromous parasitic species.

VII. EVENTS

The metamorphic interval of fish ontogeny involves two major events: initiation and climax. These events may be separated by a considerable time period or be almost simultaneous (see Section II,B.). They are characterized by a number of morphological and physiological changes that ultimately determine the metabolic and behavioral patterns of the juvenile. This section describes the changes occurring during these events.

A. Changes in Body Length and Proximate Composition

The larva period is usually typified by some growth as the animal acquires nutrition through an exogenous feeding habit (Balon, 1986) or, perhaps, through integumentary absorption of dissolved organic matter (Pfeiler, 1986). In species that have a significant and prolonged climax event, growth usually stops at metamorphosis, and this has been corroborated through examination of otoliths (Van Utrecht, 1982; Campana, 1984; Victor, 1986). In many species there is actually a decrease in body length, and this can be attributed to the cessation of feeding. The extent of decrease in body length is particularly variable among bony fishes and may be related to the degree of ossification of the skeleton that took place in earlier larval life (Ahlström and Counts, 1958): that is, it is more difficult for an animal to shrink in the presence of a hard-tissue endoskeleton. Although both lampreys (Potter, 1980b) and most elopomorphs (Harden Jones, 1968; Smith, 1971; Tesch, 1977; Pfeiler, 1986) show a reduction in length of at least a few millimeters during metamorphosis, a most spectacular reduction is seen in the bonefish *Albula* sp., where larva of 60–70 mm reduce by over 50% of their length (Hollister, 1936; Rasquin, 1955; Pfeiler and Luna, 1984). This process takes only 8–12 days (Rasquin, 1955) and is due to the resorption of a transparent, extracellular gelatinous matrix. A similar resorption of a gelatinous mass occurs during the metamorphosis of other elopomorphs (orders Anguilliformes, Elopiformes, and Notacanthiformes), but the decline in length is not always as marked as in *Albula* (Kubota, 1961; Hulet, 1978; Pfeiler, 1986). However, *Megalops atlanticus* shows a characteristic decline in length from 28 to 13 mm during its "negative growth phase" (Mercado and Ciardelli, 1972).

According to Pfeiler (1986), the leptocephali of all elopomorphs utilize the stored carbohydrates and lipids of the gelatinous matrix during their nontrophic phase of metamorphosis. However, definitive evidence has only been obtained during metamorphosis of *Albula* sp. In this bonefish there is a 50% loss of total lipid, 83% loss of carbohydrate content, and 52% decline in ash. Although protein content shows no significant change, nonprotein nitrogen declines from 70 to 58% of total nitrogen from the beginning to the end of metamorphosis (Pfeiler and Luna, 1984). Most of the proteins and carbohydrates that are lost are associated with the glycosaminoglycan fraction of the gelatinous matrix (Rasquin, 1955). In early metamorphosis keratan sulfate is the principal component of the glycosaminoglycans, but it is replaced by chondroitin sulfate at later times. Pfeiler (1984c) suggested that this change is related to the increasing importance of the glycosaminoglycans as the most suitable environment for housing morphogenetic events that are part of remodeling during metamorphosis.

Albula larvae lose 78% of their water, 83% of Na^+, and 91% of Cl^- but no K^+ during their metamorphosis (Pfeiler, 1984a), and this is the primary cause of shrinkage of the animal. The loss of ions and water is independent of environmental salinity (Pfeiler, 1984b). This confirms earlier studies on leptocephali of *Anguilla* sp. that indicated that older individuals have lower Na^+ and Cl^- concentration (Hulet *et al.*, 1972) and water content (Fontaine, 1975) than do younger animals.

From the above data it is speculated that similar mechanisms of energy utilization are likely present in metamorphosing leptocephali of many elopomorphs (Pfeiler, 1986). All lamprey species catabolize stored energy during their metamorphoses (Fig. 14), but the extent of utilization of protein and lipid may vary (Beamish and LeGrow, 1983). The storage of lipid prior to metamorphosis (14% wet body weight) and its utilization (down to 8% wet body weight) during this long

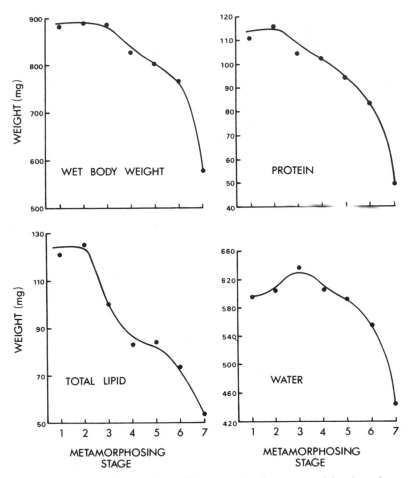

Fig. 14. The wet body weight and the weight of water, total lipid, and protein calculated for standard animals representing stages 1–7 in the metamorphosis of *Geotria australis.* [From Bird and Potter (1981).]

nontrophic phase is well documented (Lowe *et al.,* 1973; Moore and Potter, 1976; Youson *et al.,* 1979; Bird and Potter, 1981; Beamish and LeGrow, 1983). The absolute amount of total lipid declines by 57% (Bird and Potter, 1981). It seems that these stored lipids are catabolized to provide energy for protein synthesis (O'Boyle and Beamish, 1977). However, total protein declines by over 50% in *Geotria australis,* suggesting that it too may be catabolized as an energy source in this species, particularly when neutral fat has been depleted (Bird and Potter, 1981).

B. Changes in Morphology

The metamorphic interval of fish ontogeny involves three morphogenetic processes: transformation of larval tissues and organs into those of the adult, regression and eventual loss of larval structures, and the formation of new adult tissues and organs from anlagen which may have persisted since embryonic life. The complexity and breadth of this phase of the larva period in fishes are illustrated by example in the following discussion.

1. TRANSFORMATION

Presently the term transformation is used in the literature to describe most metamorphic events, when in fact it is only one of three morphogenetic processes taking place. Transformation involves both tissue regression and proliferation in functioning larval structures and is responsible for such spectacular features as eye migration in flounders, changes in the shape of the willow leaf-like leptocephalus to a glass eel, modification of the buccal funnel of ammocoetes to the suction-like disc of juvenile lampreys, and lengthening of the tail of the ribbonfish. Examples of more subtle internal changes involving tissue transformation are elongation of the intestine in logperch (Grizzle and Curd, 1978), *S. lineatus* (Bryan and Madraisau, 1977), and turbot (Cousin and Baudin Laurencin, 1985), remodeling of the blood supply to the pelvic and pectoral fins in *Auguilla* spp. (Willemse and Markus-Silvis, 1985), transformation of the brain (Fig. 15) in conger eel (Kubota, 1961), forward movement of dorsal and anal fins in many species (Castle, 1984), creation of the asymmetry in the projections of the right and left olfacatory bulbs (Rao and Finger, 1984), 90° relative displacement of the vestibular and ocular coordinates (Graf and Baker, 1983), and development of the lateral line system (Neave, 1986) in flatfishes. Transformation of ammocoete tissues and organs has received considerable attention. A few such transformations are development of intestinal folds (Youson and Connelly, 1978; Youson and Horbert, 1982; Hilliard *et al.*, 1983) and chloride cells of the gills (Peek and Youson, 1979), modification of the ammocoete endostyle to a thyroid with follicles (Wright and Youson, 1976, 1980; Wright *et al.*, 1980) and the bile duct into an endocrine pancreas (Hilliard *et al.*, 1985; Tsuneki and Ouji, 1984b; Youson, 1985; Elliott and Youson, 1987), and displacement of the larval hemopoietic sites (typhlosole and nephric fold) to the fat column (Percy and Potter, 1976, 1977; Potter *et al.*, 1978a; Ardavin *et al.*, 1984).

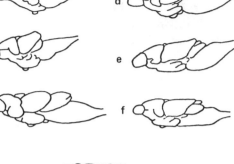

Fig. 15. Morphological changes of shape and proportion of the regions of the brain of the conger eel *Conger myriaster* during metamorphosis (a–g). From Kubota (1961).

2. REGRESSION

Regressive processes require that tissue components be either degraded by hydrolytic enzymes or catabolized as an energy source. The utilization of stored lipids by ammocoetes or of the gelatinous (glycosaminoglycans) matrix of eels (Section VII,A) illustrates regression by catabolism. In the case of ammocoetes, as lipid of the nephric fold is gradually used, the site becomes progressively occupied by newly formed kidney tissue (Ooi and Youson, 1977). Atrophy of the larval kidneys in the conger eel (Kubota, 1961) and lampreys (Ooi and Youson, 1979; Tseunki and Ouji, 1984b), loss of the pelvic appendage in gibberichthyids (de Sylva and Eschmeyer, 1977) and of the gas bladder of flounders (Richardson and Joseph, 1973), and shedding or resorption of teeth (Evseenko, 1978) and spines (Amaoka, 1973; Futch, 1977; Inui and Miwa, 1985) in many species likely all first involve an autolysis followed by some resorptive process. A particularly unusual regressive process takes place in the liver of all lamprey species in that both the gall bladder and all bile ducts disappear (Youson and Sidon, 1978; Youson, 1981c). This biliary atresia involves both hydrolytic enzymes and fibrosis (Sidon and Youson, 1983; Yamamoto *et al.*, 1986), features that are characteristic of a fatal disorder, of the same name, in humans.

3. DEVELOPMENT FROM ANLAGEN

Many processes involved in the development of definitive adult organs and tissues in fishes are initiated but arrested at an early embryo period. Metamorphosis is a phase of the larva period when these developmental processes are reinitiated. Organ or tissue rudiments (termed anlagen or primordia) and their cells are stimulated to differentiate during metamorphosis. They either may have been arrested in an advanced form of development so that they resemble, but do not function like, adult structures, or are still in an unrecognizable form. In the latter case, the anlagen may be a clump or a cord of undifferentiated cells or be represented by cells diffusely scattered among differentiated cells. Thus, the retina of ammocoetes does not complete its development until metamorphic climax (Dickson and Graves, 1982), while the definitive kidneys (Ooi and Youson, 1977; Youson and Ooi, 1979; Youson, 1984) and adult esophagus, buccal glands, and olfactory sacs (Youson, 1980) all originate from separate anlagen, which exist in an inactive but recognizable form throughout earlier larval life. This latter form of development is seen in the formation of the intermuscular bones of flounder (Hensley, 1977), the definitive kidney of eels (Kubota, 1961), and the definitive teeth of many fishes. Some of the cartilaginous skeleton associated with the new feeding apparatus of adult lampreys develops from scattered undifferentiated cells that congregate to form a blastema in early metamorphosis (Armstrong et al., 1985).

C. Physiological Change

1. BLOOD

There is little direct information on physiological or biochemical change in bony fishes during metamorphosis. Most present data are implied from observations of premetamorphic larvae and juvenile or mature adults. Blood has received the greatest attention. Hemoglobin and erythrocytes are absent in leptocephali of many eel species (Kubota, 1961; Castle, 1984), and the second hemoglobin of adult tilapia (Perez and Maclean, 1976) is not present in larvae. Near the middle of metamorphosis (stages 4 and 5) in the lamprey P. marinus there is a mixed pattern of larval and adult hemoglobin, but the typical adult pattern is present in the juvenile (Beamish and Potter, 1972). The initiation of this phase in lampreys is marked by a decrease in the hematocrit (Beamish and Potter, 1972; Macey and Potter, 1981; Beamish and Thomas, 1984). There is usually a concomitant drop in hemoglobin concentration as erythrophagocytosis takes place in the liver (Percy and Potter, 1981). The nature and quantity of serum proteins also undergo modification during lamprey metamorphosis (Filosa *et al.*, 1986). Crossed immunoelectrophoresis has shown a larval protein, AS, remains until early metamorphosis but subsequently disappears. A gradual increase in concentration of two serum proteins, SDS-1 (a glycoprotein) and CB-III (a lipoprotein), which eventually will make up 85% of the total serum protein in mature adults, marks the beginning of metamorphosis (Fig 16). The functions of these larval and adult proteins are as yet unknown. A female-specific serum protein (FSSP), likely vitellogenin, is first detectable in metamorphosing stage 4 of the nonparasitic species *Lampetra reissneri* (Fukayama *et al.*, 1986). This event coincides with the beginning of vitellogenesis in the liver (Fukayama, 1985) and with the onset of sexual maturation in females of this species. It would be of interest to establish whether the prolonged juvenile period of parasitic species results in a delay in the time of appearance of FSSP in the serum.

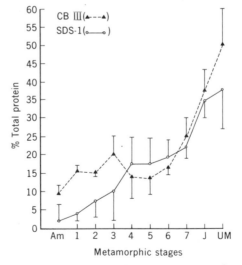

Fig. 16. The serum proteins CB-III and SDS-1 represented as a percentage of the total serum protein at various intervals of the life cycle in *Petromzon marinus*. Am, ammocoete; metamorphic stages (1–7); J, juvenile; UM, upstream migrant. [From Filosa *et al.* (1986) and J. H. Youson (unpublished data).]

Lampreys undergo a dramatic alteration in iron metabolism at metamorphosis. The plasma of larval G. *australis* has 19,000 μg Fe/dl and only 34 μg Fe/dl in mature adults (Macey *et al.*, 1982a). The major iron-binding protein of larvae is ferritin, while that of adults is more like transferrin (Macey *et al.*, 1982a, 1985). During metamorphosis the liver gradually accumulates iron (Macey *et al.*, 1982b; Youson *et al.*, 1983a; Sargent and Youson, 1986; Smalley *et al.*, 1986) and eventually becomes iron-loaded by the juvenile period (Youson *et al.*, 1983b), while at the same time plasma nonheme iron becomes much lower (Smalley *et al.*, 1986). A mechanism present in the intestinal epithelium of ammocoetes for elimination of excess iron (Macey *et al.*, 1982b) is likely lost at metamorphosis. The effects of the loss of this mechanism on body iron concentration remains to be investigated.

2. OXYGEN CONSUMPTION

There is a gradual increase in oxygen consumption during lamprey metamorphosis (29.3 to 60.4 μl g^{-1} h^{-1} at 10°C), and this accompanies extensive modification to the branchial chambers (Lewis and Potter, 1977; Lewis, 1980). The animal develops a high affinity for oxygen, which can be linked to changes in the hemoglobins at metamorphosis (Potter and Brown, 1975) but also might be somewhat related to the large number of catecholamine-secreting cells developing at this time (Epple *et al.*, 1985). A change in activity is correlated with a high rate of hemoglobin synthesis and subsequent oxygen-carrying capacity during metamorphosis in herring and plaice (de Sylva, 1974). Laurence (1975) has provided a comprehensive comparison of oxygen utilization of a larval bony fish in premetamorphic and metamorphic phases. In the larva period of winter flounder *Pseudopleuronectes americanus*, the absolute values of oxygen consumption (microliters per hour) increase until metamorphosis, at which time they decline. Oxygen consumption increases after the completion of metamorphosis. Metabolic rate [(μl μg^{-1} h^{-1}) × 10^3] decreases with increasing size from hatching through metamorphosis. Both of these parameters increase during the larva period when the temperature is raised. The change in values during the larva period are believed to reflect behavioral and morphological changes as a result of a switch from cutaneous to gill respiration (Laurence, 1975).

3. ELECTROLYTE BALANCE

There is a limited amount of evidence that metamorphosis in eels may involve an alteration in electrolyte balance (Hulet *et al.*, 1972). It seems that *Anguilla* sp. is able to modify its osmotic permeability so that net water movements are minimized in either fresh water or seawater (Evans, 1984). Such a mechanism exists in *Albula* sp. throughout metamorphosis. Although larvae of this species lose most of their water and NaCl content during metamorphosis, this net loss is not effected by subjecting animals to dilute (8‰), normal (35‰), and concentrated (48‰) seawater (Pfeiler, 1984b). The losses are due to absorption of a larval gelatinous matrix. Serum osmolality rises during early stages of lamprey metamorphosis and gradually attains the adult values ($\sim$300 mosmol kg^{-1}) by the end of the phase (Mathers and Beamish, 1974). An attempt to acclimate metamorphosing landlocked *P. marinus* to even 10‰ at stages 5 and 6 results in serum osmolality of up to 340 mosmol kg^{-1} but only 300 at later stages (cited in Beamish, 1980a).

4. HORMONES

Morphological evidence indicates that there is a change in activity of the thyroid gland during first metamorphosis in all fishes (Murr and Sklower, 1928; Leach, 1946; Miwa and Inui, 1987a). However, to date, this has only been documented in lampreys through analysis of sera concentration of the major thyroid hormones (Fig. 12). Both sera T3 and T4 concentrations decline abruptly at the very onset of metamorphosis and remain close to these levels throughout the rest of the life cycle (Wright and Youson, 1977; Lintlop and Youson, 1983a). There is little doubt that this drop in circulating levels of thyroid hormones is a significant physiological change, but it must be kept in mind in any interpretation of significance that sera levels of T3 and T4 in ammocoetes are as much as 10 times those of most vertebrates and that juvenile levels are within the normal vertebrate range. Future studies must be directed toward analyzing the relationship of sera T3 and T4 concentrations with changing metabolic activity.

Metamorphosis in eels (L'Hermite *et al.*, 1985) and that in lampreys (Elliott and Youson, 1986, 1987) share the common feature of being the phase of larval life when a specific peptide hormone is first synthesized by the endocrine pancreas. Whereas insulin appears in eel metamorphosis to accompany the somatostatin that has been present since early larval life, somatostatin appears during lamprey metamorphosis to join existing insulin. There is some suggestion from morphological (Sterba, 1955) and histochemical (Seiler *et al.*, 1981) observations that adrenocortical hormones may be significant during lamprey metamorphosis; however, no definitive evidence exists.

5. METABOLISM

In metamorphosis of the whitefish *Coregonus* sp., climax is marked by an increase in activity of enzymes of the glycolytic pathway (phosphofructokinase, pyruvate kinase, and lactic dehydrogenase) and decline in activity of enzymes of the oxidative and citric acid cycle (Fig. 17). Therefore, during metamorphosis of whitefish there may be a switch from a total utilization of amino or fatty acids to increased use of glycogenolysis (Forstner *et al.*, 1983). The biochemical pathways have not been examined in *Albula* sp. during metamorphosis, but there is strong evidence to indicate that endogenous lipid

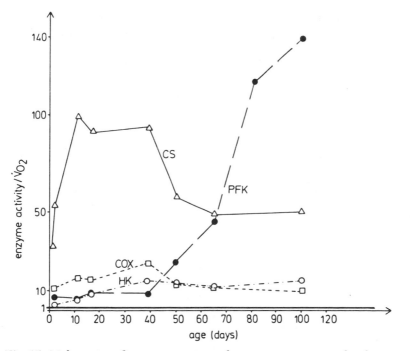

Fig. 17. Molar ratios of enzyme activity and oxygen consumption development of *Coregonus* sp. from citrate synthase (CS), phosphofructokinase (PFK), cytochrome oxidase (COX), and hexokinase (HK). The ratios were calculated on the bases of stoichiometric oxygen requirements for each enzyme-catalyzed reaction. [From Forstner *et al.* (1983).]

and carbohydrate stores are utilized as a source of energy (Pfeiler and Luna, 1984). Similarly, during the metamorphosis of lampreys, both lipids and carbohydrate stores are mobilized (O'Boyle and Beamish, 1977; Bird and Potter, 1981). A rise in serum glucose is correlated with a rapid decline in liver glycogen but a slower decline in muscle. An eventual cessation of the liver lipogenic process occurs as the activity of the relevant enzymes decline. The digestive enzyme complement changes at metamorphosis in the lake surgeon *Acipenser fulvescens* and is genetically predetermined (Buddington, 1985). Metamorphosis of this species represents a transition period when lipase, amylase, and chymotrypsin concentrations decrease and pepsin and trypsin increase.

Free amino acid composition has been examined for *Albula* sp. during their metamorphosis (Pfeiler, 1987). Although there are no changes in ninhydrin-positive substances, there is a significant decrease in a number of essential amino acids. Leucine, isoleucine, phenylalanine, histidine, valine, methionine, lysine, and arginine represent 47% of the total amino acids in early metamorphosis but are reduced to only 23% in advanced stages. Taurine is the most abundant free amino acid in whole-body extracts and accounts for 36% and 59% of the total, by weight, in early and late metamorphosing individuals, respectively. This amino acid may be important as an organic osmolyte in order to maintain intracellular volumes at a time when there is a high extracellular loss of water and salt during resorption of the gelatinous matrix. However, taurine may also increase as a necessary compensation for the decrease in total essential amino acids. The nonessential amino acids as a group generally show only small change during metamorphosis of the bonefish, but certain components show marked increases (glycine and glutamic acid) while others are reduced (tyrosine and serine).

D. Behavioral Changes

The activity levels of metamorphosing fishes usually are related to whether this phase is trophic or nontrophic and to the degree of metamorphic change. In species that continue to feed, such as *Siganus lineatus* (Bryan and Madraisau, 1977), the carnivorous larvae are active, but with early metamorphosis and increased coiling of the gut they become omnivorous, periodically active, and stop to examine benthic algae and detritus. Late metamorphosis is marked by further coiling of the gut, less activity, and a completely herbivorous diet.

Changes in feeding behavior (pelagic to benthic) have been noted for the logperch *Percina caprodes* during its subtle and inconspicuous metamorphosis (Paine and Balon, 1984), and are related to development of the stomach, elaboration of the pancreas, and the transition from a physostomous to a physoclistous swimbladder (Grizzle and Curd, 1978). Blaxter and Staines (1971) noted a decreased ability of plaice and sole to search for food during metamorphosis, that is, a drop in distance covered per minute and a decrease in time spent in feeding activity. They found a species variation in the ability to take food in the dark during eye migration. On the other hand, in many flatfishes there is a decline in ingestion rates and rate of food requirement at metamorphosis (Riley 1966; Laurence, 1977) due to their high efficiency of converting food to growth at this time in the life cycle (Laurence, 1977). T4 stimulation (Inui and Miwa, 1985) will shorten the period of time required for metamorphosing flounder to become benthic (Fig. 13). It is believed that *Taenioconger* sp. form colonies during their metamorphosis, which is a prelude to their adult fossorial behavior (Raju, 1974). The increase in density of late larvae or metamorphosing individuals may be due to transport by tidal currents during their pelagic life (Tanaka, 1985). Changes in swimming performance accompany the onset of metamorphosis and often coincide with the development of pink and red muscle fibers (Forstner *et al.*, 1983) and/or alterations to the finfold (Fukuhara, 1986).

Many behavioral changes are related to the development or alterations of the nervous system (Webb and Weihs, 1986). The presence of the lateral line system during the extension of the tail of the ribbonfish provides an antenna for the reception of water displacement and low-frequency sound. It may have acquired a head up–tail down position in order to sense predators that are below the field of view of the eyes (Rosenblatt and Butler, 1977). The flatfishes have received a considerable amount of attention because of the potential interruption to their sense of balance during eye migration and when there is no compensatory change of the vestibular system as they lose their symmetry (Platt, 1973). There are slight differences in behavior between plaice and turbot during metamorphosis, but by the end of metamorphosis the influence of light on the sense of balance is almost lost (Neave, 1985). As metamorphosis is approached in these two species, behavioral visual acuity reaches a plateau and remains unchanged (Neave, 1984). Nocturnal activity and the switch from a pelagic to benthic habitat in flounders coincides with the development of twin cones in the retina (Kawamura and Ishida, 1985).

VIII. SIGNIFICANCE

A. Ontogeny and Phylogeny

Development is of two forms, ontogenetic and phylogenetic (Balinsky, 1970; Gould, 1977), and it is clear that, as in amphibian metamorphosis (Norris, 1983), first metamorphosis in fishes often reflects on both of these. According to Gould (1977), there has been a change during evolution in the relative time of appearance and rate of development of characters that were present in the ancestor (heterochrony). This is one of the ways in which ontogeny has been involved in evolution and would explain the presence of a number of developmental strategies during the ontogeny of fishes. Metamorphosis is one developmental strategy in fishes that has been selected to permit a delay in development of definitive characters. A comparison of individuals within some groups of similar fishes suggests that variation in the duration of their larval life (age or length at metamorphosis) may reflect this evolutionary trend. For example, *Megalops atlantica* metamorphoses at an earlier age (2–3 months) and a smaller size (30 mm) than many other species of eels (Smith, 1984). Does this mean that this species is representative of a more "primitive" condition among the eels? This may not necessarily be the case, for the variation in length of larval life between North American and European varieties of *Anguilla* sp. may be essential to their distribution (Tesch, 1977) and be a reflection of the delay in the trigger to metamorphosis rather than differences in their evolutionary history. The tendency for lengthening of larval life during the phylogenetic development of a species has also been suggested both for flatfishes (Moser, 1981) and lampreys (Hardisty, 1979). In Pleuronectiformes, larvae metamorphose over a range of 5–120 mm, but most are between 10 and 25 mm. Metamorphosis at the smaller and larger sizes are examples of "derived states," and species have either shortened larval life because of limited food or habitat or have prolonged it in order to ensure dispersal (Hensley and Ahlström, 1984). However, at least in bony fishes, the developmental strategy of delayed metamorphosis does not always imply a period of increased larval growth (Victor, 1986).

The concept of "paired species" among lampreys, with a nonparasitic form derived from a closely related parasitic species (Hardisty and Potter, 1971b), has introduced many questions about the evolutionary history of lampreys. For example, there have been suggestions of the introduction of pedomorphosis among extant lampreys where sexual maturation may accompany or precede metamorphosis (Zanandrea, 1956). There is no definitive evidence to dispute a claim that fossil *Mayomyzon* (Bardack and Zangerl, 1971) may have been pedomorphic. Its size of 3–6 cm is approximately one-third the length of most extant lampreys at metamorphosis or as juveniles, implying that, if larval life was present in ancestral lampreys, it must have been of short duration (Hardisty, 1983). Zanandrea (1956, 1957) also described neotenous ammocoetes in relatively primitive nonparasitic *L. zanandreai*. However, we cannot assume that *Mayomyzon* in its small adult size was less specialized than extant adult lampreys, for the fossils may be of juveniles that had just recently metamorphosed from slow-growing ammocoetes of advanced age. Recently, Vladykov (1985) gave his assessment of previous reports of neoteny in lampreys and concluded that it does not exist. However, discovery on the west coast of Canada of newly metamorphosed individuals that are capable of feeding in the laboratory while sexually mature (Beamish, 1985) may open a further chapter to this discussion. A second major question is, are the extant nonparasitic lampreys, which undergo a prolonged larval period and a rapid sexual maturation following metamorphosis with no distinct juvenile period, representative of more recent trends in lamprey evolution? In other words, is there a repetitive trend toward the elimination of feeding in postmetamorphic lampreys (Vladykov, 1985)? Hardisty (1983) believes that lamprey evolution has been characterized by a resistance to a change in the time to sexual maturity, but the duration of larval life, and hence the time when metamorphosis is initiated, have been subjected to a great deal of adaptive modification. In the case of nonparasitic species, circumstances favored an extension of larval life and a corresponding reduction of adult life, while they favored a lengthened larval life but a retention of an extended juvenile period in parasitic species.

Szarski (1957) claims that a larval form is present in the life history of those fish species that show primitive characteristics and a larva period must be concluded by metamorphosis. It is his belief that a larva period was present in all vertebrate ancestors. A contrasting view has been provided for amphibians (Norris, 1983). As labyrinthodonts possessed large amounts of yolk in their eggs and development may have been direct, larval life and metamorphosis in extant amphibians may be a derived state. In this context, it is noteworthy that another extant agnathan and the most primitive living vertebrate, the hagfish, lacks a larva period and has large yolky eggs—that is, it has direct development. If lampreys and hagfishes share a common ancestry, then metamorphosis in lampreys may also be a derived state.

However, differences in ontogeny between the two extant agnathans can probably be explained by the fact that they followed quite different evolutionary histories (Hardisty, 1979, 1982). Taking the opposite stand, if retention of a larva period and metamorphosis are primitive and less specialized ontogenetic features among fishes, there must be selective advantage for their persistence in some species. The maintenance of a complex life cycle (including metamorphosis) in many anuran amphibians is related to the fact that tadpoles are highly specialized suspension feeders. This method of feeding is adopted in order to take advantage of rapid rises in primary production (Wassersug, 1975). Similarly, following detailed observations of metamorphosis in *Albula* sp., Pfeiler (1986) makes a claim that retention of the leptocephalous strategy in all orders of the "primitive" teleost Superorder Elopomorpha has been selected because the premetamorphic larvae live in a nutrient bath that permits absorption of dissolved nutrients across the integumentary epithelium. With this simple and efficient method of obtaining nutrients, there was likely selective pressure for extension of the larval growth phase, which was followed by a metamorphic phase. However, there is no concrete evidence that the leptocephalous strategy is primitive. As in the case of the lamprey ammocoete (Hardisty, 1982), it is unwise to place too great a phylogenetic burden on the leptocephalus. Both the ammocoete and the leptocephalus are highly specialized for their particular mode of life and, during the course of their respective evolutionary histories, the adult and larval forms have increasingly diverged in their morphology and habits of life. This divergence in both groups has resulted in specialized, and far from primitive, larvae and adults and in a radical metamorphosis.

Divergence in larvae (tadpoles) and adults also seems to be a feature of the evolutionary history of anuran amphibians (Wassersug and Hoff, 1982). In fact, there is evidence to support the view that the metamorphic process has evolved (Wassersug and Hoff, 1982), and some extant anuran species with tadpoles may have evolved from species that lacked a larva period (Wassersug and Duellman, 1984). In summary, it cannot be assumed from our present evidence that the larval developmental strategy with its terminal phase of metamorphosis is a primitive form of ontogeny in fishes.

Szarski (1957) states that the greater the differences between the larva and the adult in both fishes and amphibians, the more striking is the metamorphosis. Notable among these differences are the habitat, the behavior, and the manner in which nutrients are acquired. Therefore, metamorphosis in fishes is characterized by variable need to break down larval structures and often to construct adult structures from embryonic anlagen. Thus, the almost sedentary, blind, suspension-feeding ammocoete, which resides in a burrow of silt in a freshwater stream, undergoes a dramatic first metamorphosis to a juvenile with suctoral mouth and rasping tongue, a salivary gland secreting an anticoagulant, a well-developed eye, a new form of ventilation (tidal), and innumerable changes to the digestive, excretory, and repiratory systems, which permit, in many species, a migration to saltwater. The major changes in the body form of eel leptocephali and larval flatfishes during metamorphosis also emphasize the great differences between these larvae and their definitive forms and the specialization of individuals in both periods of the life cycle.

B. Dispersal

Northcutt and Gans (1983) and Jollie (1982) propose that ancestral larval fish were suspension feeders like modern-day lamprey ammocoetes, while Mallatt (1984) models them as raptorial feeders on phytoplankton and zooplankton much like most larvae of extant bony fishes. Regardless of the evolutionary history of feeding mechanisms in larval fishes, the premetamorphic phase of the larva period of extant fishes has been selected by many species, for it provides an opportunity for storage of energy, for growth, and for dispersal. These features of the larva period are related to pelagic feeding, a feature most agree is characteristic of the larvae of ancestral fishes (Mallatt, 1985).

Balon (1986) considers metamorphosis to be a very costly interval, in terms of energy loss, during the ontogeny of fishes. The regression, remodeling of temporary larval organs, and development of new organs consume valuable energy. Furthermore, larvae of small size are more susceptible to predation. Therefore, small larvae have to be produced in large numbers or small numbers of large larvae have to be especially equipped to ensure survival of the species. Thus, leptocephali or teleplonic larvae (Futch, 1977) are equipped by their shape for dispersal by flotation over long distances where they will not compete for food with their definitive form. This passive, but selective, transport is not a high energy consumer and is also practiced by flatfishes (Rijnsdorp et al., 1985). In such cases, the larva period is probably protracted for the above reasons (Futch, 1977). One may then take the view that the larval interval is a specialized period of ontogeny that initially evolved to take the organism through the "critical period" (Braum, 1978), that is, the shift from yolk to exogenous energy source. Success in a pelagic life resulted in a prolongation of the period. Size disparity in metamorphosing individuals of several species suggests that the larger forms may represent highly successful larval adaptation to planktonic feeding and retardation of metamorphosis (Futch, 1977). This has ultimately resulted in greater dispersal, such as in *Anguilla* sp. (Tesch, 1977) and in the landlocked form of the sea lamprey, *P. marinus* (Smith, 1971). The success of the sea lamprey in the Great Lakes of North America is not so much a result of the ability of the adult to feed and to breed but instead is a reflection of the tremendous success of the ammocoete to colonize a new habitat. Colonization is one of the objectives of the larva period in vertebrates (Norris, 1983), and the ammocoete has utilized its efficient, suspension-feeding mechanism to advantage in order to grow to metamorphosing size. Another successful case of larval dispersal is seen in eels. According to Pfeiler (1986), the thin epidermis and the thin, leaflike body of high surface to volume ratio are responsible for the successful dispersal of the eel leptocephali. Circumstantial evidence is provided suggesting that the shape of the leptocephalus is not just related to passive locomotion. Instead, the high surface to volume ratio is important for efficient epithelial uptake of dissolved organic matter from the surrounding medium. This method of acquiring nutrition during the premetamorphic phase is important to meet the energy demands of both growing leptocephali and metamorphosing larvae. The latter obtain their nutrient requirements through breakdown of a the gelatinous matrix that was formed prior to the commencement of metamorphosis.

C. Physiological and Behavioral Adaptation

There are many examples among the fishes that illustrate the significance of metamorphic change to the survival and existence of their definitive phenotype. Many of the lesser known examples, such as the lengthening of the tail in ribbonfish and of the intestine of *Siganus lineatus,* and the change in the digestive enzymes of the lake sturgeon, have been mentioned in preceding sections. Blaxter (1986) has recently summarized the development of sensory structures in fishes and how these relate to behavioral modifications. Flatfishes are constantly referred to in any discussion of adaptive change in vertebrates.

Many have found that metamorphosis in Pleuronectiformes provides an opportunity to study adaptive changes in the vestibulo–ocular reflex system (Platt, 1973; Graf and Baker, 1983). Concomitant with a 90° relative displacement of the vestibular and ocular coordinate systems during metamorphosis is the likelihood of development of pathways that permit secondary vestibular neurons of the horizontal semicircular canal to make contact with vertical eye-muscle motoneuron pools on both sides of the brain (Graf and Baker, 1983; Kawamura and Ishida, 1985). This is a necessary response to permit suitable eye movements in the new benthic environment of fully metamorphosed individuals. Similarly, the development at metamorphosis of a duplex (rods and cones) retina from a retina in pelagic leptocephali with only rods is an adaptation to a need for enhanced visual acuity in elvers (Pankhurst, 1984) and juvenile flounders (Kawamura and Ishida, 1985) in a new environment.

Multiple hemoglobins seem to be the rule for adult fishes, but these hemoglobins are not all present in larvae (Iuchi, 1985). The addition of at least one more hemoglobin at metamorphosis allows the adult to exploit warmer, more saline environments (Perez and Maclean, 1976), with less oxygen (Fontaine, 1975) than in larval niches. Pelagic behavior may account for the complete absence or low concentration of hemoglogin in some larvae. In lampreys, in addition to the change in hemoglogin, metamorphosis in anadromous species is characterized by the development of a new kidney and ion-transporting epithelia in the gills and intestine (Beamish, 1980a) to permit adaptation to a marine environment.

The adaptive value of many metamorphic alterations is not fully understood. For example, there is a loss of the gallbladder and the bile ducts in all lamprey species (Youson, 1981c, 1985). Although water conservation for a marine existence seems plausible, an aductular liver is not seen in any other marine vertebrate. In parasitic lamprey species that feed on the blood and body fluids, the ingested material may be in a form that can be readily absorbed across the intestinal wall without the emulsification properties of bile. However, lamprey species that ingest large quantities of muscle and even organs also lack a gallbladder and bile ducts. As iron is a normal component of bile, iron-loading of tissues of postmetamorphic lampreys may be a consequence of their biliary atresia, but there also seems to be little adaptive value to this manifestation of metamorphosis. A definitive explanation for the presence of the transitory larval structures in a number of bony fishes also awaits further study (de Sylva and Eschmeyer, 1977).

IX. SUMMARY AND CONCLUSIONS

Metamorphosis in fishes is found at two intervals of their ontogeny (Fig. 3). A first metamorphosis occurs in all species having a larva period and is the phase that terminates this period and that leads into a juvenile period. Bony fishes with a classical first metamorphosis during their ontogeny are Anguilliformes, Elopiformes, Notacanthiformes, and Pleuronectiformes. First metamorphosis is also present in Petromyzontiformes but not in Class Chondrichthyes or in Myxiniformes. A second metamorphosis may occur in many fish species when the juvenile undergoes sexual maturation to the definitive adult, but it is not a true metamorphosis. First metamorphosis is a true vertebrate metamorphosis: it is a phase of the larva period in which change takes place in nonreproductive structures, it results in a change of niche, and it is characterized by a dramatic and abrupt change in form that is likely triggered by internal and environmental cues. In fishes, first metamorphosis does not usually involve an increase in length, and the marked change in form assures that the larva and adult do not look alike. First metamorphosis is emphasized as a phase of the larva period with initiation and climax events, but in many species these two may occur almost simultaneously. Stages of metamorphosis can be delineated by careful evaluation of changes in external characters.

The time of first metamorphosis varies among the orders of fishes but is usually related more to body length than to age. Growth rate within a population may also be an important determinant. In species with a protracted and nontrophic metamorphic phase, physiological preparation is essential to the completion of metamorphosis. The highly synchronized timing of metamorphosis of various populations of the same species suggest environmental influence such as temperature and photoperiod.

There is increasing evidence to indicate that first metamorphosis in fishes is controlled by a contribution of both environmental (particularly temperature) and hormonal factors. Thyroid hormones are likely involved in initiating and maintaining metamorphosis in bony fishes, and regulation of secretion of the thyroid gland is controlled by thyrotropin from the pituitary. Lamprey metamorphosis may involve the interaction between several hormones and also seems to require coordination from the pituitary gland.

The duration of the metamorphic phase often reflects the magnitude of the morphological and physiological change or the importance of synchronizing the termination of change with the availability of food or with environmental conditions. During the initiation and climax events, developmental processes include resorption of stored material for energy, regression (loss) of larval structures, transformation of larval structures into adult tissues and organs, and the development of new tissues and organs from embryonic anlagen or primordia. Physiological change includes alteration of blood chemistry, electrolyte balance, hormones, and metabolism. Behavioral modifications are concomitant with morphological and physiological change.

First metamorphosis in fishes is significant in discussions of the relationship between ontogenetic and phylogenetic development, yet it is uncertain whether it reflects a primitive condition or is a more recently derived state in vertebrate evolution. In some cases the larva period provides an opportunity during ontogeny for the wide dispersal of the species at low energy expenditure. Although it is a high-energy-consuming phase and often is marked by extensive depletion of stored energy, metamorphosis has a climax event whose completion is synchronized to a time when energy can be obtained through feeding or when conditions are ideal for continuing through to a sexually mature adult. A larva period that terminates with a metamorphosis has been a highly successful developmental strategy among fishes.

It is proposed that first metamorphosis in fishes be given more attention during studies of fish ontogeny, that standard criteria for defining this phase be adopted, and that universal staging be used as much as possible at the genus or species level. If such recommendations are followed, it will ensure the successful application of this large and specialized group of vertebrates to aquaculture, to furthering our knowledge of animal diversity, and for use in many other areas of biology. Fishes are presently utilized as essential research tools in neurophysiology and pathology. The fields of developmental biology, nutrition, and taxonomy will be immediate benefactors of any new information on fish first metamorphosis.

ACKNOWLEDGMENTS

This work was supported by grants from the Natural Sciences and Engineering Research Council and the Medical Research Council of Canada. Alison Barrett and Patricia Sargent provided some technical assistance and Dr. E. Balon, Dr. M. W. Hardisty, Dr. D. J. Macey, Dr. A. H. Weatherley, and Dr. R. Winterbottom all provided valuable comments on the manuscript. The author also appreciates the constructive comments of an anonymous reviewer.

REFERENCES

Ahlström, E. H. (1968). Review of "Development of fishes of the Chesapeake Bay region, an atlas of egg, larval and juvenile stages." Part 1. *Copeia,* pp. 648–651.

Ahlström, E. H., and Counts, R. C. (1958). Development and distribution of *Vinciguerria lucetia* and related species in the eastern Pacific. *Fish. Bull.* **58,** 363–416.

Ahlström, E. H., Amaoka, K., Hensley, D. A., Moser, H. G., and Sumida, B. Y. (1984). Pleuronectiformes: Development. *In* "Ontogeny and Systematics of Fishes" (H. G. Moser *et al.,* eds.), Am. Soc. Ichthyol. Herpetol., Spec. Publ. No. 1, pp. 640–670. Allen Press, Lawrence, Kansas.

Alderdice, D. F. (1985). A pragmatic view of early life history of fishes. *Trans. Am. Fish. Soc.* **114,** 445–451.

Amaoka, K. (1972). Studies on the larvae and juveniles of the sinistral flounders. III. *Laeops kitahurae. Jpn. J. Ichthyol.* **19,** 154–165.

Amaoka, K. (1973). Studies on the larvae and juveniles of the sinistral flounders. IV. *Arnoglossus japonicus. Jpn. J. Ichthyol.* **20,** 145–156.

Ardavin, C. F., Gomariz, R. P., Barrutia, M. G., Fonfria, J., and Zapata, A. (1984). The lymphohemopoietic organs of the anadromous sea lamprey. *Petromyzon marinus.* A comparative study throughout the life span. *Acta Zool. (Stockholm)* **65,** 1–15.

Armstrong, L. A., Wright, G. M., and Youson, J. H. (1985). The development of cartilage during lamprey metamorphosis. *Anat. Rec.* **211,** 12A.

Balinsky, B. I. (1970). "An Introduction to Embryology." Saunders, Philadelphia, Pennsylvania.

Balon, E. K. (1975). Terminology of intervals in fish development. *J. Fish. Res. Board Can.* **32,** 1663–1670.

Balon, E. K. (1979). The theory of saltation and its application to the ontogeny of fishes: Steps and thresholds. *Environ. Biol. Fishes* **4,** 97–101.

Balon, E. K. (1981). Saltatory processes and altrical to precocial forms in the ontogeny of fishes. *Am. Zool.* **21,** 567–590.

Balon, E. K. (1984). Reflections on some decisive events in the early life of fishes. *Trans. Am. Fish. Soc.* **113,** 178–185.

Balon, E. K. (1985a). Saltatory ontogeny and life history models revisited. *In* "Early Life Histories of Fishes" (E. K. Balon, ed.), pp. 13–30. Martinus Nijhoff/Dr. W. Junk Publishers, Dordrecht, The Netherlands.

Balon, E. K. (1985b). Reflections on epigenetic mechanisms: Hypothesis and case histories. *In* "Early Life Histories of Fishes" (E. K. Balon, ed.), pp. 239–270. Martinus Nijhoff/Dr. W. Junk Publishers, Dordrecht, The Netherlands.

Balon, E. K. (1986). Types of feeding in the ontogeny of fishes and the life-history model. *Env. Biol. Fish* **16,** 11–24.

Bardack, D., and Zangerl, R. (1971). Lampreys in fossil record. *In* "The Biology of Lampreys" (M. W. Hardisty and I. C. Potter, eds.), Vol. 1, pp. 67–84. Academic Press, London.

Barrington, E. J. W. (1961). Metamorphic processes in fishes and lampreys. *Am. Zool.* **1,** 97–106.

Barrington, E. J. W. (1968). Metamorphosis in lower chordates. *In* "Metamorphosis: A Problem in Developmental Biology" (W. Etkin and L. E. Gilbert, eds.), pp. 223–270. Appleton, New York.

Beamish, F. W. H. (1980a). Osmoregulation in juvenile and adult lampreys. *Can. J. Fish. Aquat. Sci.* **37,** 1739–1750.

Beamish, F. W. H. (1980b). Biology of the North American anadromous sea lamprey, *Petromyzon marinus. Can. J. Fish. Aquat. Sci.* **37,** 1924–1943.

Beamish, F. W. H., and Austin, L. A. (1985). Growth of the mountain brook lamprey *Ichthyomyzon greeleyi* Hubbs and Trautman. *Copeia*, pp. 881–890.

Beamish, F. W. H., and LeGrow, M. (1983). Bioenergetics of the southern brook lamprey, *Ichthyomyzon gagei*. *J. Anim. Ecol.* **52**, 575–590.

Beamish, F. W. H., and Potter, I. C. (1972). Timing of changes in the blood, morphology, and behaviour of *Petromyzon marinus* during metamorphosis. *J. Fish. Res. Board Can.* **29**, 1271–1282.

Beamish, F. W. H., and Thomas, E. J. (1984). Metamorphosis of the southern brook lamprey, *Ichthyomyzon gagei*. *Copeia*, pp. 502–512.

Beamish, R. J. (1985). Freshwater parasitic lamprey on Vancouver Island and a theory of the evolution of the freshwater parasitic and nonparasitic life history types. *In* "Evolutionary Biology of Primitive Fishes" (R. E. Foreman, A. Gorbman, J. M. Dodd, and R. Olsson, eds.), pp. 123–140. Plenum, New York.

Beamish, R. J., and Youson, J. H. (1987). The life history and abundance of young adult *Lampetra ayresi* in the Fraser River and their possible impact on salmon and herring stocks in the Strait of Georgia. *Can. J. Fish Aquat. Sci.* **44**, 525–537.

Bird, D. J., and Potter, I. C. (1979a). Metamorphosis in the paried species of lampreys, *Lampetra fluviatilis* (L.) and *Lampetra planeri* (Bloch). 1. A description of the timing and stages. *J. Linn. Soc. London, Zool.* **65**, 127–143.

Bird, D. J., and Potter, I. C. (1979b). Metamorphosis in the paired species of lampreys, *Lampetra fluviatilis* (L.) and *Lampetra planeri* (Bloch). 2. Quantitative data for body proportions, weights, lengths and sex ratios. *J. Linn. Soc. London, Zool.* **65**, 145–160.

Bird, D. J., and Potter, I. C. (1981). Proximate body composition of the larval, metamorphosing and downstream migrant stages in the life cycle of the Southern Hemisphere lamprey, *Geotria australis* Gray. *Environ. Biol. Fishes* **6**, 285–297.

Blaxter, J. H. S. (1969). Development: Eggs and larvae. *In* "Fish Physiology" (W. S. Hoar and D. J. Randall, eds.), Vol. 2, pp. 177–252. Academic Press, New York.

Blaxter, J. H. S. (1986). Development of sense organs and behaviour of teleost larvae with special reference to feeding and predator avoidance. *Trans. Am. Fish. Soc.* **115**, 98–114.

Blaxter, J. H. S., and Staines, M. E. (1971). Food searching potential in marine fish larvae. *In* "Proceedings of the Fourth European Symposium on Marine Biology" (D. J. Crisp, ed.), pp. 467–485. Cambridge Univ. Press, London.

Braum, E. (1978). Ecological aspects of the survival of fish eggs, embryos and larvae. *In* "Ecology of Freshwater Fish Production" (S. D. Gerking, ed.), pp. 102–136. Blackwell, Oxford.

Breder, C. M. (1949). On the taxonomy and the postlarval stages of the surgeonfish, *Acanthurus hepatus*, *Copeia*, p. 296.

Bryan, P. G., and Madraisau, B. B. (1977). Larval rearing and development of *Siganus lineatus* (Pisces, Siganidae) from hatching through metamorphosis. *Aquaculture* **10**, 243–252.

Buckley, L. J. (1980). Changes in ribonucleic acid, deoxyribonucleic acid, and protein content during ontogenesis in winter flounder, *Pseudopleuronectes americanus* and effect of starvation. *Fish Bull.* **77**, 703–708.

Buckley, L. J. (1982). Nitrogen utilization by larval summer flounder (*Paralicthys dentatus*). *J. Exp. Mar. Biol. Ecol.* **59**, 243–256.

Buddington, R. K. (1985). Digestive secretions of lake sturgeon, *Acipenser fulvescens*, during early development. *J. Fish Biol.* **26**, 715–723.

Caldwell, D. K. (1962). Development and distribution of the short bigeye *Pseudo-priacanthus altus* (Gill), in the western North Atlantic. *Fish. Bull.* **62**, 103–150.

Campana, S. E. (1984). Microstructural growth patterns in the otoliths of larval and juvenile starry flounders. *Platichthys stellatus. Can. J. Zool.* **62**, 1507–1512.

Castle, P. H. J. (1970). Ergebnisse der Forschungsreisen des FFS "Walther Herwig" nach Südamerika. XI. The leptocephali. *Arch. Fischereiwiss* **21**, 1–21.

Castle, P. H. J. (1977). Leptocephalus of the muraenesocid eel *Gavialiceps taeniola. Copeia*, pp. 488–492.

Castle, P. H. J. (1979). Early life-history of the eel *Moringua edwardsi* (Pisces, Moringuidae) in the western North Atlantic. *Bull. Mar. Sci.* **29**, 1–18.

Castle, P. H. J. (1984). Notacanthiformes and Anguilliformes: Development. *In* "Ontogeny and Systematics of Fishes" (H. G. Moser *et al.*, eds.), Am. Soc. Ichthyol. Herpetol. Spec. Publ. No. 1, pp. 62–92. Allen Press, Lawrence, Kansas.

Cetta, C. M., and Capuzzo, J. M. (1982). Physiological and biochemical aspects of embryonic and larval development of the winter flounder *Pseudopleuronectes americanus. Mar. Biol. (Berlin)* **71**, 327–337.

Cohen, D. M. (1984). Ontogeny, systematics and phylogeny. *In* "Ontogeny and Systematics in Fishes" (H. G. Moser *et al.*, eds.), Am. Soc. Ichthyol. Herpetol. Spec. Publ. No. 1, pp. 7–11. Allen Press, Lawrence, Kansas.

Cole, W. C., and Youson, J. H. (1981). The effect of pinealectomy, continuous light and continuous darkness on metamorphosis of anadromous sea lampreys, *Petromyzon marinus* L. *J. Exp. Zool.* **218**, 397–404.

Cousin, J. C. B., and Baudin Laurencin, F. (1985). Morphogenese de l'appareil digestif et de la vessie gazeuse du turbot, *Scophthalmus maximus* L. *Aquaculture* **47**, 305–319.

Crawford, C. M. (1987). Development of eggs and larvae of the flounders *Rhombosolea tapirina* and *Ammotretis rostratus* (Pisces: Pleuronectidae). *J. Fish Biol.* **29**, 325–334.

de Sylva, C. (1974). Development of the respiratory system in herring and plaice. *In* "The Early Life History of Fish" (J. H. S. Blaxter, ed.), pp. 465–485. Springer-Verlag, Berlin and New York.

de Sylva, D. P., and Eschmeyer, W. N. (1977). Systematics and biology of the deep-sea fish family Gibberichthyidae, a senior synonym of the family Kasidoroidae. *Proc. Calif. Acad. Sci.* [4] **41**, 215–231.

Dickson, D. H., and Graves, D. A. (1981). The ultrastructure and development of the eye. *In* "The Biology of Lampreys" (M. W. Hardisty and I. C. Potter, eds.), Vol. 3, pp. 43–94. Academic Press, New York.

Eddy, J. M. P. (1969). Metamorphosis and the pineal complex in the brook lamprey, *Lampetra planeri. J. Endocrinol.* **44**, 451–452.

Elliott, M. W., and Youson, J. H. (1986). Immunocytochemical localization of insulin and somatostatin in the endocrine pancreas of the sea lamprey, *Petromyzon marinus* L., at various phases of its life cycle. *Cell Tissue Res.* **243**, 629–634.

Elliott, W. M., and Youson, J. H. (1987). Immunohistochemical observations of the endocrine pancreas during metamorphosis of the sea lamprey, *Petromyzon marinus* L. *Cell Tissue Res.* **247**, 351–357.

Epple, A., Hilliard, R. W., and Potter, I. C. (1985). The cardiovascular chromaffin cell system of the Southern Hemisphere lamprey, *Geotria australis* Gray. *J. Morphol.* **183**, 225–231.

Etkin, W., and Gilbert, L. I., eds. (1968). "Metamorphosis: A Problem in Developmental Biology." Appleton, New York.

Evans, D. H. (1984). The roles of gill permeability. In "Fish Physiology" (W. S. Hoar and D. J. Randall, eds.), Vol. 10, Part B, pp. 239–283. Academic Press, New York.

Evseenko, S. A. (1978). Some data on metamorphosis of larvae of the genus Bothus (Pisces, Bothidae) from the Caribbean Sea. Zool. Zh. 57, 1040–1047.

Filosa, M. F., Sargent, P. A., and Youson, J. H. (1986). An electrophoretic and immunoelectrophoretic study of serum proteins during the life cycle of the lamprey, Petromyzon marinus L. Comp. Biochem. Physiol. B 83B, 143–149.

Fontaine, M. (1975). Physiological mechanisms in the migration of marine and amphihaline fish. Adv. Mar. Biol. 13, 241–355.

Forstner, H., Hinterleitner, S., Mahr, K., and Wieser, W. (1983). Towards a better definition of "metamorphosis" in Coregonus sp.: Biochemical, histological, and physiological data. Can. J. Fish. Aquat. Sci. 40, 1224–1232.

Fox, H. (1984). "Amphibian Morphogenesis." Humana Press, Clifton, New Jersey.

Fukayama, S. (1985). Ultrastructural changes in the liver of the sand lamprey, Lampetra reissneri, during sexual maturation. Jpn. J. Ichthyol. 32, 316–323.

Fukayama, S., Takahashi, H., Matsubara, T., and Hara, A. (1986). Profiles of the female-specific serum protein in the Japanese lamprey, Lampetra japonica (Martens), and the sand lamprey, Lampetra reissneri (Dybowski), in relation to sexual maturation. Comp. Biochem. Physiol. A 84A, 45–48.

Fukuda, M., Yano, Y., and Nakano, H. (1986). Protein and nucleic acid changes during early developmental stages of cresthead flounder. Bull. Jpn. Soc. Sci. Fish 52, 951–955.

Fukuhara, O. (1986). Morphological and functional development of Japanese flounder in early life stage. Bull. Jpn. Soc. Sci. Fish 52, 81–91.

Futch, C. R. (1977). Larvae of Trichopsetta ventralis (Pisces, Bothidae), with comments on intergeneric relationships within the Bothidae. Bull. Mar. Sci. 27, 740–757.

Gilbert, L. I., and Frieden, E. (1981). "Metamorphosis: A Problem in Developmental Biology," 2nd ed. Plenum, New York.

Gordon, D. J., Markle, D. F., and Olney, J. E. (1984). Ophidiiformes: Development and relationships. In "Ontogeny and Systematics in Fishes" (H. G. Moser et al., eds.), Am. Soc. Ichthyol. Herpetol., Spec. Publ. No. 1, pp. 308–319. Allen Press, Lawrence, Kansas.

Gould, S. J. (1977). "Ontogeny and Phylogeny." Belknap Press, Cambridge, Massachusetts.

Graf, W., and Baker, R. (1983). Adaptive changes of the vestibulo–ocular reflex in flatfish are achieved by reorganization of central nervous pathways. Science 221, 777–779.

Griswold, C. A., and McKenney, T. W. (1984). Larval development of the scup, Stenotomous chrysops (Pisces: Sparidae). Fish. Bull. 82, 77–84.

Grizzle, J. M., and Curd, M. R. (1978). Posthatching histological development of the digestive system and swim bladder of logperch, Percina caprodes. Copeia, pp. 448–455.

Harden Jones, F. R. (1968). "Fish Migration." Arnold, London.

Hardisty, M. W. (1979). "Biology of the Cyclostomes." Chapman & Hall, London.

Hardisty, M. W. (1982). Lampreys and hagfishes: Analysis of cyclostome relationships. In "The Biology of Lampreys" (M. W. Hardisty and I. C. Potter, eds.), Vol. 4B, pp. 165–259. Academic Press, London.

Hardisty, M. W., and Potter, I. C. (1971a). The behaviour, ecology and growth of larval lampreys. In "The Biology of Lampreys" (M. W. Hardisty and I. C. Potter, eds.), Vol. 1, pp. 85–125. Academic Press, London.

Hardisty, M. W., and Potter, I. C. (1971b). Paired species. *In* "The Biology of Lampreys" (M. W. Hardisty and I. C. Potter, eds.), Vol. 1. pp. 249–277. Academic Press, London.

Hardy, J. D. (1978). Development of fishes in the mid-Atlantic Bight. An atlas of egg, larval, and juvenile stages, Vol. 2, pp. 1–458. US Govt. Printing Office, Washington D.C.

Hendrich, J. W., Weise, J. G., and Smith, B. R. (1980). Changes in biological characteristics of the sea lamprey (*Petromyzon marinus*) as related to lamprey abundance, prey abundance, and sea lamprey control. *Can. J. Fish. Aquat. Sci.* **37**, 1861–1871.

Hensley, D. A. (1977). Larval development of *Engyophrys senta* (Bothidae), with comments on intermuscular bones in flatfishes. *Bull. Mar. Sci.* **27**, 681–703.

Hensley, D. A., and Ahlström, E. H. (1984). Pleuronectiformes: Relationships. *In* "Ontogeny and Systematics in Fishes" (H. G. Moser *et al.*, eds.), Am. Soc. Ichthyol. Horpotol., Spoc. Publ. No 1, pp 670–687 Allen Press, Lawrence, Kansas,

Hilliard, R. W., Bird, D. J., and Potter, I. C. (1983). Metamorphic changes in the intestine of three species of lampreys. *J. Morphol.* **176**, 181–196.

Hilliard, R. W., Epple, A., and Potter, I. C. (1985). The morphology and histology of the endocrine pancreas of the Southern Hemisphere lamprey, *Geotria australis* Gray. *J. Morphol.* **184**, 253–262.

Hollister, G. (1936). A fish which grows by shrinking. *Bull. N.Y. Zool. Soc.* **39**, 104–109.

Hubbs, C. L. (1943). Terminology of early stages of fishes. *Copeia*, p. 260.

Hulet, W. H. (1978). Structure and functional development of the eel leptocephalus *Ariosoma balearicum* (De La Roche, 1809). *Philos. Trans. R. Soc. London* **282**, 107–138.

Hulet, W. H., Fischer, J., and Rietberg, B. J. (1972). Electrolyte composition of anguilliform leptocephali from the straits of Florida. *Bull. Mar. Sci.* **22**, 432–448.

Inui, Y., and Miwa, S. (1985). Thyroid hormone induces metamorphosis of flounder larvae. *Gen. Comp. Endocrinol.* **60**, 450–454.

Iuchi, I. (1985). Cellular and molecular bases of the larval-adult shift of hemoglobins in fish. *Zool. Sci,* **2**, 11–23.

Johnson, R. K. (1984). Giganturidae: Development and relationships. *In* "Ontogeny and Systematics in Fishes" (H. G. Moser *et al.*, eds.), Am. Soc. Ichthyol. Herpetol., Spec. Publ. No. 1, pp. 199–201. Allen Press, Lawrence, Kansas.

Jollie, M. (1982). What are the "Calcichordata" and the larger question of the origin of the chordates. *J. Linn. Soc. London, Zool.* **75**, 167–188.

Joss, J. M. P. (1985). Pituitary control of metamorphosis in the Southern Hemisphere lamprey, *Geotria australis. Gen. Comp. Endocrinol.* **60**, 58–62.

Just, J. J., Kraus-Just, J., and Check, D. A. (1981). Survey of chordate metamorphosis. *In* "Metamorphosis: A Problem in Developmental Biology" (L. I. Gilbert and E. Frieden, eds.), 2nd ed., pp. 265–326. Plenum, New York.

Kawamura, G., and Ishida, K. (1985). Changes in sense organ morphology and behaviour with growth in the flounder *Paralichthys olivaceus. Bull. Jpn. Soc. Sci. Fish.* **51**, 155–165.

Kendall, A. W., Ahlström, E. H., and Moser, H. G. (1984). Early life history stages of fishes and their characters. *In* "Ontogeny and Systematics of Fishes" (H. G. Moser *et al.*, eds.), Am. Soc. Ichthyol. Herpetol., Spec. Publ. No. 1, pp. 11–22. Allen Press, Lawrence, Kansas.

Krug, E. C., Honn, K. V., Battista, J., and Nicoll, C. S. (1983). Corticosteroids in serum of *Rana catesbeiana* during development and metamorphosis. *Gen. Comp. Endocrinol.* **52**, 232–241.

Kubota, S. A. (1961). Studies on the ecology, growth and metamorphosis in conger eel, *Conger myriaster* (Brevoort). *J. Fac. Fish., Prefect. Univ. Mie* **5**, 190–370 (in Japanese).

Lam, T. J. (1980). Thyroxine enhances larval development and survival in *Sarotherodon* (tilapia) *mossambicus* Ruppell. *Aquaculture* **21**, 287–291.

Lam, T. J. (1985). Role of thyroid hormone on larval growth and development in fish. *In* "Current Trends in Comparative Endocrinology" (B. Lofts and W. N. Holmes, eds.), pp. 481–485. Hong Kong Univ. Press, Hong Kong.

Lam, T. J., and Sharma, R. (1985). Effects of salinity and thyroxine on larval survival, growth and development in the carp, *Cyprinus carpio. Aquaculture* **44**, 201–212.

Lam, T. J., Juario, J. V., and Banno, J. (1985). Effect of thyroxine on growth and development in post-yolk-sac larvae of milkfish, *Chanos chanos. Aquaculture* **46**, 179–184.

Langille, R. M., and Youson, J. H. (1984a). Morphology of the intestine of prefeeding and feeding adult lampreys, *Petromyzon marinus* (L.): The mucosa of the anterior intestine, diverticulum and transition zone. *J. Morphol.* **182**, 39–62.

Langille, R. M., and Youson, J. H. (1984b). Morphology of the intestine of prefeeding and feeding adult lampreys, *Petromyzon marinus* (L.): The mucosa of the posterior intestine and hindgut. *J. Morphol.* **182**, 137–152.

Laurence, G. C. (1975). Laboratory growth and metabolism of the winter flounder *Pseudopleuronectes americanus* from hatching through metamorphosis at three temperatures. *Mar. Biol. (Berlin)* **32**, 223–229.

Laurence, G. C. (1977). A bioenergetic model for the analysis of feeding and survival potential of winter flounder, *Pseudopleuronectes americanus,* larvae during the period from hatching to metamorphosis. *Fish Bull.* **75**, 529–546.

Leach, W. J. (1946). Oxygen consumption of lampreys, with special reference to metamorphosis and phylogenetic position. *Physiol. Zool.* **19**, 365–374.

Lewis, S. V. (1980). Respiration of lampreys. *Can. J. Fish. Aquat. Sci.* **37**, 1711–1722.

Lewis, S. V., and Potter, I. C. (1977). Oxygen consumption during the metamorphosis of the parasitic lamprey, *Lampetra fluviatilis* (L.) and its non-parasitic derivative, *Lampetra planeri* (Bloch). *J. Exp. Biol.* **69**, 187–198.

L'Hermite, A., Ferrand, R., Dubois, M. P., and Andersen, A. C. (1985). Detection of endocrine cells by immunofluorescence methods in the gastroenteropancreatic system of the adult eel, glass-eel, and leptocephali larva (*Anguilla anguilla* L.). *Gen. Comp. Endocrinol.* **58**, 347–359.

Lintlop, S. P., and Youson, J. H. (1983a). Concentration of triiodothyronine in the sera of the sea lamprey, *Petromyzon marinus*, and the brook lamprey, *Lampetra lamottenii*, at various phases of their life cycle. *Gen. Comp. Endocrinol.* **49**, 187–194.

Lintlop, S. P., and Youson, J. H. (1983b). Binding of triiodothyronine to hepatocyte nuclei from sea lamprey, *Petromyzon marinus* L., at various stages of the life cycle. *Gen. Comp. Endocrinol.* **49**, 428–436.

Lowe, D. R., Beamish, F. W. H., and Potter, I. C. (1973). Changes in the proximate body composition of the landlocked sea lamprey *Petromyzon marinus* (L.) during larval life and metamorphosis. *J. Fish Biol.* **5**, 673–682.

Macey, D. J., and Potter, I. C. (1981). Measurements of various blood cell parameters during the life cycle of the Southern Hemisphere lamprey, *Geotria australis* Gray. *Comp. Biochem. Physiol. A* **69A**, 815–823.

Macey, D. J., Webb, J., and Potter, I. C. (1982a). Iron levels and major iron binding proteins in the plasma of ammocoetes and adults of the Southern Hemisphere lamprey *Geotria australis* Gray. *Comp. Biochem. Physiol. A* **72A**, 307–312.

Macey, D. J., Webb, J., and Potter, I. C. (1982b). Distribution of iron-containing granules in lampreys with particular reference to the Southern Hemisphere species *Geotria australis* Gary. *Acta Zool. (Stockholm)* **63**, 91–99.

Macey, D. J., Smalley, S. R., Potter, I. C., and Cake, M. H. (1985). The relationship between total non-haem, ferritin and haemosiderin iron in larvae of Southern Hemisphere lampreys (*Geotria australis* and *Mordacia mordax*). *J. Comp. Physiol. B* **156**, 269–276.

McKeown, B. A. (1984). "Fish Migration." Croom Helm, London.

Mallatt, J. (1984). Feeding ecology of the earliest of vertebrates. *J. Linn. Soc. London, Zool.* **82**, 261–272.

Mallett, J. (1985). Reconstructing the life cycle and the feeding of ancestral vertebrates. *In* "Evolutionary Biology of Primitive Fishes" (R. E. Foreman, A. Gorbman, J. M. Dodd, and R. Olsson, eds.), pp. 59–68. Plenum, New York.

Malmqvist, B. (1983). Growth, dynamics, and distribution of a population of the brook lamprey *Lampetra planeri* in a South Swedish stream. *Holarctic Ecol.* **6**, 404–412.

Manion, P. J., and Smith, B. R. (1978). Biology of larval and metamorphosing sea lamprey, *Petromyzon marinus*, of the 1960 year class in the Big Garlic River, Michigan. Part II. 1966–72. *Great Lakes Fish. Comm., Tech. Rep.* **30**, 1–35.

Mathers, J. S., and Beamish, F. W. H. (1974). Changes in serum osmotic and ionic concentration in landlocked *Petromyzon marinus. Comp. Biochem. Physiol. A* **49A**, 677–688.

Matsubara, K. (1942). On the metamorphosis of a clupeoid fish, *Pterothrissus gissu* Hilgendorf. *J. Imp. Fish. Inst. (Jpn.)* **35**, 1–16.

Mercado, S. J. E., and Ciardelli, A. (1972). Contribución a la morfología y organogénesis de los leptocéfalos del sábalo *Megalops atlanticus* (Pisces: Megalopidae). *Bull. Mar. Sci.* **22**, 153–184.

Miwa, S., and Inui, Y. (1987a). Histological changes in the pituitary–thyroid axis during spontaneous and artificially-induced metamorphosis of flounder larvae. *Cell Tissue Res.*, **249**, 117–123.

Miwa, S., and Inui, Y. (1987b). Effects of various doses of thyroxine and triiodothyronine on the metamorphosis of flounder (*Paralichthys olivaceus*). *Gen. Comp. Endocrinol.* **67**, 556–563.

Moore, J. W., and Potter, I. C. (1976). Aspects of feeding and lipid deposition and utilization in the lampreys, *Lampetra fluviatilis* (L.), and *Lampetra planeri* (Bloch). *J. Anim. Ecol.* **45**, 699–712.

Moriarty, C. (1978). "Eels. A Natural and Unnatural History." Universe Books, New York.

Morman, R. H. (1987). Relationship of density to growth and metamorphosis of caged larval sea lampreys, *Petromyzon marinus* Linnaeus, in Michigan streams. *J. Fish Biol.* **30**, 173–181.

Moser, H. G. (1981). Morphological and functional aspects of marine fish larvae. *In* "Marine Fish Larvae: Morphology, Ecology, and Relation to Fisheries" (R. Lasker, ed.), pp. 89–131. Univ. of Washington Press, Seattle.

Murr, E., and Sklower, A. (1926). Untersuchungen uber die Inkretorischen Organe der Fische. 1. Das Verhalten der Schilddruse in der Metamorphose des Aales. *Z. Vergl. Physiol.* **7**, 279–288.

Neave, D. A. (1984). The development of visual acuity in larval plaice (*Pleuronectes platessa* L) and turbot (*Scophthalmus maximus* L.). *J. Exp. Mar. Biol. Ecol.* **78**, 167–175.

Neave, D. A. (1985). The dorsal light reactions of larval and metamorphosing flatfish. *J. Fish Biol.* **26**, 629–640.

Neave, D. A. (1986). The development of the lateral line system in plaice (*Pleuronectes platessa*) and turbot (*Scophthalamus maximus*). *J. Mar. Biol. Assoc. U.K.* **66**, 683–693.

Norman, J. R., and Greenwood, P. H. (1963). "A History of Fishes." Ernest Benn, London.

Norris, D. O. (1983). Evolution of endocrine regulation of metamorphosis in lower vertebrates. *Am. Zool.* **23**, 709–718.

Northcutt, R. G., and Gans, C. (1983). The genesis of neural crest and epidermal placodes: A reinterpretation of vertebrate origins. *Q. Rev. Biol.* **58**, 1–28.

O'Boyle, R. N., and Beamish, F. W. H. (1977). Growth and intermediary metabolism of larval and metamorphosing stages of the landlocked sea lamprey, *Petromyzon marinus* L. *Environ. Biol. Fishes* **2**, 103–120.

Ooi, E. C., and Youson, J. H. (1976). Growth of the opisthonephric kidney during larval life in the anadromous sea lamprey, *Petromyzon marinus* L. *Can J. Zool.* **54**, 1449–1458.

Ooi, E. C., and Youson, J. H. (1977). Morphogenesis and growth of the definitive opisthonephros during metamorphosis of anadromous sea lamprey, *Petromyzon marinus* L. *J. Embryol. Exp. Morphol.* **42**, 219–235.

Ooi, E. C., and Youson, J. H. (1979). Regression of the larval opisthonephros during metamorphosis of the sea lamprey, *Petromyzon marinus* L. *Am. J. Anat.* **154**, 57–80.

Paine, M. D., and Balon, E. K. (1984). Early development of the northern logperch, *Percina caprodes semifasciata*, according to the theory of saltatory ontogeny. *Environ. Biol. Fishes* **11**, 173–190.

Pankhurst, N. W. (1984). Retinal development in larval and juvenile European eel, *Anguilla anguilla*. *Can. J. Zool.* **62**, 335–343.

Peek, W. D., and Youson, J. H. (1979). Transformation of the interlamellar epithelium of the gills of the anadromous sea lamprey, *Petromyzon marinus* L., during metamorphosis. *Can. J. Zool.* **57**, 1318–1332.

Percy, R., and Potter, I. C. (1976). Blood cell formation in the River lamprey, *Lampetra fluviatilis*. *J. Zool.* **178**, 319–340.

Percy, R., and Potter, I. C. (1977). Changes in haemopoietic sites during the metamorphosis of the lampreys, *Lampetra fluviatilis*, and *Lampetra planeri*. *J. Zool.* **183**, 111–123.

Percy, R., and Potter, I. C. (1981). Further observations on the development and destruction of lamprey blood cells. *J. Zool.* **193**, 239–251.

Perez, J. E., and Maclean, N. (1976). The haemoglobins of the fish *Sarotherodon messambicus* (Peters): Functional significance and ontogenetic change. *J. Fish Biol.* **9**, 447–455.

Pfeiler, E. (1984a). Changes in water and salt content during metamorphosis of larval bonefish (*Albula*). *Bull. Mar. Sci.* **34**, 177–184.

Pfeiler, E. (1984b). Effect of salinity on water and salt balance in metamorphosing bonefish (*Albula*) leptocephali. *J. Exp. Mar. Biol. Ecol.* **82**, 183–190.

Pfeiler, E. (1984c). Glycosaminoglycan breakdown during metamorphosis of larval bonefish *Albula*. *Mar. Biol. Lett.* **5**, 241–245.

Pfeiler, E. (1986). Towards an explanation of the developmental strategy in leptocephalous larvae of marine fishes. *Environ. Biol. Fishes* **15**, 3–13.

Pfeiler, E. (1987). Free amino acids in metamorphosing bonefish (*Albula* sp.) leptocephali. *Fish Physiol. Biochem.* **3** (in press).

Pfeiler, E., and Luna, A. (1984). Changes in biochemical composition and energy utilization during metamorphosis of leptocephalous larvae of the bonefish (*Albula*). *Environ. Biol. Fishes* **10**, 243–251.

Pietsch, T. W. (1984). Lophiformes: Development and relationships. *In* "Ontogeny and Systematics in Fishes" (H. G. Moser *et al.*, eds.), Am. Soc. Ichthyol. Herpetol. Spec. Publ. No. 1, pp. 320–325. Allen Press, Lawrence, Kansas.

Platt, C. (1973). Central control of postural orientation in flatfish. 1. Postural change dependence on central neural changes. *J. Exp. Biol.* **59**, 491–521.

Policansky, D. (1982). Influence of age, size, and temperature on metamorphosis in the starry flounder, *Platichthys stellatus. Can. J. Fish. Aquat. Sci.* **39**, 514–517.

Policansky, D. (1983). Size, age and demography of metamorphosis and sexual maturation in fishes. *Am. Zool.* **23**, 57–63.

Policansky, D., and Sieswerda, P. (1979). Early life history of the starry flounder, *Platichthys stellatus,* reared through metamorphosis in the laboratory. *Trans. Am. Fish. Soc.* **108**, 326–327.

Potter, I. C. (1970). The life cycles and ecology of Australian lampreys of the genus *Mordacia. J. Zool.* **161**, 487–511.

Potter, I. C. (1980a). The Petromyzontiformes with particular reference to paired species. *Can. J. Fish. Aquat. Sci.* **37**, 1595–1615.

Potter, I. C. (1980b). The ecology of larval and metamorphosing lampreys. *Can. J. Fish. Aquat. Sci.* **37**, 1641–1657.

Potter, I. C., and Beamish, F. W. H. (1977). The freshwater biology of adult anadromous sea lampreys *Petromyzon marinus. J. Zool.* **181**, 113–130.

Potter, I. C., and Brown, I. D. (1975). Changes in haemoglobin electropherograms during the life cycle of two closely related lampreys. *Comp. Biochem. Physiol. B* **51B**, 517–519.

Potter, I. C., and Huggins, R. J. (1973). Observations on the morphology, behaviour and salinity tolerance of downstream migrating river lampreys (*Lampetra fluviatilis*). *J. Zool.* **169**, 365–379.

Potter, I. C., Percy, R., and Youson, J. H. (1978a). A proposal for the adaptive significance of the development of the lamprey fat column. *Acta Zool.* (*Stockholm*) **59**, 63–67.

Potter, I. C., Wright, G. M., and Youson, J. H. (1978b). Metamorphosis in the anadromous sea lamprey, *Petromyzon marinus* L. *Can. J. Zool.* **56**, 561–570.

Potter, I. C., Hilliard, R. W., and Bird, D. J. (1980). Metamorphosis in the Southern Hemisphere lamprey, *Geotria australis. J. Zool.* **190**, 405–430.

Potter, I. C., Hilliard, R. W., and Bird, D. J. (1982). Stages in metamorphosis. *In* "The Biology of Lampreys" (M. W. Hardisty and I. C. Potter, eds.), Vol. 4B, pp. 137–164. Academic Press, New York.

Purvis, H. A. (1979). Variation in growth, age at transformation, and sex ratio of sea lampreys reestablished in chemically treated tributaries of the upper Great Lakes. *Great Lakes Fish. Comm., Tech. Rep.* **35**, 1–36.

Purvis, H. A. (1980). Effects of temperature on metamorphosis and the age and length at metamorphosis in sea lamprey (*Petromyzon marinus*) in the Great Lakes. *Can. J. Fish. Aquat. Sci.* **37**, 1827–1834.

Raju, S. N. (1974). Distribution, growth and metamorphosis of leptocephali of the garden eels, *Taenioconger* sp. and *Gorgasia* sp. *Copeia,* pp. 494–500.

Rao, P. D., and Finger, T. E. (1984). Asymmetry of the olfactory system in the brain of the winter flounder, *Pseudopleuronectes americanus. J. Comp. Neurol.* **225**, 492–510.

Rasquin, P. (1955). Observations of the metamorphosis of the bonefish, *Albula vulpes* (Linnaeus). *J. Morphol.* **97**, 77–118.

Richards, J. E., and Beamish, F. W. H. (1981). Initiation of feeding and salinity tolerance in the Pacific lamprey, *Lampetra tridentata. Mar. Biol.* (*Berlin*) **63**, 73–78.

Richards, W. J. (1976). Some comments on Balon's terminology of fish development intervals. *J. Fish. Res. Board Can.* **33**, 1253–1254.

Richardson, S. L., and Joseph, E. B. (1973). Larvae and young of western North Atlantic bothid flatfishes *Etropus microstomus* and *Citharichthys arctifrons* in the Chesapeake Bight. *Fish. Bull.* **71**, 735–767.

Rijnsdorp, A. D., Van Stralen, M., and Van Der Veer, H. W. (1985). Selective tidal transport of North Sea plaice larvae *Pleuronectes platessa* in coastal nursery areas. *Trans. Am. Fish. Soc.* **114**, 461–470.

Riley, J. D. (1966). Marine fish culture in Britain. VII. Plaice (*Pleuronectes platessa* L.) post-larval feeding on *Artemia salina* L. nauplii and the effects of varying feeding levels. *J. Cons., Cons. Int. Explor. Mer* **30**, 204–221.

Rosenblatt, R. H., and Butler, J. L. (1977). The ribbonfish genus *Desmodema*, with the description of a new species (Pisces, Trachipteridae). *Fish. Bull.* **75**, 843–855.

Ryland, J. S. (1966). Observations on the development of larvae of the plaice, *Pleuronectes platessa* L. in aquaria. *J. Cons., Cons. Int. Explor. Mer* **30**, 177–195.

Sargent, P. A., and Youson, J. H. (1986). Quantification of iron deposits in several body tissues of lampreys (*Petromyzon marinus* L.) throughout the life cycle. *Comp. Biochem. Physiol. A* **83A**, 575–577.

Schmidt, J. (1906). Contributions to the life-history of the eel (*Anguilla vulgaris*, Flem.). *Rapp. P.-V. Reun., Cons. Int. Explor. Mer* **5**, 137–264.

Seikai, T., Tanangonan, J. B., and Tonaka, M. (1986). Temperature influence on larval growth and metamorphosis of the Japanese flounder *Paralichthys olivaceus* in the laboratory. *Bull. Jpn. Soc. Sci. Fish.* **52**, 977–982.

Seiler, K., Seiler, R., and Claus, R. (1981). Histochemical and spectrophotometric demonstration of hydroxysteroid dehydrogenase activity in the presumed steroid producing cells of the brook lamprey (*Lampetra planeri* Bloch) during metamorphosis. *Endokrinologie* **78**, 297–300.

Sidon, E. W., and Youson, J. H. (1983). Morphological changes in the liver of the sea lamprey, *Petromyzon marinus* L., during metamorphosis. I. Atresia of the bile ducts. *J. Morphol.* **177**, 109–124.

Sklower, A. (1930). Die Bedeutung der Schilddruse fur die Metamorphose des Aales und der Plattefische. *Forsch. Fortschr.* **6**, 435–436.

Smalley, S. R., Macey, D. J., and Potter, I. C. (1986). Change in the amount of non-haem iron in the plasma, whole body and selected organs during the post-larval life of the lamprey *Geotria australis. J. Exp. Zool.* **237**, 149–157.

Smigielski, A. S. (1979). Induced spawning and larval rearing of the yellowtail flounder, *Limanda ferruginea. Fish. Bull.* **76**, 931–935.

Smith, B. R. (1971). Sea lampreys in the Great Lakes of North America. *In* "The Biology of Lampreys" (M. W. Hardisty and I. C. Potter, eds.), Vol. 1, pp. 207–247. Academic Press, London.

Smith, D. G. (1984). Elopiformes, Notacanthiformes and Anguilliformes: Relationships. *In* "Ontogeny and Systematics of Fishes" (H. G. Moser *et al.*, eds.), Am. Soc. Ichthyol. Herpetol., Spec. Publ. No. 1, pp. 94–102. Allen Press, Lawrence, Kansas.

Sterba, G. (1955). Das Adrenal-und Interrenalsystem in Lebensablauf von *Petromyzon planeri* Bloch. 1. Morphologie und Histologie einschlieblich Histogenese. *Zool. Anz.* **155**, 151–168.

Sterba, G. (1963). "Freshwater Fishes of the World." Longacre Press, London.

Szarski, H. (1957). The origin of the larva and metamorphosis in amphibia. *Am. Nat.* **91**, 283–301.

Tanaka, M. (1985). Factors affecting the inshore migration of pelagic larval and demersal juvenile red sea bream *Pagrus major* to a nursery ground. *Trans. Am. Fish. Soc.* **114**, 471–477.

Tandler, A., and Helps, S. (1985). The effects of photoperiod and water exchange rate on growth and survival of gilthead sea bream (*Sparus aurata*, Linnaeus; Sparidae) from hatching to metamorphosis in mass rearing systems. *Aquaculture* **48**, 71–82.

Tesch, F. W. (1977). "The Eel." Chapman & Hall, London.

Tsuneki, K., and Ouji, M. (1984a). Morphometric changes during growth of the brook lamprey *Lampetra reissneri*. *Jpn. J. Ichthyol.* **31**, 38–46.

Tsuneki, K., and Ouji, M. (1984b). Histological changes of several organs during growth of the brook lamprey *Lampertra reissneri*. *Jpn. J. Ichthyol.* **31**, 167–180.

Tucker, D. W. (1959). A new solution to the Atlantic eel problem. *Nature (London)* **183**, 495–501.

Van Utrecht, W. L. (1982). Some aspects of the distribution, metamorphosis and growth in *Serrivomer parabeani* Bertin, 1940 (Pisces, Apodes, Serrivomeridae) related to growth features in their otoliths. *Beaufortia* **32**, 117–124.

Victor, B. C. (1986). Delayed metamorphosis with reduced larval growth in a coral reef fish (*Thalassoma bifasciatum*) *Can. J. Fish Aquat. Sci.* **43**, 1208–1213.

Vilter, V. (1946). Action de la thyroxine sur la metamorphose larvaire de l'Anguille. *C. R. Seances Soc. Biol. Ses Fil.* **140**, 783–785.

Vladykov, V. D. (1985). "Does Neoteny Occur in Holarctic Lampreys (Petromyzontidae)?" Syllog. Ser. No. 57, pp. 1–13. Natl. Mus. Nat. Sci., Natl. Mus. Can., Ottawa.

Volk, E. C. (1986). Use of calcareous otic elements (statoliths) to determine age of sea lamprey ammocoetes (*Petromyzon marinus*). *Can. J. Fish. Aquat. Sci.* **43**, 718–722.

Wald, G. (1958). The significance of vertebrate metamorphosis. *Science* **128**, 1481–1490.

Wald, G. (1981). Metamorphosis: An overview. *In* "Metamorphosis: A Problem in Developmental Biology" (L. I. Gilbert and E. Frieden, eds.), 2nd ed., pp. 1–39. Plenum, New York.

Wassersug, R. J. (1975). The adaptive significance of the tadpole stage with comments on the maintenance of complex life cycles in anurans. *Am. Zool.* **15**, 405–417.

Wassersug, R. J., and Duellman, W. E. (1984). Oral structures and their development in egg-brooding hylid frog embryos and larvae: Evolutionary and ecological implications. *J. Morphol.* **182**, 1–37.

Wassersug, R. J., and Hoff, K. (1982). Developmental changes in the orientation of the anuran jaw suspension. A preliminary exploration into the evolution of anuran metamorphosis. *Evol. Biol.* **15**.

Webb, P. W., and Weihs, D. (1986). Functional locomotor morphology of early life history stages of fishes. *Trans. Am. Fish. Soc.* **115**, 115–127.

White, B. A., and Nicoll, C. S. (1981). Hormonal control of amphibian metamorphosis. *In* "Metamorphosis: A Problem in Developmental Biology" (L. I. Gilbert and E. Frieden, eds.), 2nd ed., pp. 363–396. Plenum, New York.

Willemse, J. J., and Markus-Silvis, L. (1985). The shifting of the aortic origin of the brachial arteries in the metamorphosing eel *Anguilla anguilla* (L.), with remarks on the shifting mechanisms in arterial junctions in general. *Acta Anat.* **121**, 216–222.

Wright, G. M., and Youson, J. H. (1976). Transformation of the endostyle of the anadromous sea lamprey, *Petromyzon marinus* L., during metamorphosis. I. Light microscopy and autoradiography with I-125. *Gen. Comp. Endocrinol.* **30**, 243–257.

Wright, G. M., and Youson, J. H. (1977). Serum thyroxine levels in larval and metamorphosing anadromous sea lamprey, *Petromyzon marinus* L. *J. Exp. Zool.* **202**, 27–32.

Wright, G. M., and Youson, J. H. (1980). Transformation of the endostyle of the anadromous sea lamprey, *Petromyzon marinus* L., during metamorphosis. II. Electron microscopy. *J. Morphol.* **160**, 231–257.

Wright, G. M., Filosa, M. F., and Youson, J. H. (1980). Immunocytochemical localization of thyroglobulin in the transforming endostyle of anadromous sea lampreys, *Petromyzon marinus* L., during metamorphosis. *Gen. Comp. Endocrinol.* **42**, 187–194.

Yamamoto, K., Sargent, P., Fisher, M. M., and Youson, J. H. (1986). Periductal fibrosis and lipocytes (fat-storing cells or Ito cells) during biliary atresia in the lamprey. *Hepatology* **6**, 54–59.

Youson, J. H. (1980). Morphology and physiology of lamprey metamorphosis. *Can. J. Fish. Aquat. Sci.* **37**, 1687–1710.

Youson, J. H. (1981a). The alimentary canal. *In* "The Biology of Lampreys" (M. W. Hardisty and I. C. Potter, eds.), Vol. 3, pp. 93–189. Academic Press, London.

Youson, J. H. (1981b). The kidneys. *In* "The Biology of Lampreys" (M. W. Hardisty and I. C. Potter, eds.), Vol. 3, pp. 191–261. Academic Press, London.

Youson, J. H. (1981c). The liver. *In* "The Biology of Lampreys" (M. W. Hardisty and I. C. Potter, eds.), Vol. 3, pp. 263–332. Academic Press, London.

Youson, J. H. (1982a). The morphology of the kidney in young adult anadromous sea lampreys, *Petromyzon marinus*, adapted to sea water. 1. General morphology and fine structure of the renal corpuscle and proximal segments. *Can. J. Zool.* **60**, 2351–2366.

Youson, J. H. (1982b). The morphology of the kidney in young adult anadromous sea lampreys, *Petromyzon marinus*, adapted to sea water. 2. Distal and collecting segments, the archinephric duct, and the intertubular tissue and blood vessels. *Can. J. Zool.* **60**, 2367–2381.

Youson, J. H. (1984). Differentiation of the segmented tubular nephron and excretory duct during lamprey metamorphosis. *Anat. Embryol.* **169**, 275–292.

Youson, J. H. (1985). Organ development and specialization in lamprey species. *In* "Evolutionary Biology of Primitive Fishes" (R. E. Foreman, A. Gorbman, J. M. Dodd, and R. Olsson, eds.), pp. 141–164. Plenum, New York.

Youson, J. H., and Connelly, K. L. (1978). Development of longitudinal mucosal folds in the intestine of the lamprey, *Petromyzon marinus* L., during metamorphosis. *Can. J. Zool.* **56**, 2364–2371.

Youson, J. H., and Horbert, W. R. (1982). Transformation of the intestinal epithelium of the larval anadromous sea lamprey, *Petromyzon marinus* L., during metamorphosis. *J. Morphol.* **171**, 89–117.

Youson, J. H., and Ooi, E. C. (1979). Development of the renal corpuscle during metamorphosis in the lamprey. *Am. J. Anat.* **155**, 201–222.

Youson, J. H., and Potter, I. C. (1979). A description of the stages of metamorphosis in the anadromous sea lamprey, *Petromyzon marinus* L. *Can. J. Zool.* **57**, 1808–1817.

Youson, J. H., and Sidon, E. W. (1978). Lamprey biliary atresia: first model for the human condition? *Experientia* **34**, 1084–1086.

Youson, J. H., Wright, G. M., and Ooi, E. C. (1977). The timing of changes in several internal organs during metamorphosis of anadromous larval lamprey, *Petromyzon marinus* L. *Can J. Zool.* **55**, 469–473.

Youson, J. H., Lee, J., and Potter, I. C. (1979). The distribution of fat in larval, metamorphosing, and adult anadromous sea lampreys, *Petromyzon marinus* L. *Can. J. Zool.* **57**, 237–246.

Youson, J. H., Sargent, P. A., and Sidon, E. W. (1983a). Iron loading in the livers of metamorphosing lampreys, *Petromyzon marinus* L. *Cell Tissue Res.* **234**, 109–124.

Youson, J. H., Sargent, P. A., and Sidon, E. W. (1983b). Iron loading in the liver of parasitic adult lampreys, *Petromyzon marinus* L. *Am. J. Anat.* **168**, 37–49.

Zanandrea, G. (1956). Neotenia in *Lampetra zanandreai* (Vladykov) e l'endocrinologia sperimentale dei Ciclostomi. *Boll. Zool.* **23**, 413–427.

Zanandrea, G. (1957). Neoteny in a lamprey. *Nature (London)* **179**, 925–926.

3

FACTORS CONTROLLING MERISTIC VARIATION

C. C. LINDSEY

Department of Zoology
University of British Columbia
Vancouver, British Columbia, Canada V6T 2A9

I. INTRODUCTION

In fish the number of meristic parts—serially repeated structures such as vertebrae or fin rays—can be greatly modified by the environment during early development. Then quite early in ontogeny the

197

meristic characters, unlike the morphometric characters, become fixed, and remain unchanged regardless of subsequent changes in the environment or in body size or in body shape. This early phenotypic pliability and later fixity can be exploited to help unravel some of the physiological processes of early development.

Because meristic characters are easily countable, they have been widely used for a century to distinguish between stocks requiring separate fishery regulation. Stock discrimination is becoming increasingly effective through the application of computer techniques that combine meristic, morphometric, and biochemical characters (Fournier *et al.*, 1984; Misra and Carscadden, 1984). The variability of meristic characters and their bilateral asymmetry have also been investigated in relation to developmental stability, to hybridization, and to environmental stress. Other topics that are of current interest but that will not be explored here include functional significance of meristic differences in fish with respect to locomotory performance (Swain and Lindsey, 1984; Swain, 1986), and the evolutionary implications of the large phenotypic variability of fish.

Although biologists have counted meristic parts in millions of wild fish, the interpretation of meristic differences is often still clouded by inability to distinguish between genotypic and phenotypic variation and by ignorance of the mechanisms whereby the environment affects the phenotype. This review will concentrate on the experimental evidence concerning modification of meristic characters and on its possible basis in embryology and physiology.

II. MERISTIC VARIATION IN WILD FISH

As background to description of experimental studies of phenotypic effects, this section outlines salient patterns of meristic variation in wild fish at both global and intraspecific levels. More information on some of these themes was given earlier in this series (Lindsey, 1978, p. 46).

A. Global Patterns

The number of segments in fishes shows a phyletic tendency to decrease. Cyclostomes, elasmobranchs, and primitive teleosts have many segments. Most but not all groups of higher teleosts have fewer, and several groups have each independently evolved very low segment numbers. Phyletic decrease in segment number has been observed in other groups of organisms, and has been solemnized as Williston's law or as Dogiel's principle of oligomerization. The description is valid only as a broad generalization. There are so many exceptions that it cannot be taken (by cladists or others) as an indication of which of two forms is ancestral.

The number of body segments in fish tends to be higher, among related forms, in those having the larger ultimate body size. This phenomenon, "pleomerism," is remarkably widespread. It occurs within (but not necessarily between) families having very different body shapes. It occurs within and between genera, between races, and between the sexes (Lindsey, 1975). Although calculations have been based on the maximum recorded size, the phenomenon probably relates more closely to size at some critical stage when selection operates most stringently (such as the free-swimming larval stage). The explanation may be related to locomotor performance (Spouge and Larkin, 1979). Recent experimental evidence supports the view that in any one body plan there is an optimal segment number for locomotion, but that optimum shifts with body length (Swain and Lindsey, 1984; Swain, 1986).

The number of vertebrae is negatively correlated, among related species, with thickness of the body. Like pleomerism, this is a widespread phenomenon observable in very differently shaped fishes and at several taxonomic levels. Calculations from hydrodynamic theory (Spouge and Larkin, 1979) also suggest that more rotund fish should have fewer vertebrae. It might be expected that thickness refers to lateral body width, and that a reduction in width (i.e., lateral compression) would be associated with more vertebral joints, both conferring greater side-to-side flexibility. In fact, within each of the four families so far examined (Clupeidae, Labridae, Scombridae, and Pleuronectidae) vertebral count is most strongly correlated negatively with the cross-sectional area, and body depth seems to contribute more than body width to the correlation. The explanation may be related to hydrodynamics but remains obscure.

The number of vertebrae tends to be higher in fish from more polar or cooler waters than in their relatives from tropical or warm waters. The phenomenon, termed Jordan's rule, holds both in Northern and Southern Hemispheres (references in Clark and Vladykov, 1960, p. 284). It occurs in many different fish groups and at different taxonomic levels. Although Jordan's rule is usually stated in terms of latitude, the operational factor is evidently temperature. In several studies where water temperature was not closely tied to latitude, the vertebral counts were more highly correlated with temperature at spawning time than with distance from the equator (Jensen, 1944; Peden, 1981; Resh *et al.*, 1976; Ritchie, 1976; Yatsu, 1980). For fine-scale comparisons within the tropical belt, latitudinal gradients are obscure both in temperature and vertebral number, and here Jordan's rule tends to break down (Casanova, 1981). In freshwater fishes, vertebral number tends also to be higher at higher altitudes; again, this is probably a reflection of lower temperatures.

Jordan's rule is not attributable to latitudinal clines in body size or in shape, even though these may influence vertebral clines. Body sizes of fish tend to be larger at higher latitudes (Lindsey, 1966), and larger fish tend to have more vertebrae (pleomerism). However, while clines in vertebral number are often thus reinforced by pleomerism, a distinct latitudinal vertebral cline persists in several families, even after the effect of body length is removed (Lindsey, 1975). Similarly, in three families tested that have significant correlations between vertebral number and latitude, these persist after the effect of body thickness is removed.

Scale count in general follows the same trends as vertebral count; among related forms, higher scale counts tend to be associated with higher latitudes, with more slender body shape, and with larger maximum body length. The numbers of rays in dorsal, anal, and pectoral fins tend also to be higher at higher latitudes (Vladykov, 1934; Berg, 1969, p. 265), but patterns are not as well documented.

The number of segments in related forms has been claimed to vary with the salinity of the habitat. The generalization that fishes in low salinity tend to have fewer vertebrae has been made by Hubbs (1926) and Nikolsky (1969a), and many supporting instances have been published. On the contrary, Jordan (1892) stated that, within groups, the freshwater forms tend to have more vertebrae than their marine relatives, and instances supporting this claim can be cited (e.g., Koli, 1969; Chernoff *et al.*, 1981). The inconsistent correlation between the salinity factor and segment number in wild populations is probably, as suggested by Vladykov (1934), because other factors such as temperature also vary between the habitats compared and may be of overriding importance.

"Space factor" was also invoked by Vladykov (1934) as affecting segment number. Space factor was loosely defined, but included the extent of the water body and its depth. Instances were cited where among related forms the segment number was higher in larger than in smaller marine areas, or in deeper than shallower waters. Experiments will be cited where the size of the rearing container has indeed affected segment number as well as growth rate. However, even more than in the case of salinity, effects of "space factor" on segment number in wild fish is likely to be confounded by other environmental influences and by associations between body size or shape in fish adapted to different habitats.

B. Evidence for Phenotypic Variation in Wild Fish

The foregoing patterns of meristic variation do not necessarily reflect any phenotypic modification by the environment. They could have a purely genetic basis [a view supported by some early work (Heincke, 1898)]. But the following evidence, from variation in wild fish, demonstrates that the environment must sometimes directly alter the phenotype. The references cited are only a few from the voluminous literature on racial variation in fishes.

Meristic counts of wild fish hatched in the same place often vary significantly between different years, in a manner associated with annual differences in water temperature (Runnström, 1941; Clark and Vladykov, 1960). Transplants to new environments can result in significant meristic differences not attributable solely to genetic selection (Ege, 1942; Svärdson, 1951). Within year classes, protracted spawning period coupled with changing temperature are sometimes associated with meristic differences between early and late hatchers. In spring-spawning species, the larger fish of a year class (developed during the early *cool* period) have *more* vertebrae (Ben-Tuvia, 1963; Komada, 1982); in autumn- and winter-spawning species, the larger fish (developed during the early *warm* period) have *fewer* segments (Ruivo and Monteiro, 1954; Komada, 1977b). This negative association observed between temperature and segment number is consistent with the commonest experimental results (see next section).

There could conceivably be hereditary meristic differences between early- and late-spawning parents, but such an explanation can scarcely be invoked in cases where detailed temperature fluctuations within as well as between years are negatively correlated with fluctuations in meristic counts (Hubbs, 1922; Ghéno and Poinsard, 1968).

The situation is less clear in instances where different stocks of the same species spawn at different temperatures and show the same negative correlation between temperature and vertebral number. DeCiechomski and deVigo (1971) point out that here genetic as well as phenotypic influences may be operating. An extended investigation of genotypic versus phenotypic meristic variation was started in 1917 by

Johannes Schmidt on *Zoarces viviparus* (references in Christiansen *et al.*, 1981). This 60-year study reveals a geographic mosaic of hereditary differences with a temporal overlay of phenotypic perturbations by the environment.

III. EXPERIMENTS ON PHENOTYPIC MERISTIC VARIATION

Although there is widespread meristic variation in wild fishes, with suggestive correlations between meristic characters and environmental variables, the effects of heredity and of the environment are hard to disentangle in nature. Laboratory rearing is required to provide direct evidence of phenotypic modification of meristic characters and to uncover the mechanisms at work. This section will review the laboratory studies, in the manner of the results section of a research paper. Discussion and synthesis come in a later section, after the range and diversity of experimental data have been laid out.

Results are assembled here from about 40 fish species in which the young have been reared under experimentally controlled conditions. The usual procedure has been to subdivide an egg batch after fertilization into lots, each of which has been reared in a different environment until after hatching. Typically the young have been fed for some further time, still in the controlled environment, until they were so large that their meristic counts were no longer subject to phenotypic modification. Meristic series were then counted, usually either on stained and cleared specimens or from radiographs.

Experiments on meristic variation will be discussed in two categories. The majority have examined meristic effects of rearing under several different levels of one environmental factor (most often temperature), each held constant throughout development. The other experiments have examined effects of *changes* in the environment during the developmental period. Responses to static conditions are better documented, but it is the responses, sometimes surprising, to environmental changes that offer the best clues to the mechanisms that underlie environmental modification of part numbers.

A. Experiments on Sustained Environmental Influences

Almost every environmental factor that has been tested has been found to produce significant meristic differences. Almost every meristic series, in almost every species tested, turns out to be subject to environmental modification. As a first step toward ordering the diversity of meristic responses, the following categories of patterns can be recognized:

Positive (P): mean meristic counts progressively higher as factor increases. Also called an acclivous response.

Negative (N): mean counts progressively lower as factor increases. Also called a declivous response.

V-Shaped (V) (in some cases actually U-shaped): mean count lowest at an intermediate factor level, and higher both at the upper and lower factor extremes.

Arched (A): mean count highest at an intermediate factor level, and lower both at the upper and lower factor extremes.

Zero (0): no clear response trend, or lack of statistically significant differences between any mean counts.

Within each category, the response curves of different species or genotypes may shift along either axis (Fig. 1), and their slopes may differ. The reliability of classifying a given set of meristic responses is affected by the number and range of environmental levels that have been tried. In most studies three or more levels were tested; when only two were tried, the responses can appear only as either positive or negative, not as arched nor V-shaped. Moreover, where the range of environmental levels tested does not span the whole range tolerable, it is possible that an apparently positive or negative response might, by extending the range, be revealed as actually an arch or V. Fortunately, most reported experiments have included three or more test levels, and have spanned most of the range tolerated by that species.

1. RESPONSES TO SUSTAINED TEMPERATURE

As an introduction to the presentation of the responses of *mean* meristic counts to environmental factors, an example is shown (Table I) of the distribution of the *individual* counts from which means can be calculated. In this example there is the highest possible genetic uniformity, as the parents were from a clone of the hermaphroditic fish *Rivulus marmoratus*. Rearing at different temperatures has produced a striking difference in vertebral counts: there is no overlap in the range of counts between those reared at low temperatures and those at medium or high temperatures. Since genetic diversity was nil and mortality was negligible, the meristic differences can be attributed to direct environmental modification of the phenotype. In other pub-

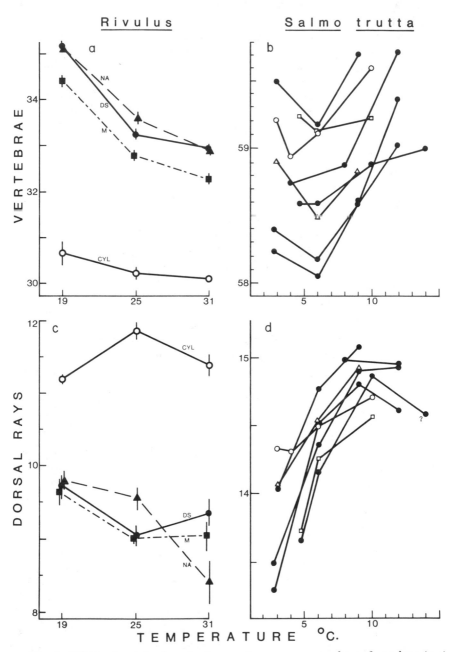

Fig. 1. Effects of sustained rearing temperature on mean numbers of vertebrae (top) and dorsal fin rays (bottom). For *Salmo trutta* (right), means shown for each of eight different crosses (Tåning, 1952). For *Rivulus* (left), NA, DS, and M are three clones of the self-fertilizing *R. marmoratus;* CYL is the outbreeding *R. cylindraceus;* vertical lines are 95% confidence intervals (Harrington and Crossman, 1976a,b).

Table I

Vertebral Frequencies in Homozygous Clone NA of *Rivulus marmoratus*
Reared at Three Temperatures[a]

| Temperature | Vertebrae | | | | | Mean | Coefficient of variation |
	32	33	34	35	36		
19°C				45	6	35.12	0.93
25°C		22	30			33.58	1.49
31°C	6	42				32.88	1.02

[a] From Harrington and Crossman (1976b).

lished experiments on meristics, the genetic diversity has been greater (using gonochoristic parents) and the mortality often higher, but arguments are presented later why most of the variation described below, within individual crosses, is nevertheless still attributable to direct modification of the phenotype by the environment.

The means of vertebral counts from experiments on one clone in Table I can be compared in Fig. 1 with means produced in experiments on other material which also tested effects of sustained rearing temperature. Means are shown, both of vertebrae and of dorsal fin rays, for three homozygous clones of the hermaphroditic *Rivulus marmoratus,* and for the gonochoristic species *R. cylindraceus* and *Salmo trutta.* The examples in Fig. 1 illustrate the following features, which are typical of meristic variation in fish:

1. In different species the same meristic series may have different response patterns.
2. Different meristic series of the same species may have different response patterns.
3. Different genotypes of one species may have response curves with similar shapes but with absolute values of the meristic character widely displaced.
4. Different genotypes of one species may even have response curves of different shapes. [This is less well illustrated in Fig. 1 than in the responses of vertebrae to temperature in *Oryzias latipes;* here some crosses showed negative responses over the whole temperature range, but most showed V-shaped responses with nadirs that varied from 24 to 32°C (Ali and Lindsey, 1974)].

Response patterns of different meristic series to sustained temperatures in 40 fish species are summarized in Table II. Abbreviations

Table II

Response Patterns of Meristic Counts to Different Sustained Developmental Temperatures[a]

Species	Vertebrae	Dorsal rays	Anal rays	Pectoral rays	Caudal rays	Others	Source
Clupea harengus	N[b]						Hempel and Blaxter (1961); Lapin et al. (1969)
Brachydanio rerio	N(V?)	N	O	N(P)	P[c]	Pelvic rays P	Dentry (1976); Dentry and Lindsey (1978)
Campostoma anomalum						Scales N	Carmichael (1983)
Cyprinus carpio	N(P)	N	A	P(N)		Pelvic rays N; Scales N(F)	Tatarko (1968)
Mylocheilus caurinus	N						C. C. Lindsey (unpublished)
Ptychocheilus oregonensis	N						C. C. Lindsey (unpublished)
Richardsonius balteatus	N(V)	V(N)	P(A)				C. C. Lindsey (unpublished)
Tribolodon hakonensis	V						Komada (1982)
Misgurnus anguillicaudatus	N						Kubota (1967)
Esox lucius	N(P)						Lubitskaya (1961); Lubitskaya and Dorofeeva (1961a)
Osmerus eperlanus	N[d]						Lubitskaya and Dorofeeva (1961b)
Plecoglossus altivelis	N						Komada (1977a)
Oncorhynchus keta	N	A,P	A	P	O	Pelvic rays O; Gill rakers N	Kubo (1950); Beacham and Murray (1986)
O. kisutch	V(P)	A(P)	N(P)				C. C. Lindsey, unpublished
O. nerka, sockeye	P[e]	A[e]	A	P[e]	N[c,e]	Pelvic rays N[e]; Scales N[e]; Branchiostegals N[e]	Canagaratnam (1959)
O. tshawytscha	V	A[e]	N[e]				Seymour (1959)
Salmo gairdneri	N	N[e]					Garside (1966); Hallam (1974); Kwain (1975); Lindsey et al. (1984);

Species						Additional characters	References
S. salar	N[f]					Gill rakers P[e]	MacGregor and MacCrimmon (1977); Mottley (1934); Orska (1962, 1963); Winter et al. (1980)
S. trutta	V	A(P)	N,P	P(A?)			Pavlov (1984); Hallam (1974); Orska (1956); Schmidt (1921); Tåning (1944, 1952)
Salvelinus fontinalis	N						Garside (1966); Hallam (1974)
S. namaycush	N						Hallam (1974)
Belone belone	V						Fonds et al. (1974)
Oryzias latipes	V(N)	A	P(A,N)	N	A(P)[g]		Ali and Lindsey (1974); Ogawa (1971)
Fundulus heteroclitus	N(V)[h]						Gabriel (1944); Linden et al. (1980)
F. majalis	N	N	P	N	P[g]		Fahy (1972, 1976, 1979, 1980, 1982, 1983)
Rivulus cylindraceus	N	A	P	A	P[g]	Pelvic rays O	Harrington and Crossman (1976a)
R. marmoratus	N	N,V	N,V,A	N	A[g]	Pelvic rays A; Scales N?[h]	Harrington and Crossman (1976b); Lindsey and Harrington (1972); Swain and Lindsey (1986a,b)
Lebistes reticulatus	V(P,A)	P					Schmidt (1919)
Poeciliopsis lucida	P	P	O	P(A?)	O(P)[c]	Predorsal scales[i] A(P); Lateral-line scales P(A)	Angus and Schultz (1983)
Leuresthes tenuis	N[b,d]						Ehrlich and Farris (1971)
Gasterosteus aculeatus	V	N[j,k]	N[j,k]	A(P)[j]	N[c]	Scutes P?; Dorsal spine basals P	Heuts (1949); Lindsey (1962a)

(continued)

Table II (*Continued*)

Species	Vertebrae	Dorsal rays	Anal rays	Pectoral rays	Caudal rays	Others	Source
Pungitius pungitius	V,P	V?	N?	N?		Scales O?	Lindsey (1962b)
Micropterus dolomieui		N	O	P?		Anal spines N Dorsal spines O	Wallace (1973)
Acerina cernua	P(N)[d]						Lubitskaya and Dorofeeva (1961a)
Etheostoma grahami × *E. lepidum*				P[l]		Dorsal spines P[l]	Strawn (1961)
E. nigrum	N[d]						Bailey and Gosline (1955)
Perca fluviatilis	O[d]						Lubitskaya and Dorofeeva (1961b)
Macropodus opercularis	V	O	V	V?	N?[g]	Anal spines N	Lindsey (1954)
Channa argus	V		V				Itazawa (1959)
Pleuronectes platessa	V	P	P				Dannevig (1950); Molander and Molander-Swedmark (1957)

[a] Patterns, as defined in text: P, positive; N, negative; V, V-shaped; A, arched; O, no clear response. Parentheses indicate inconsistency between experiments. Where different genotypes followed different patterns, all are given with the less common in parentheses. Question mark indicates marginal significance.
[b] Myomere counts.
[c] Major caudal rays.
[d] Only two temperatures tested: low and high.
[e] Only two temperatures tested: medium and high.
[f] Only two temperatures tested: low and medium.
[g] Total caudal rays (major plus minor).
[h] Gabriel (1944) found some sibs nonlabile.
[i] Median predorsal scales.
[j] Response varies between marine and freshwater populations.
[k] Rays and ray basals show same pattern.
[l] Only two temperatures tested.

refer to the categories of response described above. All responses listed involved statistically significant meristic differences unless marked otherwise.

Negative responses were commonest, occurring in 58% of the studies on vertebrae and in 41% on fin series. Other patterns of responses differed between the two types of series; for vertebrae the V-shaped response was frequent (33%) while positive or arched responses were very rare; for fin series (rays, spines, basals) positive or arched responses were fairly frequent (29% and 17%), while V-shaped was very rare. Within each species there was no correlation between the ways in which different meristic series responded to temperature. Even dorsal fin ray count and anal fin ray count more often had different than similar response patterns.

The magnitude of meristic variation induced by temperature in Fig. 1 is typical of most other species. Among all species that have been studied, the range of phenotypic variability within a cross, in the means of vertebrae and of dorsal, anal, and caudal rays, was usually about 0.3–3 units, with vertebrae rarely ranging as high as 5 (*Esox*). Of course the absolute meristic counts differed widely between species, as did the temperatures they could tolerate. To restate the above ranges in common units, the phenotypic variability within a cross was usually between 0.1 and 1.0% of total meristic count for each Centigrade degree of temperature in vertebrae, and usually between 0.3 and 3.0% in dorsal, anal, and caudal rays. Pectoral counts were more consistent in the magnitude of their phenotypic responses; absolute means showed phenotypic variability ranging from about 0.3 to 1.5 fin rays.

In the few experiments involving several crosses of one species from one place, the range of meristic differences that could be induced phenotypically within a cross was about the same as the range of genotypic differences between crosses (Fig. 1). However, when experiments have involved crosses of one species from different places, hereditary geographic differences are sometimes much larger than the differences phenotypically inducable within any one cross. For example, see data on the cyprinid *Richardsonius balteatus* in Section VI.

2. Responses to Radiation

Effects on meristic counts of exposure to radiation (usually visible light, occasionally ultraviolet or X rays) have been studied in a dozen fish species (Table III). Most have been subjected to the same daily

Table III

Response Patterns of Meristic Counts to Different Daily Radiation Dose[a,b]

Species	Vertebrae	Dorsal rays	Anal rays	Pectoral rays	Caudal rays	Others	Source
Light[c]							
Brachydanio rerio	N	O	N		O		Maginsky (1958)
Cyprinus carpio	N						Korovina et al. (1965)
Esox lucius	N						Lubitskaya (1961); Lubitskaya and Dorofeeva (1961a)
Osmerus eperlanus	N						Lubitskaya and Dorofeeva (1961b)
Oncorhynchus gorbuscha	(N)[d]	(N)[d]	(N)[d]				Canagaratnam (1959)
O. nerka, sockeye	P[e]	N[f]	N[f]		N[f,g]	Scales (N)[e]	Canagaratnam (1959)
O. nerka, kokanee	N[f]	(N)[h]	(N)[h]				Lindsey (1958); Canagaratnam (1959)
Salmo gairdneri	N (A,P,V)	A(V)	(A)	(A)		Pelvics O(A) Gill rakers (A)	Canagaratnam (1959); MacCrimmon and Kwain (1969); Kwain (1975)
S. salar	P(A)[i]						Vibert (1954)
Oryzias latipes	O	(N[d],P[j])	(P)[j]	(N)[f]	P[j]		Ali (1962); Lindsey and Ali (1965)
Leuresthes tenuis	N						McHugh (1954)
Acerina cernua	N						Lubitskaya and Dorofeeva (1961a)
Perca fluviatilis	N						Lubitskaya and Dorofeeva (1961b)

Ultraviolet[k]				
Osmerus eperlanus	N^l, P^m			Lubitskaya and Dorofeeva (1961b)
Esox lucius	$(N)^n$			Lubitskaya and Dorofeeva (1961a)
X Rays[o]				
S. gairdneri	N^p	$(N)^q$	Parr marks N^r	Welander (1954)

[a] Symbols as in Table II.
[b] Dose calculated as radiation intensity times daily hours of exposure.
[c] Daily dose same throughout development.
[d] Fewer at longer exposures.
[e] More at longer exposures, more at higher intensities
[f] Fewer at longer exposures; fewer at higher intensities.
[g] Both upper and lower minor caudal rays.
[h] Fewest at longest exposure.
[i] Gravel cover versus open trough.
[j] More at higher intensities.
[k] Single short ultraviolet light (UVL) pulse applied either at very early or at tail bud stage.
[l] In dark, early UVL pulse decreased count.
[m] In light, late UVL pulse increased count.
[n] Early or late UVL pulses usually decreased count.
[o] Single short X-ray pulses, various doses and stages.
[p] Fewer at medium or high doses.
[q] Fewer at medium or high doses, except more by low dose pulse at one-celled or late-eyed stage.
[r] Fewer than control at all doses.

light regime throughout development. Hours of exposure per day, and intensity, can be multiplied to yield a daily dosage, and this is the value plotted against mean meristic counts to determine the response patterns summarized in Table III (e.g., if higher daily dosage produced fewer parts the response is classified as negative).

The same generalizations about meristic responses to temperature are applicable to meristic responses to light or other radiation. Virtually every meristic series that has been studied can be altered by radiation. About two-thirds of all responses were negative, and one-third was arched or positive. Unlike responses to temperature, responses to light were almost never V-shaped in any series. Light usually produced somewhat less meristic response than did temperature, although vertebral count responded to light strongly in a few species.

The commonest response to light radiation, reduction in number of parts at higher exposure, is repeated by response to radiation of shorter wavelengths. Exposure to ultraviolet or X rays (which was in all experiments a single short pulse, unlike the daily exposures to light) produced strong negative responses in all meristic series tested (Table III). Some other studies have also shown X radiation to reduce the number of myotomes, but with teratological effects that preclude precise counts (Solberg, 1938).

3. Responses to Solutes

a. Oxygen. There is more consistency in meristic response to oxygen (or its lack) than to other environmental influences. *Decrease* in dissolved oxygen produced higher meristic counts in the four species that have been tested directly (Table IV). Congruently, *increase* in dissolved carbon dioxide also produced higher vertebral counts, in *S. trutta* (Tåning, 1952). Schmidt (1921) found in the same species that, at comparable temperature, the vertebral counts were higher in still than in running water. Perhaps analogously, Tåning (1944) used eggs from a female *S. trutta* that had lain dead overnight on the hatchery floor, and produced higher vertebral counts in her offspring than in others reared at comparable temperature. Meristic variation produced by different oxygen concentrations can be large, ranging from half to over three times as great as variation produced by temperature.

b. Salinity. The lack of agreement in the literature on the relation of salinity to segment number in wild populations is mirrored in experimental studies. The responses of meristic series to salinity shown in Table IV were about one-third negative, one-third positive or a mixture (between genotypes) of positive and negative, and one-third nonsignificant. None was arched or V-shaped. Meristic changes induced by salinity were not large compared with those induced by temperature (e.g., as demonstrated in the experiments on the garfish *Belone belone* by Fonds *et al.*, 1974).

The experiments on salinity in Table IV suffer from inconsistent methodologies. Hempel and Blaxter (1961) fertilized each egg batch in water of the same salinity in which it was subsequently incubated. Most or all of the other experiments consisted of fertilizing eggs at a common salinity, and then placing subdivided egg batches into different test salinities after a delay ranging from 15 min to overnight. Salinity of the water imbibed during swelling of the chorion after fertilization may have had a long-term effect on composition of the perivitelline fluid in which the embryos developed. Holliday and Blaxter (1960) found in herring that while salinity of the perivitelline fluid was partly dependent on the salinity in which eggs were reared, there were also detectable differences in perivitelline salinity between egg batches that were being incubated at the same salinity but that had been fertilized 7–8 days earlier at different salinities. Since wild eggs are likely to be fertilized at about the same salinity in which they are incubated, the protocol usually followed in salinity experiments is a poor simulation of the natural situation. In future experiments on effects of salinity, or temperature, on meristics, natural conditions would be better approximated by applying test conditions to the eggs from the moment of fertilization, or perhaps even earlier to the parents.

c. Other Dissolved Substances. Most of the preceding experiments concerned naturally occurring environmental influences (temperature, light, oxygen, salinity). A few others have examined the modification of meristic counts by artificial concentrations of chemicals in the water (Table IV). Many other recent investigations, arising from concern with environmental pollution, have examined not meristic differences but the more drastic responses of fish embryos, such as gross abnormalities or death, to toxic sustances in the environment. Nonmeristic responses to toxicants are covered in Chapter 4 of Volume XIA.

Canagaratnam (1959), in searching for the mechanisms whereby light influences vertebral number, demonstrated that pituitary primordia and a few minute thyroid follicles were already present in sockeye salmon 6–15 mm long, at stages before vertebral count was finally fixed. Histological examination at a later stage showed that thyroids were more active in fish subject to higher light dosages. Canagaratnam

Table IV

Response Patterns of Meristic Counts to Different Concentrations of Solutes[a]

Species	Vertebrae	Dorsal rays	Anal rays	Pectoral rays	Caudal rays	Source
Oxygen						
Oncorhynchus tshawytscha	N[b]	N[b]	N[b]			Seymour (1959)
Salmo gairdneri	N[c]					Garside (1966)
S. trutta	N					Tåning (1952)
Salvelinus fontinalis	N[d]					Garside (1966)
Carbon dioxide						
Salmo trutta	P					Tåning (1952)
Salinity						
Clupea harengus	P[e]					Hempel and Blaxter (1961)
Belone belone	N[f]	O	O			Fonds et al. (1974)
Oryzias latipes	P	O	O	O	O	Ali (1962)
Fundulus heteroclitus	O					Linden et al. (1980)
F. majalis	O[g]					Fahy and O'Hara (1977)
Rivulus marmoratus	N	N	N	N	N	Swain and Lindsey (1986a)
Gasterosteus aculeatus	(P)	(N,P)[h]	(N,P)[h]	(N,P)[h]	N	Heuts (1949); Lindsey (1962a)
Thyroxine						
O. latipes	(N)	O	N	N[i]	(N)	Ali (1962)
Thiourea						
O. latipes	(P)	P[f]	(P)	P[f]	(N,P)	Ali (1962)

214

Compound / Species						Reference
2,4-Dinitrophenol						
O. latipes	P^j,N^k	O	N	P	P	Ali (1962); Waterman (1939)
Urethane						
Brachydanio rerio	N	N^f	N			Battle and Hisaoka (1952)
O. latipes	P	N^f	P	O		Ali (1962)
Fuel oil						
F. heteroclitus	N^l		N			Linden et al. (1980)
HCl, phosphate, nitrate						
S. trutta	O^m					Tåning (1952)
Malachite green						
O. latipes	O	O	O	O		Ali and Lindsey (1974)
B. rerio	N	O				Battle and Hisaoka (1952)

^{a} Symbols as in Table II.
^{b} Accidental low oxygen during first 21 days, two cases
^{c} Oxygen effect greatest at high temperature.
^{d} Oxygen effect greatest at low temperature.
^{e} Myomeres.
^{f} Weak association.
^{g} Fertilized eggs overnight in full seawater before placed in test salinities.
^{h} Salinity response varies with temperature, and between marine and freshwater populations.
^{i} Very strong association.
^{j} Concentrations 1 : 800,000 to 1 : 1,000,000 (Ali, 1962).
^{k} Concentrations 1 : 40,000 to 1 : 200,000 (Waterman, 1939).
^{l} Water-soluble fraction of number 2 fuel oil.
^{m} No counts given.

suggested that influence of light on meristic count might operate by modifying metabolism via substances secreted by the thyroid. Ali (1962) found that, in medaka *O. latipes,* thyroxine tended to decrease counts in most meristic series, while thiourea tended to increase meristic counts. Two other substances that at appropriate concentrations may inhibit growth or differentiation, 2-4-dinitrophenol and urethane, produced various significant alterations in meristic counts (Table IV). Vertebral number was affected in one species by fuel oil (Linden *et al.,* 1980), but not, in another, by hydrochloric acid or phosphate or nitrate (Tåning, 1952). Malachite green, used to control fungus in many experiments, was found not to affect meristic characters (Ali and Lindsey, 1974).

B. Experiments on Inconstant Environments

Tåning (1944) attempted to determine the duration of the period when temperature could affect meristic characters, by transferring the eggs from one temperature to another at various developmental stages. To his surprise, transfer at some stages produced mean meristic counts that lay outside the range of the means produced by sustained rearing at either temperature. Such "extralimitary" responses have now been investigated in a dozen fish species. All meristic series seem capable of producing extralimitary responses. Most investigations have concerned changes in temperature, but changes in other environmental factors can also produce extralimitary meristic counts. In some cases transfers at different stages produce extralimitary responses in opposite directions. Although experiments published so far have yielded complex and sometimes apparently contradictory results, it is possible now to perceive some patterns in meristic responses to environmental change.

1. TEMPERATURE BREAKS

The term temperature "break" is applied to an experiment in which embryos are abruptly transferred, once only, from the starting temperature to a new (the final) temperature, and held at the final temperature until beyond the stages at which the meristic counts are malleable. Breaks may be applied at different times after fertilization, and they may be to either a higher or a lower temperature.

Extralimitary meristic counts, arising from transfer between two temperatures, are most usefully categorized not simply according to whether the counts are higher or lower with respect to controls (the meristic counts produced by sustained rearing at the two temperatures), but according to the *direction* in which they lie in relation to the controls. If transfer produces a change in count in the direction that would be anticipated from the known effect of sustained rearing at the final temperature, but *beyond* the latter value, this extralimitary response is called *overcompensation.* If the response is a change in the opposite direction to what would be anticipated, this extralimitary response is called a *paradox.* In *O. latipes* (Fig. 2) vertebral count is higher at sustained low, than at sustained high, rearing temperatures. *Early* embryos transferred from low to high temperatures produce even *fewer* vertebrae than those reared permanently at high temperature. Similarly, early embryos transferred from high to low produce even *more* vertebrae than those reared permanently at low temperature. Both responses are overcompensations, even though the resulting absolute values are very different, because both were in the "expected" directions. Conversely, *late* embryos transferred between temperatures may produce counts lying beyond either control count but in the "wrong" directions, and are paradoxical. In meristic series whose counts are very close at the two sustained temperatures, temperature breaks may still produce big meristic changes (e.g., dorsal rays of *R. marmoratus* in Fig. 2), but their categorization as overcompensation or paradox is dubious.

Temperature breaks were originally intended by Tåning (1944) to delimit the "thermolabile period" of meristic counts, and can be so used, but with reservations. In some studies there is supicion of undiscovered meristic lability at later or earlier stages than those tested. Also, nonlability to the particular environmental perturbation applied at a certain stage does not preclude lability then to other possible perturbation.

Despite these limitations, the period of sensitivity to temperature change can be approximated for many meristic series from transfer and from pulse experiments. The length of the period is, of course, different at different temperatures. In Table V the periods of extralimitary responses have been classified as "early" or "mid" or "late" according to their position in the estimated total time when the meristic counts are subject to phenotypic change. Different meristic series have different periods of sensitivity; vertebral count is finally fixed early (usually well before hatching), while some fin ray counts may still be labile. Information on embryogenesis of meristic series is given in the next section. Fixation of the number of parts may occur well before the total series of parts is visible and countable.

In Table V, which summarizes published results of temperature

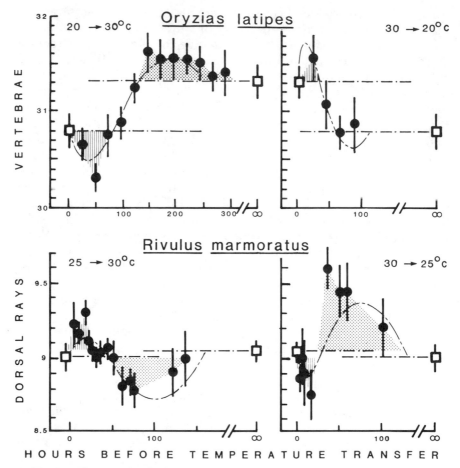

Fig. 2. Effects on vertebral and dorsal fin ray counts, in *Oryzias latipes* and *Rivulus marmoratus*, of rearing at constant high or low temperatures (squares and broken horizontal lines), or of transferring between temperatures at indicated times after fertilization (circles). Vertical lines are 95% confidence intervals. Shading indicates extralimital responses, either overcompensation (vertical hatching) or paradox (stippling). Light curves calculated from atroposic model (see text). [Modified from Lindsey and Arnason (1981) and Swain and Lindsey (1986a).]

breaks, 0 entries mean no extralimitary responses (i.e., counts produced by all breaks lay within the range of those produced by the two control temperatures). Of the 23 cases that did produce some extralimitary responses, three-quarters were either early overcompensation, or late paradox, or more often both (as in Fig. 2). In the exceptional cases, the two control values lay very close.

The commonest response pattern, an early overcompensation followed by a late paradox, was produced in some experiments by upward and in some by downward temperature breaks. This pattern was produced in some meristic series with positive, and others with negative, correlation between counts and sustained rearing temperatures. In some cases the extents by which the overcompensation and the paradox exceeded the control counts were about equal. In other cases, one of the extralimitary responses was smaller, and this pattern grades imperceptibly into cases where only a single response was detectable. Where two series of temperature breaks have been applied in opposite directions (Fig. 2), the types (but not the degree) of extralimitary response in one direction were also present in the reciprocal break response. As is implied by the term "extralimitary," the meristic responses to temperature breaks can be greater than responses to the different sustained control temperatures. In the limited number of experiments published, the magnitude of meristic variation produced by temperature breaks is roughly similar to that produced by extreme sustained rearing temperatures.

Contrary to the long-held assumption that meristic counts are not susceptible to environmental modification until a substantial time after fertilization (the beginning of the so-called "sensitive period"), both breaks (Swain and Lindsey, 1986a) and pulses (Komada, 1977a) have revealed meristic lability even within the first few hours.

2. TEMPERATURE PULSES

The term temperature "pulse" is applied to an experiment in which all development occurs at one (control) temperature except that the embryos are abruptly transferred to a new temperature for a specified time, and then returned to the control temperature for the duration of their development. Like breaks, pulses may start at various developmental stages, and may be to either a higher or lower temperature. The duration of the pulse has varied from a few hours to a few days.

Meristic responses to pulses can be categorized, like break responses, according to the direction in which they lie with respect to meristic counts produced by sustained rearing at the control temperature and at the pulse temperature. Change in count in the direction of that produced by sustained rearing at the pulse temperature is compensation; change in the opposite direction is paradox. The term overcompensation is avoided for pulse responses in Table V, as the meristic count that would be produced by continuous rearing at the pulse

Table V

Responses of Meristic Counts to Temperature Changes[a]

Species	Vertebrae	Dorsal rays	Anal rays	Pectoral rays	Caudal rays	Other	Source
Temperature breaks[b]							
Brachydanio rerio	0[c]	0[c]	?[c,d]	?[c,e]	0[c]	Pelvic rays ?[c,f]	Dentry (1976); Dentry and Lindsey (1978)
Salmo gairdneri	Early overcomp., late paradox						Hallam (1974); Lindsey et al. (1984)
S. salar	0[g]						Pavlov (1984)
S. trutta	Late paradox	Early slight overcomp., late large paradox	Early overcomp., later paradox	0			Täning (1944, 1952); Lindsey and Arnason (1981)[h]
Salvelinus fontinalis	Early overcomp., late paradox						Hallam (1974)
Oryzias latipes	Early overcomp., late paradox	Late overcomp[i]	Early overcomp., mid paradox?, late overcomp.?	Early overcomp.	Middle paradox[i]		Ali (1962); Ali and Lindsey (1974); Ogawa (1971); Lindsey and Arnason (1981)[h]
Fundulus majalis	(Late paradox?)		0				Fahy (1976, 1979, 1983)
F. heteroclitus	0[j]						Gabriel (1944)
Rivulus marmoratus	Early overcomp. (late paradox?)	Early paradox[k] late overcomp.[i]	Middle paradox[i]	Early overcomp. (late paradox)	Early overcomp.[i]		Lindsey and Harrington (1972); Swain and Lindsey (1986a)
Micropterus dolomieui		(Early overcomp., late paradox)?	0	(Early overcomp., late paradox)	(Early overcomp., late paradox)[h]	Scales late paradox	Wallace (1965, 1973)
Macropodus opercularis	Middle paradox[l]	0[i,l]	Mid paradox, late overcomp.[i,l]	0[l]		Dorsal spines late large paradox[i,l]	Lindsey (1954)
Pleuronectes platessa	0[m]	Late paradox[m]	Late paradox[m]				Dannevig (1950); Molander and Molander-Swedmark (1957)
Temperature pulses[n]							
Clupea harengus	Compens.[o]						Hempel and Blaxter (1961)
Plecoglossus altivelis	Very early paradox, later compens.						Komada (1977a); Lindsey and Arnason (1981)[h]

Species					References
S. gairdneri	(Middle compens. or paradox, later comp.)[p]	Early compens. (later paradox. or compens.)[p]	(Mid compens. or paradox., late paradox. or compens.)[p]	Late comp.	Orska (1962, 1963); Lindsey *et al.* (1984)
S. trutta	Early comp., late paradox	Late comp.	Early comp.		Orska (1956); Tåning 1950, 1952); Lindsey and Arnason (1981)[h]
Temperature alternation[q]					
O. latipes	Intermediate between controls				Lindsey and Ali (1965)[i]

---

[a] Early, middle, and late refer to segments of estimated total period of malleability of a given character. Parentheses indicate inconsistency between experiments. Question mark indicates marginal significance.

[b] One-time-only transfer from original to new temperature (which may be higher or lower). Overcompensation and paradox refer to extralimitary responses (counts beyond those produced by either control) in either expected or opposite direction respectively. Zero, mean counts of transferred lots not extralimitary.

[c] Only early transfers, at 1, 2, 4, 8, 16, or 32-cell stages.

[d] Extralimitary at eight-cell stage.

[e] Extralimitary at two- and eight-cell stage.

[f] Some extralimitary between 4- and 32-cell stage.

[g] Only one transfer, at stage 10. Total vertebrae intermediate, but abdominal and caudal counts both paradoxical.

[h] Atroposic model fitted to previously published data.

[i] Counts at control temperatures very close.

[j] Only one transfer, at 4–5 somite stage, count still unfixed.

[k] Significant only if no prefertilization effect assumed.

[l] Transfers: middle at 13 days, late at 22 days.

[m] Only two transfers, at 184 or 750 day-degrees; both after hatching.

[n] Transferred from original to new temperature (either higher or lower) for short time and then back. Compensation and paradox refer to counts different from those produced by original temperature in either expected or in opposite direction (not necessarily extralimitary) respectively.

[o] One-day cold pulse produced counts close to those at sustained cold if applied on days 7 or 8, or counts close to sustained warm if applied earlier or later.

[p] Opposite reactions in different genotypes.

[q] Alternated between low and high temperature every 12 h.

temperature may not be available, although the direction can be surmised.

Meristic responses to temperature pulses in four species (Table V) were in the direction of compensation twice as often as of paradox. Either response can occur either early or late in development, and both may occur at different developmental stages, in the same meristic series. Pulse responses can be striking; Tåning (1950), by applying a short cold pulse to one batch of S. *trutta* and a short warm pulse to another, produced a difference of 3.2 vertebrae (twice as great as any of the differences produced by sustained temperature in Fig. 1).

Evidently it is not simply the shock of the two abrupt changes that is responsible for meristic changes. Pulses must be of appreciable duration to be effective. Tåning (1952, p. 182) found that a very short pulse (2 day-degrees Fahrenheit) produced no meristic response. There is no evidence to suggest that the speed of transfer makes a difference, although experiments specifically to test the possibility are lacking.

In the only experiment involving a very early temperature pulse (Komada, 1977a), a 1-day cold pulse starting 1 h after fertilization resulted in a vertebral count paradoxically low.

3. Responses to Systematic Temperature Change

Most of the laboratory experiments described above have been poor mimics of natural rearing conditions. Eggs developing in the wild are likely to experience some diel temperature fluctuation, a regime that most experimenters have been at pains and expense to avoid. To examine the effects of diel temperature change, Lindsey and Ali (1965) compared vertebral counts of O. *latipes* produced by temperatures abruptly alternating between high and low every 12 h throughout development with those produced by sustained temperatures and by temperature breaks. Figure 2 shows how single breaks in either direction produce extralimitary counts in this species, in opposite directions depending on stage at transfer. Yet the batch subject to one upward and one downward change every day produced vertebral count intermediate between the two control counts, as though they had been reared at a sustained intermediate temperature. Survival was higher in the alternating-temperature batch than in the other batches, and malformations were negligible. These results argue against the concept of some sort of "shock effect" (Tåning, 1944) as an explanation of how breaks produce extralimitary meristic responses, since the alternating batch had been subject to the shocks of temperature change every 12 h. They also suggest that artificial constancy of rearing temperature has not importantly biassed meristic experiments.

Experiments on two other species have attempted to simulate natural temperature regimes. Seymour (1959) reared chinook salmon, which spawn in autumn, at gradually falling temperatures (1°F every 5 days). In two cases where constant-temperature controls were available, the control produced almost the same counts of vertebrae and of dorsal and anal fin rays as the experimental batch whose mean temperature during the early incubation period approximated that of the control. Fahy (1972) reared *Fundulus majalis* eggs at different daily mean temperatures, each of which followed a programmed diel cycle spanning 4°C. There were no constant-temperature controls, but the meristic counts produced by different daily mean temperatures seem to follow patterns similar to those produced in other species by different sustained temperatures.

The foregoing experiments suggest that when various temperature protocols are applied over a restricted period of development, the meristic responses seem to be dependent on total thermal units during that period, rather than on degree or direction of any temperature changes per se. Of course, the same number of thermal units applied at different developmental periods has been shown to sometimes produce very different meristic responses. The statement needs to be further qualified, since thermal units, or day-degrees, are handy in hatcheries but theoretically unsound for two reasons. First, they depend on the arbitrarily chosen temperature of "biological zero." Second, different embryonic processes do not all use the same yardstick (Hayes *et al.*, 1953), which is the essence of the models to be suggested later.

4. RESPONSES TO RADIATION CHANGE

Meristic responses to transfers between two levels of radiation have been studied in a few species. Like temperature transfers, radiation transfers can produce extralimitary responses.

Canagaratnam (1959) subjected sockeye salmon *Oncorhynchus nerka* to light "breaks" by transferring embryos between different regimes of light intensity and light duration at various stages. Some breaks produced highly significant overcompensation or paradox in vertebral count compared with counts of controls at either sustained regime. Unlike responses to temperature breaks, the light breaks tested did not suggest that paradox or overcompensation was usually associated with early or late transfer, nor were the results of reciprocal breaks always mirror images.

Light pulses of various intensities and starting at various stages were applied by Eisler (1961) to chinook salmon O. tshawytscha. Bright light pulses (5 days duration) if applied early produced a rise in vertebral count (paradox), but if applied later produced a drop in count (compensation).

Short pulses of X rays were applied by Welander (1954) to rainbow trout S. gairdneri at various dosages and various stages. Usually higher dosages reduced the number of dorsal and anal rays and parr marks (Table III). However, anal fin ray count increased significantly (paradox) if exposed to a low X-ray dose in the one-cell stage, and a similar response occurred to exposure at the prehatch stage.

Lubllskaya and Dorofeeva (1961a,b) applied pulses of ultraviolet radiation at one of two developmental stages, to four European species (Table III). Short ultraviolet exposure produced inconsistent but sometimes large changes in vertebral count.

5. Responses to Other Changes

Changes in temperature or radiation during the course of development, either as breaks or pulses, have been shown above to sometimes produce large meristic differences. Other environmental factors that at different sustained levels can affect counts may similarly produce extralimital differences if altered during development.

Five meristic series in R. marmoratus have a negative correlation with salinity (Table IV). A "salinity break" from saline to fresh water 45 days after fertilization produced an extralimital caudal ray count higher than counts produced by sustained rearing in either saline or freshwater (Swain and Lindsey, 1986a).

An accidental "low oxygen pulse" may have occurred when Hallam (1974, p. 83) left a transferred egg batch of S. gairdneri in its cup poorly aerated for 3 days. The result was a marked decrease in vertebral count, extralimital with respect to the control temperatures being tested. Since Garside (1966) had demonstrated in this species that sustained low oxygen increases vertebral count, the response to Hallam's short pulse of low oxygen (or high CO_2?) can be classed as paradoxical.

C. Responses to Other Postfertilization Influences

Mechanical disturbances seem usually to have little influence on meristic counts, except for anal rays (and sometimes caudal rays), which often yield puzzling results and may be susceptible to influences that have not been properly controlled. Different embryos within one experimental batch occasionally show surprising differences in developmental rates even in *R. marmoratus*. Dr. E. K. Balon has voiced to me the suspicion that noises in the laboratory (e.g., conversation, or slamming of doors) may affect developing fish embryos, so developmental noise *sensu* Waddington may arise in part from noise *sensu* Balon.

Violent shaking of eggs of *Oryzias latipes* for 4 min each day had no effect on vertebral, dorsal, pectoral, and caudal ray counts, but anal ray counts were significantly higher in eggs subjected to this treatment starting 4 days after fertilization (Ali and Lindsey, 1974). Pricking of the chorion (to test its permeability to experimental chemicals) produced no difference in vertebral count in *Brachydanio rerio* (Battle and Hisaoka, 1952). Similar pricking of the chorion in *O. latipes* (Ali, 1962) had no effect on vertebrae nor on pectoral or caudal rays, but produced fewer dorsal fin rays in one of two tests, and fewer anal fin rays in both tests. Subjecting *Brachydanio rerio* eggs to two atmospheric pressures produced slight retardation of somite formation (Goff, 1940), but final meristic counts were not recorded.

Meristic counts can be affected by crowding the developing embryos. Eggs of *O. latipes* were reared at four densities (from 25 to 200) in screen baskets suspended in the same bath (Ali and Lindsey, 1974). In crowded conditions there were significantly fewer anal and caudal rays, slightly fewer pectoral and dorsal rays, and no difference in vertebrae. Mortality was slight in all baskets. Time to 50% hatch was not correlated with density. In a similar experiment, *F. majalis* eggs were reared at six densities (44 to 154 per container). The most crowded batch hatched earliest and produced significantly fewer vertebrae (Fahy, 1978).

In these experiments on crowded conditions, low oxygen tension or high CO_2 might be suspected to be operative, but they seem to be exonerated since these factors increase rather than decrease meristic counts in other species (Table IV). Moreover, concentrations of these or other solutes were probably similar between egg batches, since all baskets were suspended in agitated and aerated water. As a further test, three batches each of 40 eggs of *O. latipes* were reared, one in a basket suspended in a tank of aerated and flushed water, one in a jar with continuous aeration, and one in a jar with the same amount of stagnant water (Ali and Lindsey, 1974). Hatching was much faster in the stagnant jar, and mortality was slightly higher, but there was no difference between the three batches in numbers of vertebrae or of dorsal, anal, pectoral, or caudal rays. (Densities were lower in these than in the more crowded conditions of the earlier experiment.)

The factor that affects meristic counts in crowded conditions remains mysterious. Experimental results, while not conclusive, tend to rule out buildup or deficiency of solutes, hatching rate, or selective mortality. Experimental crowding shows a parallel, perhaps coincidental, to Vladykov's 1934 "space factor" observed in wild fish; in both, confined conditions are associated with lower segment number.

D. Prefertilization Influences

Until recently, meristic variation has been assumed by most fishery biologists to be attributable to only two sources: genetic variation, and environmental influences on the developing embryos after fertilization. Now, experiments have demonstrated that events impinging on the gametes *before* fertilization can also affect meristic counts produced by the subsequent zygotes. The demonstration casts doubt on the interpretation of some previous experimental meristic studies, because prefertilization conditions of the parents have not usually been controlled nor even reported. It also rouses curiosity about the mechanisms involved. Latitudinal clines, and seasonal races, may be attributable in part to environmental effects on parents prior to reproduction. There may even be interesting evolutionary implications, with echoes of Lamarck, although there is as yet no reason to believe that the prefertilization effects reach into the genome.

Prefertilization influences fall in two categories: external environmental influences that impinge on the parents, and "internally generated" factors such as sequence of egg formation, parental age, and egg size.

1. MERISTIC RESPONSES TO ENVIRONMENTAL HISTORY OF PARENTS

Influence of parental holding temperature on offspring meristic count was first studied experimentally in zebrafish *B. rerio*. Dentry and Lindsey (1978) compared pairs of samples of young reared at the same temperature and from the same parents, and found the vertebral counts were higher in the sample laid when the parents had been held at a higher temperature. Four other meristic series also responded to parental holding temperature, but in the opposite direction: dorsal, pectoral, pelvic, and caudal ray counts were lower in samples from high parental holding temperature. Anal ray count showed no prefertilization effect. Comparison of results from eggs laid soon or long after the parental temperature change suggested that a new parental holding temperature begins to have an effect quickly, but that eggs laid shortly after a parental shift may already have incorporated some temperature effects that predate the shift.

Although the experiments with *B. rerio* were strongly suggestive, they were complicated by differences in meristic response patterns of the different breeding pairs. There was also the possibility that meristic differences among survivors might be attributable to differential survival among genetically diverse embryos. Therefore, prefertilization effects were explored further by Swain and Lindsey (1986a,b) using a genetically uniform clone of the self-fertilizing fish *R. marmoratus*. Comparison of counts in fish reared at the same temperature but arising from parents held at two different temperatures again showed a marked prefertilization effect. Counts of vertebrae and of pectoral and caudal rays were all significantly lower in samples laid when the parents were held at a high temperature than when parents had been moved recently (about 5 days ago) from a low temperature. Counts of dorsal and anal rays displayed the same trend only weakly. For vertebrae and caudal rays, the effect of a new parental temperature was apparent only in offspring produced more than 10 days after parental transfer. For pectoral rays, the effect was already significant in offspring produced less than 10 days after parental transfer, but was maximal in offspring of parents with longer parental exposure to the new temperature.

A few other studies on fish have suggested that parental holding temperature can influence other characters of the offspring. Hubbs and Bryan (1974) found in *Menidia audens* that a difference in parental acclimation temperature produced a difference in thermal tolerance of the eggs. Tsukuda (1960) found in *Lebistes reticulatus* that offspring from parents held at 25°C were decidedly more tolerant of temperature extremes than were offspring from parents held at 20 or 30°C. Another instance of prefertilization effect cited earlier concerns the very high vertebral counts produced by eggs from an *S. trutta* female who had already died before being stripped (Tåning, 1944).

Mechanisms whereby parental temperature history can affect meristic counts in offspring are conjectural. Temperature acclimation of several fish species affects the composition of their body fats and enzymes (Hoar, 1983, p. 669) and perhaps, therefore, the composition of their eggs. In fish and other organisms, development after fertilization may be controlled up to the high blastula stage by messenger RNA that has been synthesized and in some cases translated before fertilization, during oogenesis (references in Swain and Lindsey, 1986a). Therefore, prefertilization influences that affect fats or enzymes during oogenesis might be expected to modify early postfertilization development. Enzymes synthesized on maternal RNA templates are only gradually substituted by embryonic (zygotic) proteins. Products of some alleles coming from the male parent can be detected in fish embryos only at late blastula, early gastrula, or even later (Kirpichnikov, 1981, p. 200). Although the female parent is more likely to be involved, the relative roles of the two sexes in prefertilization temperature affect have yet to be investigated. The possible relevance of the atroposic model to prefertilization influence is referred to later.

2. Meristic Responses to Parental Reproductive History

Another influence on meristic variation that operates before fertilization is the reproductive history of the parents. The term parental reproductive history (PRH) refers to the position of an individual in its mother's sequence of oogenesis. The position that an egg occupies in the egg-laying sequence tends usually to be correlated closely with the age of the mother, and the two influences cannot be distinguished in many experiments.

In a clone of R. marmoratus held under constant conditions, the offspring produced soon after their parents had begun to lay eggs had fewer parts than did those produced longer after the onset of oviposition (Swain and Lindsey, 1986b). Characters affected (in decreasing order of response) were anal rays, dorsal rays, vertebrae, pectoral rays, and caudal rays. Most meristic differences were due to low counts produced by the early embryos, laid within 8 days of first oviposition. In some characters the embryos laid longest after first oviposition (average 155 days) showed a slight decrease in mean compared with those laid near the middle of the oviposition sequence. The curve of meristic count against PRH could therefore sometimes be fitted as well by a parabolic as an asymptotic model. Analysis that included some parents who began laying eggs at an unusually old age suggested that the causal agent of meristic differences was not the age of the parent, but rather the time since the parent first laid eggs.

In O. latipes, two batches of eggs laid by the same pair 15 days apart yielded significantly higher numbers of anal and caudal rays (but not of vertebrae or pectoral or dorsal rays) in the earlier batch (Ali and Lindsey, 1974).

In the viviparous fish Poecilia reticulata, the heritability of caudal fin ray number was low using very young parents and uniformly high using parents of medium or old age (Beardmore and Shami, 1976); heritability of lateral line scale number was greatest using parents of medium age (Shami and Beardmore, 1978).

Parental age in fishes has also been shown to affect characters other than meristic ones: viability of eggs and young in *Cyprinus carpio* (Nikolsky, 1969b) and in hybrid splake (Ayles and Berst, 1973), and fingerling growth in *S. gairdneri* (Gall, 1974). Parental age in *Drosophila* and other invertebrates can also affect morphology and life history characteristics (references in Swain and Lindsey, 1986b).

3. EGG SIZE AND MERISTIC COUNT

Among related species or races, those with larger adult size tend to have more vertebrae (described in Section II,A as pleomerism). Those with larger adult size also tend, often but not invariably, to lay larger eggs (Mann and Mills, 1985). Very old fish may show a secondary decline in egg size, in *Clupea* (Hempel, 1965) or *Acipenser* (Krivobok and Storozhuk, 1970). The question arises whether egg size may exert a direct influence on the number of vertebrae formed in the larva. Kyle (1926) believed this, but based his opinion on observed correlations in the wild. Eggs of the same species laid at different times in the season are commonly different in size (Hiemstra, 1962; Southward and Demir, 1974), and the larvae hatched at different times are commonly different also in meristic count (Section II,B). But environmental factors also vary seasonally. Meristic differences within a year class could arise from genetically different seasonal races or from direct environmental modification, or from differences in egg size, which in turn arose either from environmental sources (Dauolas and Economou, 1986; Marsh, 1984) or from heredity. Another complication is that in some species the older females tend to lay larger eggs (Peters, 1963; Mann and Mills, 1985), and the age distribution of spawners may vary seasonally. The problem, as always when trying to interpret meristic variation in wild populations, is to separate environmentally induced from genotypic components.

When control of meristic counts by egg size has been tested experimentally, the results have been mostly negative. There was no correlation between egg size and vertebral number in 36 reciprocal crosses of hatchery stock of *S. gairdneri* (C. C. Lindsey and G. B. Ayles, unpublished data). In another experiment on a pair of large anadromous *S. gairdneri*, the fertilized eggs were sorted mechanically into three size classes and reared separately. There were no significant differences between the three batches in vertebrae, dorsal or anal rays, or median pterygiophores. Tåning (1952) similarly concluded that in *S. trutta* the vertebral count was not influenced by egg diameter, nor by whether fry hatched first or last within an egg batch. Among laboratory-reared fish that have been hatched from common parents and reared in uniform environment, a correlation might be expected between fry size and vertebral number if egg size controls vertebral count. In almost all experiments, *no* such correlation has been found (references in Lindsey and Ali, 1971).

In each of two sets of experiments with *O. latipes,* two females that laid different-sized eggs were crossed reciprocally with two males (Lindsey and Ali, 1971). There was no correlation between egg size and the resulting numbers of vertebrae or dorsal, pectoral, or caudal rays. Only in numbers of anal rays did the larger eggs in all four pairs of crosses yield higher ray counts than did the smaller eggs mated with the same male. This correlation is of doubtful significance, as anal ray number in the species can vary between different egg batches from the same parents, and is sensitive to mechanical and perhaps other uncontrolled influences.

Indirect evidence concerning control of meristic count by egg size comes from observations on reduced segment number in partly twinned salmonid embryos (Garside and Fry, 1959), and from vertebral counts of hybrids between species having different egg sizes. The evidence, reviewed by Lindsey and Ali (1971), suggests that morphogenic control of vertebral count does not reside in initial egg size. The relevance of this conclusion to building a model of embryonic development and meristic determination will be discussed in Section V.

IV. EMBRYOGENESIS OF MERISTIC SERIES

In order to represent the action of environmental influences on segment number by a realistic model, there must be information on the geomtry and sequence of segment embryogenesis. The following summary of available information on the early stages of segment formation (the important period in determining ultimate segment number) is based on scattered literature, much of it old. Documentation and illustration of the ontogeny of later stages, particularly of the skeletal elements that respond to staining, have recently improved greatly with an upsurge of interest in identification of larval fishes (Shaw, 1980; Dunn, 1983, 1984). The sequence of ossification does not, however, always follow order of earliest formation of the primordia, and could be misleading in constructing causal relations concerning environmental modification, as well as in establishing ancestral and derived states for cladistic analysis.

A. Trunk Segmentation

The first meristic structures to appear in a fish embryo are the somites, closely followed by the vertebrae. In the neurula stage, the embryo axis is a raised ridge, with nerve chord lying externally along its crest. Beneath it lies the notochord, flanked on either side by strips of mesodermal tissue without visible structure. As the germ ring progressively encircles the yolk, the posterior of the embryo axis continually elongates to follow its advancing edge. The first pairs of somites appear, behind the site of the auditory vesicles, as roughly rectangular blocks separated by transverse fissures that cleave the mesoderm on either side of the notochord (Fig. 3a). More somites are then formed in succession backward (Fig. 3b, 3b'). The embryo axis continues to elongate posteriorly after blastopore closure, to form a tail bud free from the yolk (Fig. 3c). When the backward progression of segmentation reaches close to the notochord tip, somite formation is complete. The tip flexes dorsally at the site of the caudal fin base and the first hypural elements appear (Fig. 3d). The proportion of axial segments that comprise the tail bud varies widely in different species, as does the resultant ratio of abdominal to caudal vertebrae.

The skeleton of the vertebrae and ribs, and probably also of the internal median fin supports, is formed by mesenchymal cells that migrate inward from the sclerotome or inner surface of each somite and cluster around the notochord and nerve cord (Gabriel, 1944, p. 133). Vertebral spacing seems to be dictated by somite spacing rather than vice versa. In fact, if teleosts follow the same sequence as amphibians, somite spacing probably controls spacing of the neural-crest cells that settle to form segmental ganglia, and these in turn govern the spacing of the neural arches. The somites, after giving off the sclerotome, split into an inner myotome from which arises msucle, and an outer dermatome that joins the ectoderm and gives rise (as "somatopleure") to the body wall.

Visible appearance of cartilaginous centra, and of vertebral arches, and their subsequent ossification do not always follow the same direction as the appearance of the somites. In different taxa, the appearance of vertebral elements may originate at more than one point and may proceed in either direction (Itazawa, 1963; Nagiec, 1977; Dunn,

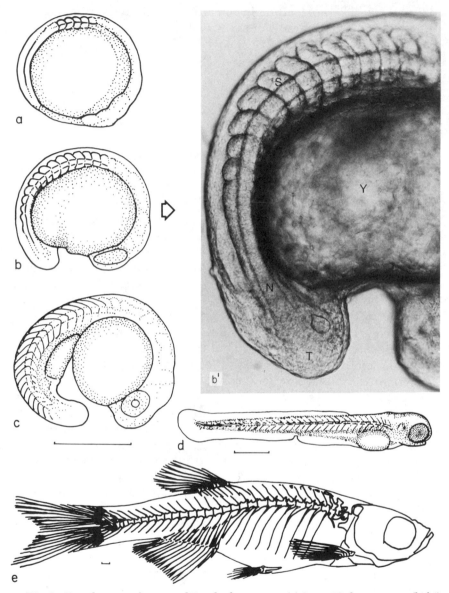

Fig. 3. Developmental stages of *Brachydanio rerio*. (a) Stage 18, five somites. (b, b′) Stage 19, 15 somites. (c) Stage 21, 30 somites. (d) Newly hatched, 34 somites. (e) Camera lucida drawing of cleared and stained juvenile. Scale bars are 0.5 mm. T, Tail bud; N, notochord; S, somite; Y, yolk. [(a)–(d) After Hisaoka and Battle (1958); (b′) courtesy of Dr. H. W. Laale.]

1984). The series of median neural and haemal spines is complete and countable before all the centra are ossified. Counting of trunk segments is usually possible shortly after hatching; fixation of the count usually occurs well before hatching.

Counts of myomeres and of vertebrae are not interchangeable; herring may have three or four more of the former than of the latter (Blaxter, 1957; Lapin, 1975). Techniques for counting myomeres or vertebrae include use of toluidine, alizarin, or other stains (Potthoff, 1984; Taylor and van Dyke, 1985), polarized light (Calkina, 1969), phase contrast (Linden et al., 1980), or radiography (Miller and Tucker, 1979). Distinction between abdominal and caudal vertebrae is often possible (Gordon and Benzer, 1945). Irregularities are common in vertebrae, particularly near the caudal flexure; in any investigation, criteria for consistent enumeration of irregular structures must be established (Ford, 1933; Tåning, 1944; Gabriel, 1944; Landrum, 1966).

Of all meristic series in fish, the somites seem to follow the most straightforward geometry of development: linear appearance along an elongating strip of apparently homogeneous tissue, without interference from adjacent meristic series. Although the mechanisms determining the internal spacing of the somites are as yet unknown, no other meristic series offers better material for modeling embryonic responses to environmental influences.

B. Dermal Structures

1. SCALES

Scale rows first become visible in bony fishes well after the pattern of myomeres has been established. Initial spacing of the scale precursors is dictated by spacing of the lateral line sense organs, dictated in turn by spacing of the myomeres. In salmonids, these sense organs are present in a single series along the flank at or about the time of hatching (Neave, 1943). There is at first one sense organ per body segment, but in some species new sense organs then arise alternating with the original members, so that the number becomes approximately doubled. Cellular accumulations called scale papillae appear beneath each sense organ, and additional papillae may develop between the primary members of the lateral line series. Oblique lines of papillae spread dorsally and ventrally, with sometimes an increase in number as they diverge from the lateral line. The doubling of papillae is omitted in those nonsalmonid species that maintain a 1:1 ratio of scale rows to myomeres. In forms such as clupeids having reduced lateral line systems, some of the primary papillae must arise independent of neuromasts.

Each papilla gives rise to one scale, with a fibrous basal plate and ossified disc (Waterman, 1970). Scales enlarge and overlap, to form a continuous cover from head to tail, like roof shingles. The order of appearance of scale ossification varies between taxa, and may spread from more than one center (White, 1977; Fukuhara and Fushimi, 1984). The full complement of scales may not be visible until well into the juvenile phase. Scale count then remains constant in most species, but a few such as *Phoxinus phoxinus* are known or suspected to increase their scale count throughout life (Repa, 1974).

The series of body segments develops first without interference from other meristic series (although subject to early environmental influences); the scales probably pick up their initial spacing from the former, and then may interpolate extra members and be subject to modification by other diverse influences. Hence the pathways whereby the environment can modify meristic count are probably more complex for scales than for vertebrae.

2. SCUTES

Bony plates or scutes form countable series along the flanks in some fishes, and have received an immense amount of attention in sticklebacks, family Gasterosteidae (Coad, 1981). Stickleback scutes are segmentally arranged, each on a separate myomere. Each first forms as a thickening containing scleroblasts, in the corium in the neighborhood of a lateral-line sense organ (Roth, 1920). The primordium appears as a plate with several openings or with a ridge, and a ring-shaped opening for the sense organ. Each plate enlarges, particularly dorso–ventrally, and may develop a midlateral projection slightly overlapping its posterior neighbour.

The order of visible appearance is complex; scutes first develop at the pectoral girdle, and then on the caudal peduncle, and from these two centres they then develop in two directions (Bertin, 1925). Some populations maintain a gap between the anterior and posterior scute series, and in others there may be scutes only at the anterior, or only on the peduncle. In *Gasterosteus*, scute number seems to be strongly heritable, largely from the mother (Heuts, 1947; Hagen and Gilbertson, 1973), but may be subject to slight phenotypic modification (Lindsey, 1962a). Certainly in those forms that are incompletely plated, and perhaps even in those fully plated, the scutes should be treated as more than one meristic series. The first scutes are not visible until after hatching. Scutes are fully formed in *G. aculeatus* by a length of 2.8 cm.

Ontogenetically, scutes resemble scales in that their initial spacing is superimposed on spacing of the myomeres (and because of the 1 : 1 ratio the maximum number of scutes is limited by the myomere number). Scutes differ from scales in that the latter, if present at all, almost always form in the adult one complete series from head to tail, while scutes may not.

3. GILL RAKERS

The gill rakers, in rows on the inner or outer borders of the gill arches, are considered by Nelson (1969) to be part of the dermal skeleton of the arches and are little more than modified tooth plates. Each gill raker arises as a papilla of the pharyngeal mucosa, within which develops a cartilaginous core which later ossifies. The first ossified members appear near the angle between upper and lower limbs of the arch. Gill rakers become visible relatively late in the larval period (Beacham and Murray, 1986). Often the full complement is not reached until well into the juvenile stage, and some species may even continue throughout life to add members to the ends of the series. Gill-raker number has a large hereditary component, but is subject to some environmental modification (Lindsey, 1981; Beacham and Murray, 1986). Unlike the other meristic series described so far, the rows of rakers resemble the paired fins in that they develop perpendicular to and mechanically independent of the body axis.

C. Median Fins

1. DORSAL AND ANAL FINS

Each fin comprises several rows of structures that are tightly in step with each other along most of the fin. Each fin ray or spine articulates at its base with one set of median internal supports, the pterygiophores, and is operated by up to three sets of muscles. At the extremities of each fin, there may be departure, in the adult, from strict serial conformity: at the front, more than one small ray or spine on an enlarged internal pterygiophore, and at the back two rays articulating with a single internal base. The base that supports two external elements in the adult starts ontogenetically as two separate cartilages [usually (Kohno and Taki, 1983), but not always (Kinoshita, 1984)].

The number and spacing of skeletal supports of the dorsal and anal fins are usually independent of the trunk segmentation (Fig. 3e). Although in the adult the pterygiophores may interdigitate deeply with processes from the vertebral column, in the embryo they do not. At their inception, the median fin series give the impression of developing as parallel but autonomous organisms. Only later the fin series and body segments become intimately bound by nervous, circulatory, and other connections. In only a few higher teleosts have the elements of the median fins increased in spacing so as to correspond 1 : 1 with the underlying vertebrae over part but not all of the body.

Descriptions of median fin ontogeny are available for few species. Most are fragmentary, and many of the best are old. Key references to the literature may be found in Dunn (1984), Eaton (1945), François (1957), Lindsey (1955), and Wood and Thorogood (1984). François (1958) presents an entré to the earlier literature, much of which is German or French. For a general review and bibliography on fin anatomy, see Lindsey (1978). The following outlines the pattern of development of dorsal and anal fins, based largely on S. *gairdneri* as described by François (1957, 1958).

In most fish embryos, a median fold of epidermis runs continuously from behind the head back around the tail and forward ventrally to the anus. Into this fold thickenings, arising probably from three main sources, are insinuated at the sites of the future dorsal, caudal, and anal fins. Farthest out in the fold, a double layer of cells becomes applied to the inner faces of the dermis. In S. *gairdneri*, this "primary mesenchyme" of cells with large nuclei probably arises from loose proliferations from the dorsal edges of adjacent somites. It later gives rise to the fin rays.

Proximally, a median mass of cells with small nuclei, the "secondary mesenchyme," enters the fin base; this mesenchyme is at first continuous with the loose cells, probably sclerotomal in origin, which occupy most of the intersomitic spaces in the young embryo (and which invest the notochord and form the neural and haemal arches). Within this median cell mass appear aggregations from which develop cartilaginous rods, the "basals" or proximal members of the pterygiophores.

Alternating with the developing pterygiophores, sets of fin muscles form along either side of the base of the finfold. Cells giving rise to fin muscles have large nuclei and are derived from proliferations of the distal extremities of adjacent myotomes. In some fishes, the proliferations are distinct drop-shaped muscle buds from each myotome involved (Harrison, 1895). In others, particularly in advanced teleosts, the muscle-producing tissue is derived partially or wholly from amorphous proliferations of loose cells from the myotome edges, which aggregate along the fin base and bear no visible trace of the somite segmentation.

Mechanical support for the finfold comes initially from small, very numerous horny rods (actinotrichia) imbedded in the outer margin, which are scarcely countable and are not considered here as meristic series. They are superceded by the rays or spines, which appear in the finfold after the internal supports and fin muscles are in place. Strips of primary mesenchyme cells form dense bands in opposing pairs on the inner surfaces of the epidermis of the finfold. Along these pairs of bands, thickenings develop in the basement membrane of the dermis, which subsequently fuse distally to form the dermal fin rays. Each ray is proximally continous with erector muscles.

The direction of successive appearance and ossification of rays or spines may be from the front backward, or both forward and backward from part way along a fin. The last median fin elements may be quite late in appearing, and there has been a tendency to preserve experimental fish before the definitive ray count has become visible (Lindsey et al., 1984; Perlmutter and Antopol, 1963).

The three sets of median fin components (muscles, and internal and external skeletal supports) are in serial correspondence from their earliest appearance. Our knowledge of the inductive mechanisms among these series is meager. François (1957) concluded from experimental fin ablations in S. gairdneri that it is the muscle buds that determine the metameric disposition of the skeletal elements, and not vice versa. Support for the view that metamerism of the rays is determined by the other series comes from the late appearance of rays, after the muscles and pterygiophores. Furthermore, there are often pterygiophores without rays, but never rays without pterygiophores. On the other hand, the "primary mesenchyme" that produces dermal fin rays appears farthest out in the finfold as a homogenous band before the muscle buds develop, even though its segmental pattern emerges last. Also, experimental removal of the primary mesenchyme strip prevents the formation of muscles and pterygiophores (François, 1958).

These observations could be encompassed by the hypothesis that the total length of the meristic series to be produced in the median fin is governed by the antero–posterior extent of the strip of primary mesenchyme, while the spacing of segments within that strip is con-

trolled by the extent of contributions from adjacent muscle proliferations and sclerotomal tissue. In any event, the embryonic sequences
are complex. There seem to be more possible pathways for environmental modification of the ultimate meristic count in the median fin
series than in the body somites. Concomitantly, counts of anal fin rays
seem particularly susceptible to extraneous environmental influences
and show less consistent environmental responses than do other
meristic series.

2. CAUDAL FIN

The internal supports of the caudal fin are more complex than of
the dorsal and anal fin, and vary widely between fish groups. The
ventral surface of the upturned notochordal tip bears from two to nine
hypural elements, variously fused into a plate that supports the external caudal fin rays. These elements form a continuation of the median
series of haemal arches, but there is not serial correspondence between centra and hypurals. Nor is there one-to-one correspondence
between hypurals and fin rays; in the caudal region, rays exceed hypurals, which exceed centra. Smaller epural elements may lie dorsal
to the notochord between the last neural spine and the ossified uro-
neural. Although serial relationships between the caudal elements
tend to be conservative and to some extent define families or orders,
there can be some variation, within species, in number of parts in each
series.

Counts of the caudal fin rays can be divided into the major rays,
and a dorsal and a ventral row of minor unbranched rays running
anteriorly along the peduncle ahead of the caudal base. The count of
major caudal rays (the upper and lower members of which may be
stout and unbranched) can further be subdivided into those arising
from the dorsal and from the ventral set of hypural supports.

Ontogeny of the caudal fin elements, like that of the dorsal and
anal fins, involves material from more than one source. The primordium of the hypural supports is at first continuous with mesenchyme
investing the notochord, but the epurals may arise from primordia
distinct from the axis. The caudal fin muscles evidently arise from
myotomal proliferations. Rays of the caudal fin arise, like those of the
other median fins, as paired dermal thickenings in the finfold.

The sequence of appearance of the caudal fin supports varies and
is probably related to early function. In sucker larvae (Catostomidae),
which are free-swimming early, the first caudal rays to calcify are in
the ventral lobe, and are already supported by the first three hypurals.

In the live-bearing guppy, calcification occurs before the larva is released; here the first caudal rays appear before the hypurals, in the center of the fin (Weisel, 1967).

The geometry of development of the series of major caudal fin rays differs from the dorsal and anal rays in that it develops perpendicular to the axis of other meristic series in the body. The available space within which parts can form is in this case limited by the dorsal and ventral contours of the finfold. Another difference from the preceding series is that here the major elements merge at each end of the series into minor ones (the small unjointed unbranched rays). Distinction between the two becomes visible rather late in development. Environment might affect ultimate count of the major elements by altering the fate of members lying at the junction of the two series.

D. Paired Fins

Paired fins resemble the median fins in having internal skeletal supports, external supporting fin rays, and sets of muscles. Unlike the dorsal and anal fins, but like the caudal, the meristic series of the pectorals and pelvics do not lie parallel to the axial segmentation. Also, there is no serial correspondence between the rays and the internal supports. The number of elements in the internal supports is small and exhibits scarcely any variation within species. The number of fin rays is higher, and shows environmentally induced and genotypic variability, particularly in the pectoral fin. Counts in left and right fins are not always identical, and the question of bilateral asymmetry in fishes has received increased attention.

1. PECTORAL FINS

The "wrist" of the pectoral fin in most teleosts is supported by a row of hourglass-shaped radial bones or actinosts. There are four radials in all but a few groups of teleosts. They articulate proximally with the pectoral girdle, and are clasped distally by the split bases of the fin rays or lepidotrichia.

Ontogeny of the pectoral fins has been studied most thoroughly in *S. trutta*, by Bouvet (1968, 1974, 1978) and by Lubitskaya and Svetlov (1935). At the pectoral fin sites on each side, the ectoderm develops a fold, through cell division both in the ectoderm and in the underlying mesoderm. Each tissue contributes through alternating phases of greater and lesser proliferation. The apical fold elongates into a paddle, like the toe of a flattened sock, into which mesenchyme insinuates. An oval plate of precartilaginous cells comes to occupy the paddle, within which the skeletal elements appear successively. The four basal radials and the scapula appear in a row across the paddle base distal to the girdle elements. Next, immediately distal to these, there forms a second row of cartilaginous nodules. Unlike the basal radials, these distal radials never ossify, and may become indistinguishable in the adult. Their significance from a meristic viewpoint is that they are serially independent of the adjacent basal radials, and their number (about 14) corresponds exactly to the ultimate fin ray number. Until this stage, the fin paddle has been supported only by large numbers of actinotrichia, which originated from ectodermal cells of the fin bud apical ridge At hatching, the lepidotrichia begin to differentiate. As in the median fins, each ray is laid down as a pair of thickenings on adjacent faces of the dermis; these are bundles of collagen on which hydroxyapatite crystals are deposited. Each ray base is in line with one distal radial. The first appears opposite the largest distal radial behind the scapula at the top or front of the fin. The mesodermal osteoblasts that produce lepidotrichia can be seen microscopically to be in immediate proximity to the proximate radials. Bouvet (1968) developed fate maps of the presumptive areas of the pectoral bud, and showed that if some distal radials are excised the corresponding lepidotrichia fail to develop.

Migration of cells during pectoral ontogeny has been studied in the killifish *Aphyosemion scheeli* by Wood (1982) and Wood and Thorogood (1984). In isolated paddles, somatopleural mesenchyme cells can be seen by time-lapse video recording to be moving parallel to the actinotrichia. The source and timing of those cells that give rise to the endoskeletal elements and of those that give rise to the lepidotrichia are unclear. They may be comparable to the primary and secondary mesenchyme described in the median fins.

For reasons outlined in Section VII, the pectoral fins offer particularly promising material for meristic studies. Although pectoral buds are the first to form in most fishes, the full complement of fin rays is usually visible later in the pectoral than in the dorsal and anal fins (Beacham and Murray, 1986). There is danger in experimental studies on meristic variation that the premature preservation of young fish may produce incomplete pectoral ray counts (Valentine and Soulé 1973; Lindsey *et al.*, 1984).

2. PELVIC FINS

Skeletal supports are simpler in the pelvic than in the pectoral fins. Lower teleosts may have from one to three skeletal nodules or radials at the junction of the ray bases with the pelvic girdle and usually six or more soft rays. Higher teleosts have no radials, and often one spine and five or fewer soft rays that articulate directly on the girdle.

Ontogeny of the pelvic fins in *Salmo* has been studied by Géraudie and François (1973) and Géraudie and Landis (1982). Histogenesis and chondrogenesis follow similar patterns in pectorals and pelvics (Bouvet, 1968). A primary mesenchyme (local proliferations of somatopleura) enters the developing fin buds and will eventually produce the lepidotrichia. Secondary mesenchyme, dispersed from ventral processes of four adjacent somites, penetrates the primary mesenchyme blastema, and will give rise to the endoskeletal supports and to fin muscles. Unlike the sequence in *Salmo* pectoral fins, in the pelvics the lepidotrichia begin to appear before the cartilaginous radials. The first pelvic ray appears in the fin center in *Salmo,* but in some fishes ossification is in an inward direction and in others it is nearly simultaneous in all pelvic rays (Dunn, 1983). It is complete in the pelvics at about the same time as in the pectorals, or often somewhat earlier.

E. Other Countable Structures

1. PYLORIC CECA

The cecal outgrowths from the intestine close to the pyloric end of the stomach differ from meristic characters considered so far in that they are usually not arranged in linear series, but in a clump. The number of pyloric ceca can range from 1 to over 1000 (Suyehiro, 1942). In some species, individual ceca branch, particularly in older fish. Usually the tips are counted. Ceca develop relatively late, and may attain their ultimate count only in juveniles (Northcote and Paterson, 1960). Cecal number has been shown to have a large genetic component in trout (Bergot *et al.*, 1976).

2. RIBS AND INTERMUSCULAR BONES

Ribs (pleural, epipleural, and epineural) form countable series that extend for variable distances along the body, and within each there is usually one-to-one correspondence with the vertebrae. There are also series of paired intermuscular bones, sometimes very numerous and complex in shape (Lieder, 1961; Lindsey, 1978, p. 50). The ribs sometimes develop as cartilaginous bones, which subsequently ossify, and sometimes develop directly from bone cells. Intermuscular bones can develop in the myocommata separate from the vertebrae, presumably from wandering sclerotomal cells. First visible appearance of the rib series is commonly at the anterior end. In carp, the ribs and intermuscular bones together form unbroken series, in serial correspondence with the vertebrae, but the numbers of dorsal and of ventral intermusculars behave as two independent characters (Moav *et al.*, 1975). The number of intermuscular bones in carp may be subject to some environmental influence.

3. BRANCHIOSTEGAL RAYS

In teleosts, the branchiostegal rays usually number from three to 20; there are up to 50 in a few groups. Branchiostegal rays are strap-shaped dermal bones formed in a fold of skin, the branchiostegal membrane, which extends from the hyoid as a posteriorly directed flap. The first rays appear at the posterior. Branchiostegal rays often form their adult number quite early. They can be counted separately on the right and left side, and further subdivided as arising from either the ceratohyal or epihyal bones. Asymmetry between the sides is common in several forms (Hubbs and Hubbs, 1945). By comparing two year classes of *E. masquinongy*, Crossman (1960) concluded that total count and also asymmetry may be subject to some environmental modification.

V. POSSIBLE EMBRYONIC MECHANISMS FOR MERISTIC MODIFICATION BY THE ENVIRONMENT

A. Growth versus Differentiation

The most promising hypothesis to encompass the welter of observation and experiment on phenotypic meristic variation in fish is that the environment modifies the number of segments formed in embryos by differentially affecting the processes of growth (elongation) and of differentiation (segment formation). If, for example, low temperature inhibits differentiation more than it inhibits growth, the embryo axis will be longer at the time of differentiation, and more (approximately equal-sized) segments will be laid down.

This hypothesis has emerged gradually. Kyle (1923, 1926) observed that the number of vertebrae is higher in species whose early postlarvae are longer. Kyle also suggested that high temperature produces low vertebral counts because more yolk is absorbed and the larvae are therefore shorter when vertebrae appear. He then digressed to state that it is the movements of the young fish, coupled with their balance (the proportion between head and body), that determines the number of vertebrae.

Hubbs (1926) made the distinction between modification of growth and of differentiation, but he stressed the role of developmental *rate* in controlling segment number. Environmental factors such as temperature or salinity that accelerate rates generally result, he said, in fewer segments. Many subsequent authors have generalized that accelerated development produces fewer meristic parts, although there are in fact many cases where higher temperature produces *more* parts. In any event, acceleration per se cannot be the whole story; acceleration of all embryonic processes to the same extent would not produce a different number, but simply the same number faster.

In 1927 Hubbs recorded that parasitized fish showed retarded development and ultimately produced abnormally high scale count; he suggested that meristic parts are laid down at approximately constant absolute sizes, so their number is roughly proportional to the available space. Clear enunciations of the hypothesis of meristic control by differential environmental alteration of growth and of differentiation were provided by Gabriel (1944, p. 122) and by Hubbs and Hubbs (1945, p. 268). Turner (1942), studying the effects of hormone concentrations on the number of gonopodial segments in *Gambusia affinis*, suggested that growth and differentiation are antagonistic, and that differentiation terminates growth. The essential feature, dissimilar response to environmental influence, has been called "dissociability" or "out-of-gearishness" (Needham, 1933), or "uncoupling of embryonic processes" (Hayes, 1949).

The hypothesis that growth and differentiation may be dissociated is consistent with many recent observations on experimentally reared fish. In several species, higher rearing temperature (either sustained or varied) has produced shorter body length at comparable developmental stage (Murray and Beacham 1986; Pavlov, 1984, 1985; Peterson *et al.*, 1977). Low oxygen concentration can produce shorter fry (Silver *et al.*, 1963; Braum, 1973). Salinity can alter fry length at hatching (Kinne and Kinne, 1962; Fonds *et al.*, 1974). Different phases of salmon embryogenesis, some dominated by growth and others by organogenesis, have very different temperature coefficients (Lubitskaya, 1935; Trifinova *et al.*, 1939). Dissociability has been demonstrated clearly in *S. salar*, in which temperature so alters relative rates of development of various structures as to change the order of their first visible appearance (Hayes, 1949; Hayes *et al.*, 1953). (Hence the concept of development proceeding through an immutable sequence of "embryonic stages" is misleading.) These observations, although many were made without reference to meristic variation, suggest a physical reality to the hypothesis that number of parts might be environmentally altered via the interplay of growth and differentiation.

The general hypothesis that meristic variation depends on dissociability is promisingly versatile. All patterns of meristic response (positive, negative, V-shaped, or arched) could be simulated by selecting appropriate temperature coefficients of growth and of differentiation (Barlow, 1961). Any other environmental factor that differentially affected the rates of these two processes could be similarly effective. Moreover, a variety of extreme meristic responses to temperature *change* might be simulated by moving between low and high temperature-response curves with appropriate shape so as to maximize the dissociation between growth and differentiation. The possibility of expressing these qualitative statements quantitatively will now be examined.

B. The Atroposic Model

1. Description of Model

A first attempt to quantify the foregoing hypothesis has been called the "atroposic" model. This model turns out to be surprisingly effective in assimilating the disparate experimental data, considering its oversimplified assumptions. Features of the model are summarized here; more detailed discussions, particularly of the mathematical aspects, are given by Lindsey and Arnason (1981), Lindsey *et al.* (1984), and Swain and Lindsey (1986a).

The model (Fig. 4) supposes that the quantitative outcome of embryogenesis depends on two independent processes each building up with time. The final number of parts is fixed by the level that one process has attained when it is suddenly terminated; termination is triggered at the moment when the other process has risen to some critical level. Varying the conditions affects differentially the time courses of the two processes, and hence alters the outcome.

A classical terminology, based on the metaphor of the three Fates of Greek mythology, is adopted to stress the general nature of the model and to avoid restricting the possible nature of its components. One sister, Clotho, was said to spin the thread of life; a second, Lache-

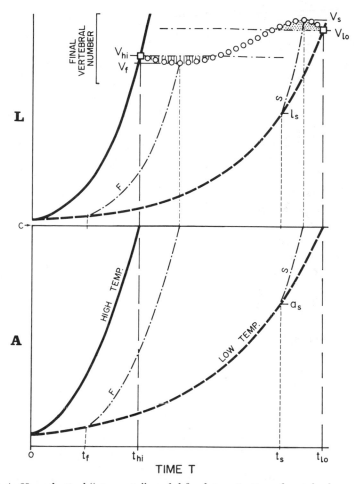

Fig. 4. Hypothetical "atroposic" model for determination of vertebral number. Development of L process and of A process follows heavy solid curves at constant high temperature, heavy broken curves at constant low temperature; F and S are portions of developmental curves followed after transfer from low to high temperature at times t_f or t_s, respectively. Resulting vertebral numbers shown by squares (constant temperature) or circles (transfers from low to high temperature). [Modified from Lindsey and Arnason (1981).]

sis, drew off the thread and controlled its course; the third, Atropos, then cut off the thread, rendering her sisters' work irreversible. The class of model is named "atroposic" because of its feature of termination at a critical moment. In Fig. 4, that process that controls the moment of termination is labeled A for Atropos; the other, whose level measures off the outcome, is labeled L for Lachesis. The L can be thought of as Length of elongating tissue (i.e., "growth"), and the A as Allócation (i.e., "differentiation"). Clotho's role is so far unassigned, unless it be the conditions (temperature, light, etc.) that control the rates of L and A processes.

On the graphical version of the atroposic model in Fig. 4, the outcomes of rearing at sustained low or sustained high temperatures, or of breaks from low to high temperature, can be followed without recourse to mathematics. Lindsey and Arnason (1981) describe the procedures as follows.

The heavy lines on Fig. 4 show the time curves of the L process and A process in an embryo reared either at sustained high or low temperature. At high temperature, A rises steeply and reaches the threshold level c at the time t_{hi}; the resulting number of vertebrae is determined by the level V_{hi} to which L has risen at time t_{hi}. At low temperature, A reaches the threshold later, at time t_{lo}, by which time L has risen to the level V_{lo}. In Fig. 4, the curves are so arranged that fewer vertebrae are formed at high than at low temperature, which conforms to the commonest empirically observed results. Slightly different curves can be chosen so as to produce fewer vertebrae at low temperature, or at an intermediate temperature. Indeed, so long as the model is applied only to effects of sustained temperatures, it is so flexible as to have little predictive value.

The value of the atroposic model emerges when it is applied to effects of temperature changes during development. The outcome of shifting from low to high temperature at either of two developmental times can be traced in Fig. 4. In the first example, both L and A processes proceed slowly following their respective low-temperature curves, until the embryo is shifted to high temperature at t_f. Each process then proceeds faster, parallel to its high-temperature curve, along the segment labeled F. The final vertebral number is determined by the height V_f, reached by the L curve at the moment when the A curve reaches the threshold c. In the second example, a later shift from low to high temperature, at time t_s, causes L and A to follow the segments labeled S after transfer, and produce a final number V_s. For the particular curves shown here, a series of shifts from low to high temperature at intermediate times will produce counts shown as open circles in Fig. 4. Early transfers produce counts below those resulting from either sustained temperature, but late transfer produces counts above.

Temperature shifts in the opposite direction, not shown in Fig. 4, can be simulated in the same way, with L and A starting along the high temperature curves until transfer, and then paralleling the low temperature curves. The resulting counts show a reversed pattern, as in the right-hand side of Fig. 2. Temperature pulses can also be simulated by switching back and forth between the appropriate curve sections. Extralimitary responses in both directions can be so generated (Lindsey and Arnason, 1981, p. 342).

The atroposic model has been tested by computerized fit to published experimental data. The curves for the A and L process are taken to follow the exponential form:

$$A = \alpha_1[\exp(\alpha_2 t) - 1]$$

$$L = \lambda_1[\exp(\lambda_2 t) - 1]$$

where t is time in days since fertilization, α_1 and α_2 are the two parameters governing the shape of the A curve, and λ_1 and λ_2 are the two parameters governing the L curve.

Inputs to calculating these equations for each species, which must be obtained experimentally, are the meristic means resulting from temperature breaks or pulses, and the means as well as completion times at sustained temperatures. Completion time (when the A process reaches the threshold c in Fig. 4, set arbitrarily at 10) is estimated from series of breaks or pulses as that time after which the meristic count could no longer be influenced. Outputs of the calculations are, for each temperature, the parameters α_1 and λ_1 (plus α_2 and λ_2, which are constrained once the other variables are fixed). The best shapes of the A and L curves to fit any set of experimental transfer data are arrived at by iterative trials, which reveal the best combination of the parameters α_1, α_2, λ_1, and λ_2. Once the parameters have been derived from one type of experiment, they can be used to predict the meristic response of that genotype to any other type of experiment involving the same two temperatures.

2. STRENGTHS OF THE MODEL

The model has been found to be capable of moderately good fit to almost all available experimental data on vertebrae and to some on fin rays (Tables II and V and Fig. 2). For *S. trutta,* a good model fit was obtained from published results of temperature-pulse experiments; using these parameters, the results of six temperature-break experiments, published by the same and by another author, were predicted and found to agree well with the observed results (Lindsey and Arnason, 1981). As the type of experiment used to produce the results to be predicted differed in kind from the type used for fitting, the prediction was a model extrapolation that depended critically on the model structure. Some further corroboration of the model was provided by agreement in overall pattern (although not in absolute values or details) between experiments on *S. gairdneri* by Hallam (1974) and by Lindsey *et al.* (1984). There was also internal consistency between the model fit to temperature breaks and the results of experiments on prefertilization influences in *R. marmoratus* (Swain and Lindsey, 1986a).

Fitting the model to complex experimental results requires few parameters. For each temperature, meristic count and completion count are based on observation, so only α_1 and λ_1 need be calculated. For breaks and/or pulses between two temperatures, only two pairs of parameters need be calculated that generate all the curves to fit all combinations of experiments.

Where experimental data involve more than two temperatures, the parameters of the A and L curves seem to vary with temperature in a regular fashion. Each of the four parameters at eight temperatures for *S. trutta* falls along an approximately straight line (slightly inflected above 10°C) when plotted on log–log axes against temperature (Lindsey and Arnason, 1981, p. 339).

Even when the vertebral response to different sustained temperatures is V-shaped, the calculated parameters of A and L curves form a smooth family of curves (Lindsey and Arnason, 1981, pp. 340, 342). The atroposic hypothesis for V-shaped responses is therefore simpler than previously published suggestions, which postulated two unspecified mechanisms, each dominant at one temperature extreme (Lindsey, 1954), or a "most economical metabolism" at intermediate temperature (Marckmann, 1958), or maximum specific gravity corresponding to lowest vertebral count (Johnsen, 1936), or inflections in temperature coefficients (Barlow, 1961), or differential mortality, or miscounting complex vertebrae at temperature extremes (Garside, 1966, 1970).

Moreover, while different species, or even different genotypes within one species (*Oryzias latipes*: Ali and Lindsey, 1974), may display either negative or V-shaped vertebral responses, all can be fitted by the same basic model, sometimes with only modest parameter changes. In all these cases, the atroposic curves have an underlying regularity not apparent in the experimental data.

The puzzling extralimitary responses of meristic counts to some temperature breaks or pulses were at first thought of as reactions to temperature change per se (sometimes to "metabolic upset"). They were described as "shock effects," and the stages at which they were observed were called "supersensitive periods" (Tåning, 1944), or "phenocritical periods" or "periods of paradoxical reaction" (Orska, 1956). In the atroposic model, no shock from temperature change is invoked; all meristic responses, including extralimitary ones, are attributed to cumulative effects of the temperature regime on the progress of the L and A curves. There seems no profit in designating supersensitive periods when the timing and direction of these responses are now known to vary according to the environmental change administered.

Another feature of the atroposic model is that it encompasses the large extralimitary meristic response occasionally evoked by very early environmental perturbations (Komada, 1977a). Even fin-ray counts in *R. marmoratus* can be affected by temperature breaks well before the site of the fins are visible (Swain and Lindsey, 1986a). These very early responses have probably been rarely reported because few researchers have thought to look for them. The concept of a sensitive period *before* and after which meristic counts are not malleable should be abandoned. Perhaps all meristic series are potentially labile from (or possibly before) fertilization and remain so until the precursors of the last-formed end of the series have been irrevocably committed.

3. WEAKNESSES OF THE MODEL

A problem in testing the atroposic (or any other) meristic model is the lack of good experimental data. Without more closely spaced transfer times and smaller variance in counts, the calculated parameters in the model are only loosely constrained. Moreover, it is impossible without better data to choose between some alternative versions of the model, such as equations other than exponential for the A and L processes.

The model in its present form predicts that larger eggs, which produce longer embryos, should yield higher vertebral count. Since they do not (as described in an earlier section), there may be at an early stage some type of "regulation" (adjustment of internal segment size in accordance with space available). Possibly, in teleosts, an early regulation in somite size is dictated by egg size, while the ultimate number is controlled according to the atroposic model. Amphibian embryos seem to follow this pattern; somite formation in *Xenopus* is regulative in early stages but not in later stages (Cooke, 1981).

An alternative explanation of why egg size does not control somite number, suggested by D. P. Swain, is that the L process represents not the length of undifferentiated tissue but the number of cells. Perhaps all early somites have the same number of cells, but the cells are smaller if they arise from smaller eggs. The atroposic model works equally well whether L stands for absolute length or cell number. The size and number of cells in early fish somites have not yet been adequately documented.

The atroposic model assumes that the number of segments is directly controlled by the absolute value of L, but makes no reference to the mechanism governing the size of the individual segments. The model may of course be valid without this information, provided the environment does not affect initial segment size. The topic of "pattern formation" is currently under intense scrutiny (Cooke, 1981; Pate and Othmer, 1984; Russell, 1985) in invertebrates, amphibians, and birds, but not in fish.

On two scores, the mathematical neatness of the model departs from reality. Differentiation of the whole row of segments is assumed to occur instantaneously (when the A process reaches its critical level), and all segments are assumed to be of equal size (i.e., segment number is directly proportional to L). In fact, the somites and other meristic series appear sequentially over an appreciable time, and the last-formed members are usually the smallest. However, the L axis can be interpreted not as physical distance along the embryo but as a scale giving the segment number, which will be formed if completion occurs at that time. The scale can be nonlinear. When, in the future, schedules become available for times and lengths of segments appearing at different temperatures, the model can be refined; at present there is probably some error in assuming, in analyzing temperature transfers, that the same increment in L, regardless of when it occurs, will have equal effect on ultimate segment number.

Although the model can be fitted quite well to almost all available experimental data on vertebrae, it is effective for only some of the data on fin rays. A few of the fin-ray series show asymmetry in their responses to temperature breaks applied in opposite directions, and they resist model fitting. Several explanations are possible. Some published data on fin-ray counts, particularly on salmonid species, are suspect because the young may have been preserved before all rays were countable (Lindsey et al., 1984). Ray counts, particularly of the anal fin, seem sometimes to be subject to extraneous environmental influences, which may have obscured the effects of the experimental variable being tested. Finally, it may be deficiencies in the model rather than in the data. Embryogenesis seems to follow more complex pathways in the fin series than in the body segments, as described earlier. The comparatively simple atroposic model may provide only an incomplete simulation of how the environment affects developing fin rays.

In sum, the atroposic model in its current form has imperfections, but it does provide a fairly simple framework to many apparently disparate experimental findings. Perhaps its principle advantage is that it can make quantified predictions that are capable of disproof.

C. Alternative Interpretations of Experimental Results

1. RELATIVE DURATION OF DEVELOPMENTAL PERIOD

The common generalization that slow absolute rate of early development leads to high meristic counts is unwarranted. Developmental rate alone cannot be a simple correlate with the diversity of meristic responses described here. The supposition that the high meristic counts often found in hybrids arise from reduced developmental rate (Leary et al., 1983) has had to be forsaken; interspecific and interstrain salmonid hybrids often do not have slower development than both parents (Leary et al., 1985a; Ferguson and Danzmann, 1987).

A negative correlation has also been suggested to occur between vertebral number and the reciprocal of the "period of vertebral formation" expressed as a proportion of the whole developmental period until hatching (Garside, 1966; Hallam, 1974). The data, confined to salmonid species, do show such a correlation, generally although not invariably. Within experiments, temperature or oxygen regimes (either constant or changed) that produce relatively long periods of vertebral formation also produce more vertebrae. In these calculations, the period of formation has been arbitrarily defined as extending between two particular developmental stages (from one-quarter epiboly to posterior flexure of the notochord). This relationship is not necessarily inconsistent with the atroposic model, since with certain combinations of exponential A and L curves a relatively long period between two intermediate levels on the A axis might be associated with a higher segment number. No reason has been advanced for the relationship, which considers only times to reach developmental stages, and not embryo dimensions. While objections can be raised to the methods of calculation used in these studies, their results warrant further investigation, and should be harmonized with any general model of meristic variation. However, the concept of a "period of formation" *before* which the environment is considered to have no effect should be abandoned for reasons stated earlier.

2. SELECTIVE MORTALITY

Arguments why meristic differences in fish reared in different experimental conditions cannot be due solely to selective mortality between genotypes have been summarized by Lindsey *et al.* (1984). Evidence includes (a) production of meristic differences by different rearing temperatures within a genetically uniform clone (Harrington and Crossman, 1976b); (b) experimentally induced meristic variation in gonochoristic species even when mortalities were extremely low; (c) absence of correlation between meristic means and mortality rates; (d) absence of correlation between meristic means and their variance (Heuts, 1949); and (e) the impossibility of explaining by moderate mortality all the various patterns of environmental responses in different meristic series.

Nevertheless, although selective mortality cannot be the cause of all experimentally induced meristic variation, there remains the possibility that whatever mortality did occur in early development may not have been selectively neutral with respect to meristic count. Selection might operate on early physiological processes that are pleiotropically associated with meristic characters. Selective mortality with respect to meristic counts can also occur after the young are free-swimming. Different vertebral numbers (or ratios of abdominal to caudal vertebrae) have been shown in young *Gasterosteus aculeatus* to be associated with locomotory ability, and with ability to avoid predation (Swain, 1986). The optimal counts shift as body size increases. Whether this posthatching selective mortality is operating on environmentally or genetically determined meristic variation, it needs to be considered as a complicating factor when interpreting meristic variation in older fish.

3. PREFERTILIZATION INFLUENCES

Discovery of a prefertilization temperature effect raises the possibility that in some experiments the pattern of meristic variation attributed to postfertilization environmental influences may actually have been an artifact of the temperature history of the parents. It is notable that the response curve of vertebral counts to different laboratory rearing temperatures has often been V-shaped, and yet in wild populations living in gradients of latitude (and hence of temperature) V-shaped clines in vertebrae are almost never observed. A speculative explanation of how this paradox may arise from prefertilization influences is suggested in Fig. 5.

The term "unacclimated parents" can be applied in Fig. 5 when the prefertilization temperature to which the parents were exposed differed from the rearing temperature of the offspring. "Acclimated parents" applies when temperatures were the same before and after fertilization. ["Acclimatized parents" would be more appropriate if one wished to imply, for a wild parent, the whole history of exposure to the total environmental complex throughout its life up to the time of the test (Fry, 1971, p. 14.] Almost all laboratory experiments have used unacclimated parents (except where one of the incubation temperatures has happened to coincide with the parental temperature). In contrast, wild embryos typically arise from acclimated parents whose recent prefertilization temperature was close to the early postfertilization developmental temperature of the embryos.

Two assumptions are made in Fig. 5: (a) that there is a prefertilization parental temperature influence on offspring vertebral count (such as has been described in Section III,D,1) with a direction and degree proportional to the difference between temperatures of parent accli-

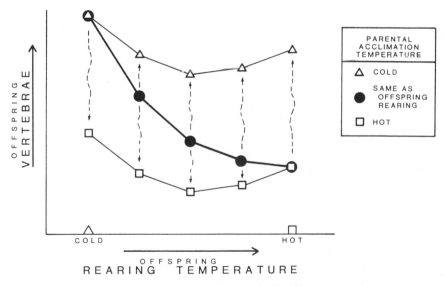

Fig. 5. Hypothetical relationship whereby vertebral response to incubation temperatures could be declivous (heavy curve) when parents were acclimated to each incubation temperature, but V-shaped when parents were acclimated to a single temperature. Broken lines with arrowheads indicate prefertilization effect, which is proportional to difference between temperatures of parental acclimation and offspring incubation.

mation and offspring rearing, and (b) that the phenotypic response of vertebral count in offspring of *acclimated* parents is negative with decreasing steepness at higher temperatures (corresponding in shape to response curves in Fig. 1A). Given these assumptions, a prefertilization effect of appropriate magnitude can be chosen that will generate V-shaped vertebral response curves rather than a negative curve, when parents are unacclimated rather than acclimated. Parental acclimation temperature is seen in Fig. 5 not to affect the shapes of the response curves of the offspring, but it does affect their position along the vertical axis. Wild populations would therefore display the commonly observed negative correlation between temperature and vertebral count, while eggs taken from parents at any one temperature and incubated at several temperatures would yield the commonly observed V-shaped vertebral response to temperatures in the laboratory.

Prefertilization effects different from those of Fig. 5, or response curves (using acclimated parents) different from those in Fig. 5, generate vertebral response curves other than V-shaped, but still with vertical axes displaced depending on the particular acclimation temperature of the parents. Regardless of the response shape, use of genetically similar parents with different acclimation histories (e.g., from different dates within the spawning season) could therefore give the illusion of genetic differences. If the response curve of acclimated parents is flat (i.e., the same meristic count is produced at all temperatures), then the response curves using unacclimated parents will slope either up or down if there is a prefertilization effect working in one direction or the other.

The scheme in Fig. 5 is highly conjectural. Even if it does mirror reality, it need not drastically modify most previous conclusions, except in species with meristic series that respond strongly to prefertilization influences. Evidence is so far based mostly on *R. marmoratus*, which is highly atypical in its extreme homozygosity (Vrijenhoek, 1985). Further speculations along these lines are unwarranted until more data are available on the extent and nature of prefertilization influences.

Eventually it may be possible to marry the atroposic model, or its descendent, to a model of prefertilization phenonena. The latter may reflect a carry-over from before fertilization of influences that preferentially affect either growth or development in the very early embryo, giving one of the processes a head start that is manifest later in the ultimate number of meristic parts. In conformity, the three meristic series in *R. marmoratus* that show clear responses to prefertilization temperature (vertebrae, pectoral rays, and caudal rays) have steeply rising A curves for which a slight difference in early timing would greatly affect completion time; the two series (dorsal and anal rays) that show little or no response have A curves with very flat early portions along which slight time shifts would have negligible meristic effects (Swain and Lindsey, 1986a, Fig. 6).

4. OTHER INTERPRETATIONS

Garside (1966) suggested that the apparent V-shaped response of mean vertebral counts to rearing temperature might be in part an artifact of the methods of enumerating abnormal vertebrae. Abnormal vertebrae are often commonest at temperature extremes. However, abnormalities have been too few to account for the distinct V-shaped responses to temperature found in many species (Table II). Their method of treatment (counting irregular vertebrae as 1, 1.5, or 2.0, or omitting them) alters absolute values but seldom alters the overall pattern of environmental response.

Irregularities, such as double hemal or neural arches on single centra, bilateral irregularities, and spiral sutures, do pose difficulties in counting (Ford, 1933). Systems of classifying and quantifying irregularities have received much attention because of their bearing on use of meristic characters for racial distinctions (Gabriel, 1944; Tåning, 1944; Blackburn, 1950). In experimental as well as taxonomic studies, data on meristic irregularities should be recorded, even though in subsequent computations the partial counts may be lowered or raised (preferably the latter), as they can provide additional insight into the embryonic processes underlying meristic variation.

Kyle (1926) suggested that movements of the larval fish, both before and after hatching, influenced the formation and ultimate number of body segments. Ford (1933, p. 221) and Barsukov (1955) echoed this possibility. Although it now appears that vertebral number is fixed before hatching, the possible influence of movement within the chorion cannot be wholly dismissed. Van Raamsdonk et al. (1979) demonstrated that when body movements in developing B. rerio were prevented (by anesthesia, agar, or a glass rod in the neural tube), the shapes of the somites were modified. Urethane, an anesthetic that may inhibit muscular movements, lowers somite number in B. rerio (Battle and Hisaoka, 1952). Urethane raises vertebral and pectoral ray number, and lowers dorsal and anal ray number, in O. latipes (Ali, 1962). On the other hand, the mechanical disturbances described in Section III,C had little effect on meristic counts in either species. Conceivably, the influence of temperature, light, and oxygen on segment number may operate in part via contractile responses in the embryo.

VI. HEREDITARY MERISTIC VARIATION

Meristic characters seem to be threshold traits (Falconer, 1981) whose number depends on an underlying continuity with thresholds that impose discontinuities on their visible expression. In the realm of statistics, discontinuous variables are also known as "meristic" or discrete variables, and "have only certain fixed numerical values, with no intermediate values possible" (Sokal and Rohlf, 1981). The real world of morphology is less tidy. Although most meristic series in most individual fish yield unambiguous whole counts, examination of large samples reveals various conditions of intermediacy, ranging from tiny additional members to incompletely separated (or slightly fused?) pairs of full-sized members. Hubbs and Hubbs (1945, p. 267), in an extensive study of bilateral asymmetry in paired fins, concluded:

Convention and convenience demand that rays be enumerated as whole numbers, but it is important to keep in mind that the classes of ray counts could be indefinitely divided, according to the degree of development of the last ray. Not only for the pectoral fins, but also for many other meristic series, we are finding that there are many gradations between the successive counts; that is, that there is no real distinction between continuous and discontinuous (or discrete) variations.

Genetic analysis of meristic characters shows them to be multifactorial. Kirpichnikov (1981) reviews the estimates of the number of genes controlling meristic counts in fish; in carp, vertebral number is controlled by "dozens of genes," fin-ray number by "a rather high number of genes." In some other fish, fin-ray number is controlled by "not less than 10 genes." The number of genes controlling gill-raker count in whitefish (Svärdson, 1979), lateral plates in G. aculeatus (Hagen and Gilbertson, 1973), and dorsal spines in Apeltes quadracus (Hagen and Blouw, 1983) may not be very high, but in these and other meristic characters inheritance is polygenic.

Grüneberg (1952) suggests that despite the discontinuous phenotypic values of meristic characters, their mode of inheritance is likely that of a continuously varying character, and, further, that "the additiveness of gene effects and their sensitivity to the environment may trace back to the common cause that the so-called multiple genes of quantitative genetics are in fact remote gene effects." In keeping with this view, according to the atroposic model, genes would control meristic number only indirectly via changes in the parameters of the curves of growth and of differentiation; indeed, their influence would likely be still more remote since the forms of these curves are themselves the result of many interacting processes (Powsner, 1935). Falconer (1981) writes, "Threshold characters do not provide ideal material for the study of quantitative genetics, because the genetic analyses to which they can be subjected are limited in scope and subject to assumptions that one would be unwilling to make except under the force of necessity."

Meristic counts of hybrid fish may be intermediate between the counts of their parents, or they may lie close to or even beyond the count of one parent. Leary et al. (1983) found that of 10 meristic characters in hybrid *Salvelinus* the hybrids had counts higher than either parent in two characters and close to the higher parent in seven characters. On the other hand, Ross and Cavender (1981) found in four intergeneric cyprinid hybrid combinations that anal fin-ray counts in all hybrids were the same as that of the parent with the lower count. The presence of numerous patterns of phenotypic expression that were *not* intermediate led Ross and Cavender to conclude that many characters are probably controlled by a complex interaction of genetic factors.

The relative masking in the hybrid of one parent phenotype by the other may involve regulatory or modifer genes that strongly influence character expression toward that typical of one parent species. The presence of pleiotropic genes could lead to phenotypic dominance by a parent species of several seemingly unrelated characters. Regardless of the mechanisms involved, phenotypic expression of fish hybrids may be controlled by additive inheritance to a much lesser degree than is generally assumed.

This opinion on the prevalence of additive gene effects differs from that of Grüneberg (1952). Reported differences in inheritance of meristic characters in hybrids may arise in part because in some studies the parents were from the same population, while in others they were from different species or even genera.

Despite the obscurity of the genetic mechanisms, high heritabilities have been reported: 0.65–0.90 for vertebrae, and usually in the range of 0.40–0.93 for fin rays, lateral plates, gill rakers, and pyloric ceca (Kirpichnikov 1981; Leary et al., 1985a). These calculated heritabilities of meristic characters in fish are, however, heavily influenced by the rearing environment. When response curves of different genotypes diverge or cross (as in Fig. 1), calculated heritabilities will depend on the developmental temperature. Hagen (1972) computed heritabilities for plate count in G. aculeatus as 0.50 or as 0.83, depending on the temperature at which the offspring were incubated. A large error may be introduced if meristic counts of parents were laid down at a different development temperature from those of their offspring. Lindsey (1962a) found vertebral counts of G. aculeatus parents (particularly of mothers) positively correlated with those of offspring reared at low temperature, but *negatively* correlated with those of offspring reared at high temperature (due to a combination of V-shaped phenotypic response curves, variation between genotypes in the temperatures producing the fewest vertebrae, and presumed low developmental temperatures that had influenced the wild parents).

Differences in meristic counts between the sexes have been reported in many species. Counts differ between males and females in the gonochoristic *R. cylindraceus*, whose phenotypic sex is determined by a genetic switch mechanism, but counts do not differ between primary males and hermaphrodites in the self-fertilizing *R. marmoratus*, whose sex phenotype is determined by a thermal switch mechanism (Harrington and Crossman, 1976a). A maternal effect in inheritance of vertebral counts of carp, particularly in distant crosses, is reported by Kirpichnikov (1981, p. 130), who suggests it is because the anterior part of the vertebral column forms early in embryogenesis when the father's genotype has little effect.

Fluctuating asymmetry (Van Valen, 1962) between meristic counts on left and right sides has been predicted to be greater in homozygous than in heterozygous individuals, because of supposed reduced "developmental stability" (Waddington, 1957). To test this hypothesis, many studies have compared the frequency of asymmetry in counts of paired structures with the heterozygosity of protein loci, or with some measure of environmental stress (Valentine and Soulé, 1973; Jagoe and Haines, 1985). In some instances the hypothesis has been supported, but in many it has been refuted. Particularly where heterozygosity is due to hybridization between unlike genomes, asymmetry may actually be greater in the more heterozygous individuals (references in Beacham and Withler, 1985; Graham and Felley, 1985; King, 1985; Leary *et al.*, 1985a,b).

Another attempted measure of developmental stability, the use of a coefficient of variation (CV) in meristic (or other) characters, is questionable. Leary *et al.* (1985a) found only a weak negative correlation between CV and heritability, and concluded that CV has a large additive genetic component as well as a component due to random developmental accidents.

Another source of possible error in using either CV or asymmetry to measure developmental stability has been overlooked until recently (Swain, 1987). Since meristic counts fall into a few discrete classes, variation in the precise position that the mean count occupies between adjacent whole values will affect the variance, particularly when the range in count is small. In the example in Table I, based on genetically uniform fish, the CV is less in the two samples reared under conditions producing mean meristic counts close to whole numbers than in the sample reared under conditions producing a mean count in between. Frequencies of asymmetry in paired characters are similarly affected. Swain (1987) shows that pectoral fin rays of *R. marmoratus* display a wide variation in asymmetry of samples reared in different environments, which has nothing to do with differences between samples in genetically based developmental stability (since all individuals are genetically the same) but is explained simply by shifts in the mean count between whole values. In order to assess this source of variation, future publications should include frequency distributions of the meristic characters examined.

Meristic differences between wild fish populations might arise from heredity, or from environmental modification of the phenotype, or from both. Whenever embryos from different populations have been reared in the same controlled environment, their meristic differences have shown a large genetic component. Two geographically distinct races of *Clupea harengus*, three homing races of kokanee *Oncorhynchus nerka* in one large lake, inland and coastal races of *O. kisutch*, and early- and late-spawning stocks of *O. keta* have all retained meristic differences when laboratory-reared (Hempel and Blaxter, 1961; Vernon, 1957; C. C. Lindsey, unpublished; Beacham and Murray, 1986). Similarly, latitudinal clines in vertebral counts of sockeye salmon *O. nerka* and of *O. tshawytscha* are mirrored by differences between laboratory-reared samples (Beacham, 1985; Seymour, 1959).

One of the most spectacular latitudinal meristic clines, in anal fin-ray counts of the redside shiner *Richardsonius balteatus*, is also evidently largely genetic. While rearing temperature has a strong positive phenotypic influence on counts in embryos from one locality (producing mean differences as great as three rays), there is an even stronger tendency, probably genetic, for embryos from central British Columbia to produce, at comparable rearing temperature, mean counts *eight* rays higher than embryos from 1300 km farther south in Nevada (Lindsey, 1953, and unpublished). Here genotypic variation and environmentally induced variation in ultimate count seem to depend on different mechanisms, judging from the relationships between body length and the number of developing anal fin rays that are visible. In fish from the same locality reared at two different temperatures, rays start appearing at a shorter body length at higher temperature, and the ultimate number if greater. In contrast, in fish from two localities reared at the same temperature, there is no obvious correlation between the body length when rays start appearing and the ultimate number formed. This observation is merely suggestive, as the visible appearance of fin rays may be well after their embryonic fixation. It does provide further circumstantial support to the hypothesis that mechanisms controlling meristic variation involve changes in the relationships between developmental stage and the size of the embryo.

VII. FUTURE RESEARCH

This review indicates that there has been a moderate number of experimental studies on meristic variation in fishes, but these have been largely unfocused and exploratory, and have seldom been designed for testing hypotheses. Embryological studies on development of meristic structures have been scattered, and usually have lacked the quantification useful for model building. The few explanations that have been suggested for the striking reactions of meristic counts to environmental influences have been highly speculative, and have not been adequately tested.

The sequence of investigations in the past should now be reversed. Sufficient background exists for the construction of firm quantitative hypotheses, of which the atroposic model described here is only a first step. If necessary, embryological descriptions should be undertaken to assure that components of prospective models have a basis in reality. Experiments should then be designed in order to test specific hypotheses. Based on their results, the cycle of hypothesis building and testing can then be repeated.

The ideal material for blocks of meristic experiments would be large numbers of eggs with maximum genetic uniformity. Because of the demonstrated genetic differences in meristic responses between different parents from one population, the use of mixed batches of eggs from several parents is likely to blur the pattern of meristic response. For the same reason, series of temperature transfers at various developmental stages cannot satisfactorily incorporate offspring from different parents (Orska, 1956, 1962; Ogawa, 1971). The hermaphrodite cyprinodont *R. marmoratus* has provided embryos that are isogenic and homozygous, but they have the disadvantage of being retained within the parent for variable and imprecisely known times, and only one or a few eggs are laid per day. Perhaps large numbers of genetically identical newly fertilized eggs may be provided in the near future by artificial cloning (Donaldson and Hunter, 1982). In the meantime, the best experimental subjects are large species that lay many eggs at one time, or iteroparous species that lay successive egg batches at short intervals.

Because of possible prefertilization influences, the temperature and other environmental history of the parents should in future be recorded and if possible controlled. Eggs should be transferred to their experimental conditions immediately after fertilization, and the precise time delays and conditions of transfer should be recorded. In tests of the affects of dissolved salts or other substances, fertilization should occur in each test solution. Young must be reared in experimental environments long enough to ensure that fixation of their meristic parts has occurred, and they must be grown long enough beyond that point to ensure that counts will yield the ultimate numbers of parts.

Experiments are needed to compare meristic responses to pulses (in temperature or other factors) of various durations, and of different magnitude, and of different abruptness. Effects of very early pulses particularly need investigation. The best data for fitting curves in models like the atroposic one will probably come from pulse rather than break experiments, because then each treatment is confined to one short segment of the curve being constructed. The best test of such a model will probably be prediction, and subsequent test, of the meristic response to one type of experiment (such as temperature breaks), using data from a different type of experiment (such as temperature pulses). Experiments are needed on the interaction of two or more environmental influences on meristic count (e.g., are the responses additive, or are there meristic limits beyond which further environmental pressure produces no additional change?).

The prefertilization phenomenon begs for investigation. To examine the meristic effects of different temperature histories on the same parental genotype, one needs either a supply of genetically uniform fish (see above), or iteroparous species where the same parents can produce young after different exposures. Here the possible complications of parental age or egg-laying sequence must be considered. Questions to be tackled include (a) does environmental exposure of the father have a meristic result in the offspring? (b) what are the earliest and the latest times preceding fertilization that the effect is manifest? (c) are environmental influences on the parents other than temperature also effective? and (d) to what extent have the published responses of meristic characters in laboratory experiments been influenced by prefertilization history of the parents?

Surgical manipulation has contributed much to embryological studies in amphibians, birds, and some invertebrates, and should be tried more extensively on fish. Although the chorion poses mechanical problems to teleost surgery, they are not insuperable. In vitro culture of embryos even for restricted periods (Laale, 1985) might allow useful observation.

The pectoral fins are more accessible to surgical intervention and to in vivo observation than are other series. The phenotypic responses of pectoral ray counts to environmental influences are relatively consistent. The geometry of their early development is free from interference from other series. Hence the pectoral fin-ray series seems particularly promising for investigation, and may lend itself to modeling the processes producing meristic variation.

Although environmentally induced meristic variation has been studied mostly in fishes, it has also been demonstrated experimentally in amphibians (Peabody and Brodie, 1975; Pawlowska-Indyk, 1985), in reptiles (Osgood, 1978), in birds (Lindsey and Moodie, 1967; Orska et al., 1973), and in mammals (Barnett, 1965). The physiological responses of the processes controlling meristic number (growth versus differentiation?) can perhaps best be teased apart in fishes because of their malleability. The mechanisms, once they are understood, may well be found to apply also to other groups of organisms.

REFERENCES

Ali, M. Y. (1962). Meristic variation in the medaka (*Oryzias latipes*) produced by temperature, and by chemicals affecting metabolism. Ph.D. Thesis, University of British Columbia, Vancouver, B.C.

Ali, M. Y., and Lindsey, C. C. (1974). Heritable and temperature-induced meristic variation in the medaka, *Oryzias latipes. Can. J. Zool.* **52**, 959–976.

Angus, R. A., and Schultz, R. J. (1983). Meristic variation in homozygous and heterozygous fish. *Copeia*, pp. 287–299.

Ayles, G. B., and Berst, A. H. (1973). Parental age and survival of progeny in splake hybrids (*Salvelinus fontinalis* × *S. namaycush*). *J. Fish. Res. Board Can.* **30**, 570–582.

Bailey, R. M., and Gosline, W. A. (1955). Variation and systematic significance of vertebral counts in the American fishes of the family Percidae. *Misc. Publ. Mus. Zool., Univ. Mich.* **93**, 5–44.

Barlow, G. W. (1961). Causes and significance of morphological variation in fishes. *Syst. Zool.* **10**, 105–117.

Barnett, S. A. (1965). Genotype and environment in tail length in mice. *Q. J. Exp. Physiol.* **50**, 417–429.

Barsukov, V. V. (1955). On the variability in the number of rays in the fins of *Anarhichas lupus lupus* L. *Dokl. Akad. Nauk SSSR, Biol. Sci.* **103**, 715–716 (cited in Orska, 1963).

Battle, H. I., and Hisaoka, K. K. (1952). Effects of ethyl carbamate (Urethan) on the early development of the teleost (*Brachydanio rerio*). *Cancer Res.* **12**, 334–340.

Beacham, T. D. (1985). Meristic and morphometric variation in pink salmon (*Oncorhynchus gorbuscha*) in southern British Columbia and Puget Sound. *Can. J. Zool.* **63**, 366–372.

Beacham, T. D., and Murray, C. B. (1986). The effect of spawning time and incubation temperature on meristic variation in chum salmon (*Oncorhynchus keta*). *Can. J. Zool.* **64**, 45–48.

Beacham, T. D., and Withler, R. E. (1985). Heterozygosity and morphological variability of chum salmon (*Oncorhynchus keta*) in southern British Columbia. *Heredity* **54**, 313–322.

Beardmore, J. A., and Shami, S. A. (1976). Parental age, genetic variation and selection. *In* "Population Genetics and Ecology" (S. Karlin and E. Nevo, eds.), pp. 3–22. Academic Press, New York.

Ben-Tuvia, A. (1963). Influence of temperature on the vertebral number of *Sardinella aurita* from the eastern Mediterranean. *Isr. J. Zool.* **12**, 59–66.

Berg, L. S. (1969). "Nomogenesis or Evolution Determined by Law" (transl. from Russian). MIT Press, Cambridge, Massachusetts.

Bergot, P., Chevassus, B., and Blanc, J.-M. (1976). Déterminisme génétique du nombre de caeca pyloriques chez la truite fario (*Salmo trutta* Linné) et la truite arc-en-ciel (*Salmo gairdneri* Richardson). I. Distribution du caractère et variabilité phénotypique intra et interfamilles. *Ann. Hydrobiol.* **7**, 105–114.

Bertin, L. (1925). Recherches bionomiques, biométriques et systématiques sur les epinoches (Gastérostéides). *Ann. Inst. Oceanograph. Monaco* [N.S.] **2**,(1), 1–204.

Blackburn, M. (1950). The Tasmanian whitebait, *Lovettia seali* (Johnston), and the whitebait fishery. *Aust. J. Mar. Fish. Res.* **1**, 155–198.

Blaxter, J. H. S. (1957). Herring rearing. III. The effect of temperature and other factors on myotome counts. *Scott. Home Dep., Mar. Res., 1957*, No. 1, pp. 1–16.

Bouvet, J. (1968). Histogenèse précoce et morphogenèse du squelette cartilagineux des ceintures primaires et des nageoires paires chez la truite (*Salmo trutta fario* L.). *Arch. Anat. Microsc. Morphol. Exp.* **57**, 35–52.

Bouvet, J. (1974). Différenciation et ultrastructure du squelette distal de la nageoire pectorale chez la truite indigène (*Salmo trutta fario* L.). II. Différenciation et ultrastructure des lepidotriches. *Arch. Anat. Microsc. Morphol. Exp.* **63**, 323–335.

Bouvet, J. (1978). Cell proliferation and morphogenesis of the apical ectodermal ridge in the pectoral fin bud of the trout embryo (*Salmo trutta fario* L.). *Wilhelm Roux's Arch. Dev. Biol.* **185**, 137–154.

Braum, E. (1973). Einflusse chronischen exogenen Sauerstoffmangels auf die Embryogenese des Herings (*Clupea harengus*). *Neth. J. Sea Res.* **7**, 363–375.

Canagaratnam, P. (1959). The influence of light intensities and durations during early development on meristic variation in some salmonids. Ph.D. Thesis, University of British Columbia, Vancouver, B.C.

Carmichael, G. J. (1983). Scale-number differences in Central stonerollers incubated and reared at different temperatures. *Trans. Am. Fish. Soc.* **112**, 441–444.

Casanova, J.-P. (1981). Amendments of the ecological rules known as Bergmann's and Jordan's rules. *J. Plankton Res.* **3**, 509–530.

Chernoff, B., Conner, J. V., and Bryan, C. F. (1981). Systematics of the *Menidia beryllina* complex (Pisces: Atherinidae) from the Gulf of Mexico and its tributaries. *Copeia*, pp. 319–336.

Christiansen, F. B., Neilsen, B. V., and Simonsen, V. (1981). Genetic and morphological variation in the eelpout *Zoarces viviparus*. *Can. J. Genet. Cytol.* **23**, 163–172.

Clark, J. C., and Vladykov, V. D. (1960). Definition of haddock stocks of the northwestern Atlantic. *Fish. Bull.* **169**, 283–296.

Coad, B. W. (1981). A bibliography of the sticklebacks (Gasterosteidae: Osteichthyes). *Syllogeus No. 35*, National Museums of Canada.

Cooke, J. (1981). The problem of periodic patterns in embryos. *Philos. Trans. R. Soc. London, Ser. B* **295**, 509–524.

Crossman, E. J. (1960). Variation in number and asymmetry in branchiostegal rays in the family Esocidae. *Can. J. Zool.* **38**, 363–375.

Dannevig, A. (1950). The influence of the environment on number of vertebrae in plaice. *Rep. Norw. Fish. Mar. Invest.* **9**, 3–6.

Dauolas, C., and Economou, A. N. (1986). Seasonal variation of egg size in the sardine, *Sardina pilchardus* Walb., of the Saronikos Gulf: Causes and a probable explanation. *J. Fish Biol.* **28**, 449–457.

de Ciechomski, J. Dz. and de Vigo, G. W. (1971). The influence of the temperature on the number of vertebrae in the Argentine anchovy, *Engraulis anchoita* (Hubbs and Marini). *J. Cons., Cons. Int. Explor. Mer* **34**, 37–42.

Dentry, W. (1976). Effect of parental temperature experience and sustained offspring rearing temperature on some meristic series in zebra fish (*Brachydanio rerio*). M.Sc. Thesis, University of Manitoba, Winnipeg.

Dentry, W., and Lindsey, C. C. (1978). Vertebral variation in zebrafish (*Brachydanio rerio*) related to the prefertilization temperature history of their parents. *Can. J. Zool.* **56**, 280–283.

Donaldson, E. M., and Hunter, G. A. (1982). Sex control in fish with particular reference to salmonids. *Can. J. Fish. Aquat. Sci.* **39**, 99–110.

Dunn, J. R. (1983). The utility of developmental osteology in taxonomic and systematic studies of teleost larvae: A review. *NOAA* (*Natl. Oceanic Atmos. Adm.*) *Tech. Rep. Circ.* **450**, 1–19.

Dunn, J. R. (1984). Developmental osteology. *In* "Ontogeny and Systematics of Fishes" (H. G. Moser *et al.*, eds.), Am. Soc. Ichthyol. Herpetol., Spec. Publ. No. 1, pp. 48–50. Allen Press, Lawrence, Kansas.

Eaton, T. H. (1945). Skeletal supports of the median fins of fishes. *J. Morphol.* **76**, 193–212.

Ege, V. (1942). Johs. Schmidt: Racial Investigations XI. A transplantation experiment with *Zoarces viviparus* L. *C. R. Lab. Carlsberg, Ser. Physiol.* **23**, 271–384.

Eisler, R. (1961). Effects of visible radiation on salmonoid embryos and larvae. *Growth* **25**, 281–346.

Ehrlich, K. F., and Farris, D. A. (1971). Some influences of temperature on the development of the grunion *Leuresthes tenuis* (Ayres). *Calif. Fish Game* **57**, 58–68.

Fahy, W. E. (1972). Influence of temperature change on number of vertebrae and caudal fin rays in *Fundulus majalis* (Walbaum). *J. Cons., Cons. Int. Explor. Mer* **34**, 217–231.

Fahy, W. E. (1976). The morphological time of fixation of the total number of vertebrae in *Fundulus majalis* (Walbaum). *J. Cons., Cons. Int. Explor. Mer* **36**, 243–250.

Fahy, W. E. (1978). The influence of crowding upon the total number of vertebrae developing in *Fundulus majalis* (Walbaum). *J. Cons., Cons. Int. Explor. Mer* **38**, 252–256.

Fahy, W. E. (1979). The influence of temperature change on number of anal fin rays developing in *Fundulus majalis* (Walbaum). *J. Cons., Cons. Int. Explor. Mer* **38**, 280–285.

Fahy, W. E. (1980). The influence of temperature change on number of dorsal fin rays developing in *Fundulus majalis* (Walbaum). *J. Cons., Cons. Int. Explor. Mer* **39**, 104–109.

Fahy W. E. (1982). The influence of temperature change on number of pectoral fin rays developing in *Fundulus majalis* (Walbaum). *J. Cons., Cons. Int. Explor. Mer* **40**, 21–26.

Fahy, W. F. (1983). The morphological time of fixation of the total number of caudal fin rays in *Fundulus majalis* (Walbaum). *J. Cons., Cons. Int. Explor. Mer* **41**, 37–45.

Fahy, W. E., and O'Hara, R. K. (1977). Does salinity influence the number of vertebrae in developing fishes? *J. Cons., Cons. Int. Explor. Mer* **37**, 156–161.

Falconer, D. S. (1981). "Introduction to Quantitative Genetics." Longmans, Green, New York.

Ferguson, M. M., and Danzmann, R. G. (1987). Deviation from morphological intermediacy in interstrain hybrids of rainbow trout, *Salmo gairdneri. Environ. Biol. Fishes* **18**, 249–256.

Fonds, M., Rosenthal, H., and Alderdice, D. F. (1974). Influence of temperature and salinity on embryonic development, larval growth and number of vertebrae of the garfish, *Belone belone. In* "The Early Life History of Fish" (J. H. S. Blaxter, ed.), pp. 509–525. Springer-Verlag, Berlin and New York.

Ford, E. (1933). The "Number of Vertebrae" in the herring and its variation. *J. Cons., Cons. Int. Explor. Mer* **8**, 211–222.

Fournier, D. A., Beacham, T. D., Riddell, B. E., and Busack, C. A. (1984). Estimating stock composition in mixed stock fisheries using morphometric, meristic and electrophoretic characters. *Can. J. Fish. Aquat. Sci.* **41**, 400–408.

François, Y. (1957). Quelques problèmes relatifs à la nageoire dorsale des Téléostéens. *Annee Biol.* **33**, 413–426.

François, Y. (1958). Recherches sur l'anatomie et le développement de la nageoire dorsale des Téléostéens. Thèses, Fac. Sci. Univ. Paris, Ser. A 3206, No. 4078, 108P.

Fry, F. E. J. (1971). The effect of environmental factors on the physiology of fish. *In* "Fish Physiology" (W. S. Hoar and D. J. Randall, eds.), Vol. 6, pp. 1–98. Academic Press, New York.

Fukuhara, O., and Fushimi, T. (1984). Squamation of larval greenling, *Hexagrammos otakii* (Pisces: Hexagrammidae) reared in the laboratory. *Bull. Jpn. Soc. Sci. Fish.* **50**, 759–762.

Gabriel, M. L. (1944). Factors affecting the number and form of vertebrae in *Fundulus heteroclitus. J. Exp. Zool.* **95**, 105–147.

Galkina, I. A. (1960). Method of determining the number of myomeres in herring larvae and the change in their number during development. *Dokl. Akad. Nauk SSSR, Biol. Sci.* **184**, 723–726. (pp. 194–196 in transl.)

Gall, G. A. E. (1974). Influence of size of eggs and age of female on hatchability and growth in rainbow trout. *Calif. Fish Game* **60**, 26–35.

Garside, E. T. (1966). Developmental rate and vertebral number in salmonids. *J. Fish. Res. Board Can.* **23**, 1537–1551.

Garside, E. T. (1970). Structural responses. *In* "Marine Ecology" (O. Kinne, ed.), Vol. 1, Part 1, pp. 561–616. Wiley (Interscience), London.

Garside, E. T., and Fry, F. E. J. (1959). A possible relationship between yolk size and differentiation in trout embryos. *Can. J. Zool.* **37**, 383–386.

Géraudie, J., and François, Y. (1973). Les premiers stades de la formation de l'ébauche de nageoire pelvienne de truite (*Salmo fario* et *Salmo gairdneri*). I. Etude anatomique. *J. Embryol. Exp. Morphol.* **29**, 221–237.

Géraudie, J., and Landis, W. J. (1982). The fine structure of the developing pelvic fin dermal skeleton in the trout *Salmo gairdneri. Am. J. Anat.* **163**, 141–156.

Ghéno, Y., and Poinsard, F. (1968). Observations sur les jeunes sardinelles de la baie de Pointe-Noire (Congo). *Cah. O.R.S.T.O.M., Ser. Oceanogr.* **6**, 53–67.

Goff, R. A. (1940). The effects of increased atmospheric pressure on the developing embryo of the zebra fish, *Brachydanio rerio* (Hamilton). *Trans. Kans. Acad. Sci.* **43**, 401–410.

Gordon, M., and Benzer, P. (1945). Sexual dimorphism in the skeletal elements of the gonopodial suspensoria in xiphophorin fishes. *Zoologica (N.Y.)* **30**, 57–72.

Graham, J. H., and Felley, J. D. (1985). Genomic coadaptation and developmental stability within introgressed populations of *Enneacanthus gloriosus* and *E. obesus* (Pisces, Centrarchidae). *Evolution (Lawrence, Kans.)* **39**, 104–114.

Grüneberg, H. (1952). Genetical studies on the skeleton of the mouse. IV. Quasi-continuous variations. *J. Genet.* **51**, 95–114.

Hagen, D. W. (1972). Inheritance of numbers of lateral plates and gill rakers in *Gasterosteus aculeatus*. *Heredity* **30**, 301–312.

Hagen, D. W., and Blouw, D. M. (1983). Heritability of dorsal spines in the fourspine stickleback (*Apeltes quadracus*). *Heredity* **50**, 275–281.

Hagen, D. W., and Gilbertson, L. G. (1973). The genetics of plate morphs in freshwater threespine sticklebacks. *Heredity* **31**, 75–84.

Hallam, J, C, (1974). The modification of vertebral number in Salmonidae. Ph.D. Thesis, University of Toronto, Toronto, Ontario, Canada.

Harrington, R. W., Jr., and Crossman, R. A., Jr. (1976a). Effects of temperature and sex genotype on meristic counts of the gonochoristic cyprinodontid fish, *Rivulus cylindraceus* (Poey). *Can. J. Zool.* **54**, 245–254.

Harrington, R. W., Jr., and Crossman, R. A., Jr. (1976b). Temperature induced meristic variation among three homozygous genotypes (clones) of the self-fertilizing fish *Rivulus marmoratus*. *Can. J. Zool.* **54**, 1143–1155.

Harrison, R. G. (1895). Die Entwicklung der unpaaren und paarigen Flossen der Teleostier. *Arch. Mikrosk. Anat.* **46**, 500–578.

Hayes, F. R. (1949). The growth, general chemistry, and temperature relations of salmonid eggs. *Q. Rev. Biol.* **24**, 281–308.

Hayes, F. R., Pelluet, D., and Gorham, E. (1953). Some effects of temperature on the embryonic development of the salmon (*Salmo salar*). *Can. J. Zool.* **31**, 42–51.

Heincke, F. (1898). Naturgeschichte des Herings. I. Die Lokalformen und die Wanderungen des Herings in den europäischen Meeren. *Abh. Dtsch. Seefischerei-Ver.* **2**, Part 1, 1–128.

Hempel, G. (1965). Fecundity and egg size in relation to the environment. *Int. Comm. N.W. Atl. Fish., Spec. Publ.* **6**, 687–690.

Hempel, G., and Blaxter, J. H. S. (1961). The experimental modification of meristic characters in herring (*Clupea harengus* L.). *J. Cons., Cons. Int. Explor. Mer* **26**, 336–346.

Heuts, M. J. (1947). Experimental studies on adaptive evolution in *Gasterosteus aculeatus* L. *Evolution (Lawrence, Kans.)* **1**, 89–102.

Heuts, M. J. (1949). Racial divergence in fin ray variation patterns in *Gasterosteus aculeatus*. *J. Genet.* **49**, 183–191.

Hiemstra, W. H. (1962). A correlation table as an aid for identifying pelagic fish eggs in plankton samples. *J. Cons., Cons. Int. Explor. Mer* **27**, 100–108.

Hisaoka, K. K., and Battle, H. I. (1958). The normal developmental stages of the zebrafish, *Brachydanio rerio* (Hamilton-Buchanan). *J. Morphol.* **102**, 311–327.

Hoar, W. S. (1983). "General and Comparative Physiology," 3rd ed. Prentice-Hall, Englewood Cliffs, New Jersey.

Holliday, F. G. T., and Blaxter, J. H. S. (1960). The effects of salinity on the developing eggs and larvae of the herring (*Clupea harengus*). *Mar. Biol. Assoc. U.K.* **39**, 591–603.

Hubbs, C. L. (1922). Variations in the number of vertebrae and other meristic characters of fishes correlated with the temperature of water during development. *Am. Nat.* **56**, 360–372.

Hubbs, C. L. (1926). The structural consequences of modifications of the developmental rate in fishes, considered in reference to certain problems of evolution. *Am. Nat.* **60**, 57–81.

Hubbs, C. L. (1927). The related effects of parasites on a fish. *J. Parasitol.* **14**, 75–84.

Hubbs, C., and Bryan, C. (1974). Effect of parental temperature experience on thermal tolerance of eggs of *Menidia audens*. In "The Early Life History of Fish" (J. H. S. Blaxter, ed.), pp. 431–435. Springer-Verlag, Berlin and New York.

Hubbs, C. L., and Hubbs, L. C. (1945). Bilateral asymmetry and bilateral variation in fishes. *Pap. Mich. Acad. Sci., Arts Lett.* **30**, 229–310.

Itazawa, Y. (1959). Influence of temperature on the number of vertebrae in fish. *Nature (London)* **183**, 1408–1409.

Itazawa, Y. (1963). The ossification sequences of the vertebral column in the carp and the snake-head fish. *Bull. Jpn. Soc. Sci. Fish.* **29**, 667–674.

Jagoe, C. H., and Haines, T. A. (1985). Fluctuating asymmetry in fishes inhabiting acidified and unacidified lakes. *Can. J. Zool.* **63**, 130–138.

Jensen, A. S. (1944). On specific constancy and segregation into races in sea-fishes. *K. Dan. Videnskab. Selsk. Biol. Medd.* **19**(8), 1–19.

Johnsen, S. (1936). On the variation of fishes in relation to environment (Preliminary account). *Bergens Mus. Årbok* 1936, Natur. Rekke 4, 1–26.

Jordan, D. S. (1892). Relations of temperature to vertebrae among fishes. *Proc. U.S. Nat. Mus.* **14**, 107–120.

King, D. P. F. (1985). Enzyme heterozygosity associated with anatomical character variance and growth in the herring (*Clupea harengus* L.). *Heredity* **54**, 289–296.

Kinne, O., and Kinne, E. M. (1962). Rates of development in embryos of a cyprinodont fish exposed to different temperature–salinity–oxygen combinations. *Can. J. Zool.* **40**, 231–253.

Kinoshita, T. (1984). Dorsal and anal fin rays of the Japanese anchovy, *Engraulis japonica*, and their pterygiophores. *Bull. Fac. Fish., Hokkaido Univ.* **35**, 66–82.

Kirpichnikov, V. S. (1981). "Genetic Bases of Fish Selection." Springer-Verlag, Berlin and New York.

Kohno, H., and Taki, Y. (1983). Comments on the development of fin-supports in fishes. *Jpn. J. Ichthyol.* **30**, 284–290.

Koli, L. (1969). Geographical variation of *Cottus gobio* L. (Pisces, Cottidae) in northern Europe. *Ann. Zool. Fenn.* **6**, 353–390.

Komada, N. (1977a). Influence of temperature on the vertebral number of the ayu *Plecoglossus altivelis*. *Copeia*, pp. 572–573.

Komada, N. (1977b). The number of segments and body length of *Plecoglossus altivelis* fry in the Nagara River, Japan. *Copeia*, pp. 573–574.

Komada, N. (1982). Vertebral anomalies in the cyprinid fish, *Tribolodon hakonensis*. *Jpn. J. Ichthyol.* **29**, 185–192.

Korovina, V. M., Lubitskaya, A. I., and Dorofeeva, E. A. (1965). Effect of visible light and darkness on the rate of formation of cartilaginous elements in the skeletons of teleost fishes. *Vopr. Ikhtiol.* **5**, 403–410.

Krivobok, M. N., and Storozhuk, A. Y. (1970). The effect of the size and age of Volga sturgeon on the weight and chemical composition of mature eggs. *J. Ichthyol. (Engl. Transl.)* **10**, 761–765.

Kubo, T. (1950). A preliminary report of the study on the groups of *Oncorhynchus keta* (Walbaum) (dog salmon) and the numbers of their segments. *Bull. Fac. Fish., Hokkaido Univ.* **1**, 1–11.

Kubota, Z. (1967). Morphology of the Japanese loach, *Misgurnus anguillicaudatus*

(Cantor). VI. Vertebral number of the loach hatched at various water temperature. *J. Shimonoseki Univ. Fish.* **15**, 273–278.

Kwain, W.-H. (1975). Embryonic development, early growth, and meristic variation in rainbow trout (*Salmo gairdneri*) exposed to combinations of light intensity and temperature. *J. Fish. Res. Board Can.* **32**, 397–402.

Kyle, H. M. (1923). The factors determining the number of vertebrae in fishes. *Dtsch. Wiss. Komm. Meeresforsch., Abt. Helgol. Heft.* **1**,(9), 1–8.

Kyle, H. M. (1926). "The Biology of Fishes." Sidgwick and Jackson, London.

Laale, H. W. (1985). Kupffer's vesicle in *Brachydanio rerio:* Multivesicular origin and proposed function *in vitro. Can. J. Zool.* **63**, 2408–2415.

Landrum, B. J. (1966). Bilateral asymmetry in paired meristic characters of Pacific salmon. *Pac. Sci.* **20**, 193–202.

Lapin, Y. Y. (1975). Variability in the number of vertebrae and larval myomeres in the White Sea herring (*Clupea harengus pallasi* natio *maris-albi*) in connection with its ecology. *J. Ichthyol (Engl Transl.)* **14**, 466–476.

Lapin, Y. Y., Belmakov, V. S., and Stepanenko, A. V. (1969). Relationship between number of segments in the White Sea herring and temperature regime during incubation. *Probl. Ichthyol.* **9**, 725–728.

Leary, R. F., Allendorf, F. W., and Knudsen, K. L. (1983). Consistently high meristic counts in natural hybrids between brook trout and bull trout. *Syst. Zool.* **32**, 369–376.

Leary, R. F., Allendorf, F. W., and Knudsen, K. L. (1985a). Inheritance of meristic variation and the evolution of developmental stability in rainbow trout. *Evolution (Lawrence, Kans.)* **39**, 308–314.

Leary, R. F., Allendorf, F. W., and Knudsen, K. L. (1985b). Developmental instability and high meristic counts in interspecific hybrids of salmonid fishes. *Evolution (Lawrence, Kans.)* **39**, 1318–1326.

Lieder, V. (1961). Untersuchungsergebnisse über die Grätenzahlen bei 17 Susswasser-Fischarten. *Z. Fisch.* **10**, 329–350.

Linden, O., Laughlin, R., Jr., Sharp, J. R., and Neff, J. M. (1980). The combined effect of salinity, temperature and oil on the growth pattern of embryos of the killifish, *Fundulus heteroclitus* Walbaum. *Mar. Environ. Res.* **3**, 129–144.

Lindsey, C. C. (1953). Variation in anal ray count of the redside shiner *Richardsonius balteatus* (Richardson). *Can. J. Zool.* **31**, 211–225.

Lindsey, C. C. (1954). Temperature-controlled meristic variation in the paradise fish *Macropodus opercularis* (L.). *Can. J. Zool.* **32**, 87–98.

Lindsey, C. C. (1955). Evolution of meristic relations in the dorsal and anal fins of teleost fishes. *Trans. R. Soc. Can.* **49**, Sect. 5, 35–49.

Lindsey, C. C. (1958). Modification of meristic characters by light duration in kokanee, *Oncorhynchus nerka. Copeia*, pp. 134–136.

Lindsey, C. C. (1962a). Experimental study of meristic variation in a population of threespine sticklebacks, *Gasterosteus aculeatus. Can. J. Zool.* **40**, 271–312.

Lindsey, C. C. (1962b). Observations on meristic variation in ninespine sticklebacks, *Pungitius pungitius*, reared at different temperatures. *Can. J. Zool.* **40**, 1237–1247.

Lindsey, C. C. (1966). Body sizes of poikilotherm vertebrates at different latitudes. *Evolution (Lawrence, Kans.)* **20**, 456–465.

Lindsey, C. C. (1975). Pleomerism, the widespread tendency among related fish species for vertebral number to be correlated with maximum body length. *J. Fish. Res. Board Can.* **32**, 2453–2469.

Lindsey, C. C. (1978). Form, function, and locomotory habits in fish. *In* "Fish Physiology" (W. S. Hoar and D. J. Randall, eds.), Vol. 7, pp. 1–100. Academic Press, New York.

Lindsey, C. C. (1981). Stocks are chameleons: Plasticity in gill rakers of coregonid fishes. *Can. J. Fish. Aquat. Sci.* **38**, 1497–1506.

Lindsey, C. C., and Ali, M. Y. (1965). The effect of alternating temperature on vertebral count in the medaka (*Oryzias latipes*). *Can J. Zool.* **43**, 99–104.

Lindsey, C. C., and Ali, M. Y. (1971). An experiment with medaka, *Oryzias latipes,* and a critique of the hypothesis that teleost egg size controls vertebral count. *J. Fish. Res. Board. Can.* **28**, 1235–1240.

Lindsey, C. C., and Arnason, A. N. (1981). A model for responses of vertebral numbers in fish to environmental influences during development. *Can. J. Fish. Aquat. Sci.* **38**, 334–347.

Lindsey, C. C., and Harrington, R. W., Jr. (1972). Extreme vertebral variation induced by temperature in a homozygous clone of the self-fertilizing cyprinodontid fish *Rivulus marmoratus. Can. J. Zool.* **50**, 733–744.

Lindsey, C. C., and Moodie, G. E. E. (1967). The effect of incubation temperature on vertebral count in the chicken. *Can. J. Zool.* **45**, 891–892.

Lindsey, C. C., Brett, A. M., Swain, D. P., and Arnason, A. N. (1984). Responses of vertebral numbers in rainbow trout to temperature changes during development. *Can. J. Zool.* **62**, 391–396.

Lubitskaya, A. I. (1935). Zur Erforschung des Temperatureffekts in der Morphogenese. II. Einfluss der Temperatur auf die Entwicklungsgeschwindigkeit und Wachstum des Embryos von *Salmo trutta* L. morpha *fario. Zool. Jahrb., Abt. Allg. Zool. Physiol.* **54**, 405–422.

Lubitskaya, A. I. (1961). Effect of the visible light, ultraviolet rays and temperature on the metamery of the fish body. Part I. Action of different portions of visible spectrum, of darkness and temperature on the survival and body metamery of fishes. *Zool. Zh.* **40**, 397–407.

Lubitskaya, A. I., and Dorofeeva, E. A. (1961a). Effect of the visible light, ultraviolet rays and temperature on the metamery of the fish body. Part 2. Effect of ultraviolet rays on the survival and metamery of the body of *Esox lucius* L. and *Acerina cernua* L. *Zool. Zh.* **40**, 1046–1057; *Referat. Zhur., Biol.,* 1962, No. 17114 (Translation).

Lubitskaya, A. I., and Dorofeeva, E. A. (1961b). Effect of the visible light, ultraviolet rays and temperature on the metamery of the fish body. Part 3. Effect of ultraviolet rays on the survival and metamery of the body of *Osmerus eperlanus eperlanus* (L.) and *Perca fluviatilis* L. *Vopr. Ikhtiol* **1**, 497–509.

Lubitskaya, A. I., and Svetlov, P. (1935). Differential beschleunigungen der Entwicklungsstadien der Brustflossen bei der Bachforelle unter Temperatureinwirkung. *Biol. Zentralbl.* **54**, 195–210.

MacCrimmon, H. R., and Kwain, W.-H. (1969). Influence of light on early development and meristic characters in the rainbow trout, *Salmo gairdneri* Richardson. *Can. J. Zool.* **47**, 631–637.

MacGregor, R. B., and MacCrimmon, H. R. (1977). Evidence of genetic and environmental influences on meristic variation in the rainbow trout, *Salmo gairdneri* (Richardson). *Environ. Biol. Fishes* **2**, 25–33.

McHugh, J. L. (1954). The influence of light on the number of vertebrae in the grunion, *Leuresthes tenuis. Copeia,* pp. 23–25.

Maginsky, R. B. (1958). A study of the effects of light on the meristic characters of the zebra fish *Brachydanio rerio.* M.Sc. Thesis, University of Pittsburgh, Pittsburgh.

Mann, R. H. K., and Mills, C. A. (1985). Variations in the sizes of gonads, eggs and larvae of the dace, *Leuciscus leuciscus. Environ. Biol. Fishes* **13**, 277–287.

Marckmann, K. (1958). The influence of the temperature on the respiratory metabolism during the development of the sea-trout. *Medd. Danm. Fisk. Havundersøg* [N.S.] 2, 21, 1–20.

Marsh, E. (1984). Egg size variation in Central Texas populations of *Etheostoma spectabile* (Pisces: Percidae). *Copeia*, pp. 291–301.

Miller, J. M., and Tucker, R. W., Jr. (1979). X-Radiography of larval and juvenile fishes. *Copeia*, pp. 535–538.

Misra, R. K., and Carscadden, J. E. (1984). Stock discrimination of capelin (*Mallotus mallotus*) in the Northwest Atlantic. *J. Northwest. Atl. Fish. Sci.* 5, 199–205.

Moav, R., Finkel, A., and Wohlfarth, G. (1975). Variability of intermuscular bones, vertebrae, ribs, dorsal fin rays and skeletal disorders in the common carp. *Theor. Appl. Genet.* 46, 33–43.

Molander, A. R., and Molander-Swedmark, M. (1957). Experimental investigations on variation in plaice (*Plouronectes platessa* L.), *Inst. Mar. Res., Lysekil, Ser. Biol., Rept.* 7, Fish, Board Sweden, 1–45.

Mottley, C. McC. (1934). The effects of temperature during development on the number of scales in the Kamloops trout, *Salmo kamloops* Jordan. *Contrib. Can. Biol.* 8, 253–263.

Murray, C. B., and Beacham, T. D. (1986). Effect of varying temperature regimes on the development of pink salmon (*Oncorhynchus gorbuscha*) eggs and alevins. *Can. J. Zool.* 64, 670–676.

Nagiec, C. (1977). Ossification of the axial skeleton and fins in the whitefish, *Coregonus lavaretus* L. *Acta Biol. Cracov. Ser. Zool.* 20, 155–180.

Neave, F. (1943). Scale pattern and scale counting methods in relation to certain trout and other salmonids. *Trans. R. Soc. Can.* 37, Sect. 5., 79–91.

Needham, J. (1933). On the dissociability of the fundamental processes in ontogenesis. *Biol. Rev. Cambridge Philos. Soc.* 8, 181–223.

Nelson, G. J. (1969). Gill arches and the phylogeny of fishes, with notes on the classification of vertebrates. *Bull. Am. Mus. Nat. Hist.* 141, Art. 4, 475–552.

Nikolsky, G. V. (1969a). Parallel intraspecific variation in fishes. *Probl. Ichthyol.* 9, 4–8.

Nikolsky, G. V. (1969b). "Theory of Fish Population Dynamics." Oliver & Boyd, Edinburgh.

Northcote, T. G., and Paterson, R. J. (1960). Relationship between number of pyloric caeca and length of juvenile rainbow trout. *Copeia*, pp. 248–250.

Ogawa, N. M. (1971). Effect of temperature on the number of vertebrae with special reference to temperature-effective period in the medaka (*Oryzias latipes*). *Annot. Zool. Jpn.* 44, 125–132.

Orska, J. (1956). The influence of temperature on the development of the skeleton in teleosts. *Zool. Pol.* 7, 271–326.

Orska, J. (1962). The influence of temperature on the development of meristic characters of the skeleton in Salmonidae. Part I. Temperature-controlled variations of the number of vertebrae in *Salmo irideus* Gibb. *Zool. Pol.* 12, 309–339.

Orska, J. (1963). The influence of temperature on the development of meristic characters of the skeleton in Salmonidae. Part II. Variations in dorsal and anal fin ray count correlated with temperature during development of *Salmo irideus* Gibb. *Zool. Pol.* 13, 49–76.

Orska, J., Lecyk, M., and Indyk, F. (1973). The effect of high temperatures of incubation on the structure of vertebral column in chickens. *Zool. Pol.* 23, 269–278.

Osgood, D. W. (1978). Effects of temperature on the development of meristic characters in *Natrix fasciata*. *Copeia*, pp. 33–47.

Pate, E., and Othmer, H. G. (1984). Applications of a model for scale-invariant pattern formation in developing systems. *Differentiation* **28**, 1–8.

Pavlov, D. A. (1984). Effect of temperature during early ontogeny of Atlantic salmon, *Salmo salar*. 1. Variability of morphological characters and duration of development under different temperatures. *J. Ichthyol. (Engl. Transl.)* **24**, 30–38.

Pavlov, D. A. (1985). Effect of temperature during early ontogeny of the Atlantic salmon, *Salmo salar*. 2. Growth of the embryo and consumption of yolk during development at various temperatures. *J. Ichthyol (Engl. Transl.)* **25**, 41–52.

Pawlowska-Indyk, A. (1985). The effect of temperature during the embryonic development on the structure of the vertebral column of *Xenopus laevis* (Daudin). *Zool. Pol.* **32**, 211–222.

Peabody, R. B., and Brodie, E. D., Jr. (1975). Effect of temperature, salinity and photoperiod on the number of trunk vertebrae in *Ambystoma maculatum*. *Copeia*, pp. 741–746.

Peden, A. E. (1981). Meristic variation of four fish species exhibiting lowest median counts in Georgia Strait, British Columbia. *Can. J. Zool.* **59**, 679–683.

Perlmutter, A., and Antopol, W. (1963). The effect of neotetrazolium on the newly fertilized eggs of the brook trout, *Salvelinus fontinalis* (Mitchill). *Copeia*, pp. 166–168.

Peters, H. M. (1963). Eizahl, Eigewicht und Gelegeentwicklung in der Gattung *Tilapia* (Cichlidae, Teleostei). *Int. Rev. Ges. Hydrobiol. Hydrogr.* **48**, 547–576.

Peterson, R. H., Spinney, H. C. E., and Sreddharan, A. (1977). Development of Atlantic salmon (*Salmo salar*) eggs and alevins under various temperature regimes. *J. Fish. Res. Board Can.* **34**, 31–43.

Potthoff, T. (1984). Clearing and staining techniques. In "Ontogeny and Systematics of Fishes" (H. G. Moser et al., eds.), Am. Soc. Ichthyol. Herpetol. Spec. Publ. No. 1, pp. 35–37. Allen Press, Lawrence, Kansas.

Powsner, L. (1935). The effects of temperature on the durations of developmental stages of *Drosophila melanogaster*. *Physiol. Zool.* **8**, 474–520.

Repa, P. (1974). Zunahme der Schuppenzahl an der Seitenlinie bei einigen Susswasserfischarten in der Abhängigkeit von deren Körperlänge. *Vestn. Čsl. Spol. Zool.* **38**, 295–308.

Resh, V. H., White, D. S., Elbert, S. A., Jennings, D. E., and Krumholz, L. A. (1976). Vertebral variation in the emerald shiner *Notropis atherinoides* from the Ohio River: An apparent contradiction to "Jordan's Rule." *Bull. South. Calif. Acad. Sci.* **75**, 76–84.

Ritchie, L. D. (1976). Systematics and meristic variation in the butterfish [*Odax pullus* (Forster)]. *Fish. Tech. Rep. N.Z. Minist. Agric. Fish.* **145**, 1–46.

Ross, M. R., and Cavender, T. M. (1981). Morphological analyses of four experimental intergeneric cyprinid hybrid crosses. *Copeia*, pp. 377–387.

Roth, F. (1920). Über den Bau und die Entwicklung des Hautpanzers von *Gasterosteus aculeatus*. *Anat. Anz.*, **52**, 513–534.

Ruivo, M., and Monteiro, R. (1954). Influence du facteur température sur le déterminisme de la composition vertébrale chez *Sardina pilchardus* (Walb.) des eaux de Banyuls. *C.R. Hebd. Seances Acad. Sci.* **239**, 1875–1877.

Runnström, S. (1941). Racial analysis of the herring in Norwegian waters. *Rep. Norw. Fish. Mar. Invest.* **6**, 1–110.

Russell, M. A. (1985). Positional information in insect segments. *Dev. Biol.* **108**, 269–283.

Schmidt, J. (1919). Racial studies in fishes. 2. Experimental investigations with *Lebistes reticulatus* (Peters) Regan. *J. Genet.* **8**, 147–153.

Schmidt, J. (1921). Racial investigations. 7. Annual fluctuations of racial characters in *Zoarces viviparus* L. *C.R. Trav. Lab. Carlsberg* **14**, 1–24.

Seymour, A. H. (1959). Effects of temperature upon the formation of vertebrae and fin rays in young chinook salmon. *Trans. Am. Fish. Soc.* **88**, 58–69.

Shami, S. A., and Beardmore, J. A. (1978). Stabilizing selection and parental age effects on lateral line scale number in the guppy, *Poecilia reticulata* (Peters). *Pak. J. Zool.* **10**, 1–16.

Shaw, R. F. (1980). A bibliography of the eggs, larval and juvenile stages of fishes, including other pertinent references. *Main Sea Grant, Tech. Rep.* **61**, 1–266.

Silver, S. J., Warren, C. E., and Doudoroff, P. (1963). Dissolved oxygen requirements of developing steelhead trout and chinook salmon embryos at different water velocities. *Trans. Am. Fish. Soc.* **92**, 327–343.

Sokal, R. R., and Rohlf, F. J. (1981). "Biometry: The Principles and Practices of Statistics in Biological Research," 2nd ed. Freeman, New York.

Solberg, N. A. (1938). The susceptibility of *Fundulus heteroclitus* embryos to X-radiation. *J. Exp. Zool.* **78**, 441–469.

Southward, A. J., and Demir, N. (1974). Seasonal changes in dimensions and viability of the developing eggs of the Cornish pilchard (*Sardina pilchardus* Walbaum) off Plymouth. *In* "The Early Life History of Fish" (J. H. S. Blaxter, ed.), pp. 53–68. Springer-Verlag, Berlin and New York.

Spouge, J. L., and Larkin, P. A. (1979). A reason for pleomerism. *J. Fish. Res. Board Can.* **36**, 255–269.

Strawn, K. (1961). A comparison of meristic means and variances of wild and laboratory-raised samples of the fishes, *Etheostoma grahami* and *E. lepidum* (Percidae). *Tex. J. Sci.* **13**, 127–159.

Suyehiro, Y. (1942). A study of the digestive system and feeding habits of fish. *Jpn. J. Zool.* **10**, 1–303.

Svärdson, G. (1951). The Coregonid problem. 3. Whitefish from the Baltic successfully introduced into fresh waters in the north of Sweden. *Rep. Inst. Freshwater Res., Drottningholm* **32**, 79–125.

Svärdson, G. (1979). Speciation of Scandinavian *Coregonus*. *Rep. Inst. Freshwater Res., Drottningholm* **57**, 1–95.

Swain, D. P. (1986). Adaptive significance of variation in vertebral number in fishes: Evidence in *Gasterosteus aculeatus* and *Mylocheilus caurinus*. Ph.D. diss., Univ. British Columbia, Vancouver.

Swain, D. P. (1987). A problem with the use of meristic characters to estimate developmental stability. *Am. Nat.*, **129**, 761–768.

Swain, D. P., and Lindsey, C. C. (1984). Selective predation for vertebral number of young sticklebacks, *Gasterosteus aculeatus*. *Can. J. Fish. Aquat. Sci.* **41**, 1231–1233.

Swain, D. P., and Lindsey, C. C. (1986a). Meristic variation in a clone of the cyprinodont fish *Rivulus marmoratus* related to temperature history of the parents and of the embryos. *Can. J. Zool.* **64**, 1444–1455.

Swain, D. P., and Lindsey, C. C. (1986b). Influence of reproductive history of parents on meristic variation in offspring in the cyprinodont fish *Rivulus marmoratus*. *Can. J. Zool.* **64**, 1456–1459.

Tåning, Å. V. (1944). Experiments on meristic and other characters in fishes. 1. On the influence of temperature on some meristic characters in sea-trout and the fixation period of these characters. *Medd. Dan. Fisk. Havunders. Ser. Fisk.* **11**, 1–66.

Tåning, Å, V. (1950). Influence of the environment on number of vertebrae in teleostean fishes. *Nature (London)* **165**, 28.

Tåning, Å. V. (1952). Experimental study of meristic characters in fishes. *Biol. Rev. Cambridge Philos. Soc.* **27**, 169–193.

Tatarko, K. I. (1968). Effect of temperature on the meristic characters of fish. *Probl. Ichthyol.* **8**, 339–350.

Taylor, W. R., and van Dyke, G. C. (1985). Revised procedures for staining and clearing small fishes and other vertebrates for bone and cartilage study. *Cybium* **9**, 107–119.

Trifinova, A. N., Vernidoube, M. F., and Phillipov, N. D. (1939). La physiologie de la différentiation et de la croissance. 2. Les périodes critiques dans le développment des salmonides et leur base physiologique. *Acta Zool. (Stockholm)* **20**, 239–267.

Tsukuda, H. (1960). Temperature adaptation in fishes. 3. Temperature tolerance of the guppy, *Lebistes reticulatus*, in relation to the rearing temperature before and after birth. *Biol. J. Nara Women's Univ.* **10**, 11–14.

Turner, C. L. (1942). A quantitative study of the effects of different concentrations of ethynyl testosterone and methyl testosterone in the production of gonopodia in females of *Gambusia affinis*. *Physiol. Zool.* **15**, 263–280.

Valentine, D. W., and Soulé, M. (1973). Effect of *p,p'*-DDT on developmental stability of pectoral fin rays in the grunion, *Leuresthes tenuis*. *Fish. Bull.* **71**, 921–926.

van Raamsdonk, W., Mos, W., Tekrounie, G., Pool, W. W., and Mijzen, P. (1979). Differentiation of the musculature of the teleost *Brachydanio rerio*. II. Effects of immobilization on the shape and structure of somites. *Acta Morphol. Neerl.-Scand.* **17**, 259–274.

Van Valen, L. (1962). A study of fluctuating asymmetry. *Evolution (Lawrence, Kans,)* **16**, 125–142.

Vernon, E. H. (1957). Morphometric comparison of three races of kokanee (*Oncorhynchus nerka*) within a large British Columbia lake. *J. Fish. Res. Board Can.* **14**, 573–598.

Vibert, R. (1954). Effect of solar radiation and loss of gravel cover on development, growth, and loss by predation in salmon and trout. *Trans. Am. Fish. Soc.* **83**, 194–201.

Vladykov, V. D. (1934). Environmental and taxonomic characters of fishes. *Trans. R. Can. Inst.* **20**, 99–140.

Vrijenhoek, R. C. (1985). Homozygosity and interstrain variation in the self-fertilizing hermaphroditic fish, *Rivulus marmoratus*. *J. Hered.* **76**, 82–84.

Waddington, C. H. (1957). "The Strategy of the Genes." Allen & Unwin, London.

Wallace, C. R. (1965). The embryology of smallmouth bass and the effect of temperature on meristic characters. M.S. Thesis University of Arkansas, Fayetteville.

Wallace, C. R. (1973). Effects of temperature on developing meristic structures of smallmouth bass *Micropterus dolomieui* (Lacépède). *Trans. Am. Fish. Soc.* **102**, 142–144.

Waterman, A. J. (1939). Effects of 2,4-dinitrophenol on the early development of the teleost, *Oryzias latipes*. *Biol. Bull. (Woods Hole, Mass.)* **76**, 162–170.

Waterman, R. E. (1970). Fine structure of scale development in the teleost, *Brachydanio rerio*. *Anat. Rec.* **168**, 361–380.

Weisel, G. F. (1967). Early ossification in the skeleton of the sucker (*Catostomus macrocheilus*) and the guppy (*Poecilia reticulata*). *J. Morph.* **121**, 1–18.

Welander, A. D. (1954). Some effects of X-irradiation of different embryonic stages of the trout (*Salmo gairdnerii*). *Growth* **18**, 227–255.

White, D. S. (1977). Early development and pattern of scale formation in the spotted sucker, *Minytrema melanops* (Catostomidae). *Copeia*, pp. 400–403.

Winter, G. W., Schreck, C. B., and McIntyre, J. D. (1980). Meristic comparison of four stocks of steelhead trout (*Salmo gairdneri*). *Copeia*, pp. 160–162.

Wood, A. (1982). Early pectoral fin development and morphogenesis of the apical ectodermal ridge in the killifish, *Aphyosemion scheeli*. *Anat. Rec.* **204**, 349–356.

Wood, A., and Thorogood, P. (1984). An analysis of *in vivo* cell migration during teleost fin morphogenesis. *J. Cell Sci.* **66**, 205–222.

Yatsu, A. (1980). Geographic variation in vertebral numbers in two Pholidid fishes, *Enedrias nebulosa* and *E. crassispina* around Japan. *Jpn. J. Ichthyol.* **27**, 115–121.

4

THE PHYSIOLOGY OF SMOLTING SALMONIDS

W. S. HOAR

Department of Zoology
University of British Columbia
Vancouver, British Columbia, Canada V6T 2A9

I. INTRODUCTION

Many salmonids, fish of the genera *Oncorhynchus, Salmo,* and *Salvelinus,* are anadromous and undergo a distinct transformation prior to seaward migration. Typically, the cryptically colored, stream-dwelling juvenile (usually called a *parr*) changes into a more stream-lined, silvery and active pelagic individual referred to as a *smolt,* physiologically adapted for life in ocean waters. The prototype to which the term smolt was first applied is the Atlantic salmon *Salmo salar.* In this species, the gay markings of the stream-dwelling parr,

FISH PHYSIOLOGY, VOL. XIB

aged 1, 2, or several years, are covered with a silvery layer of purines (guanine and hypoxanthine), while the body form becomes slender in relation to that of the parr with a decline in the weight per unit length (condition factor); in addition, the fins—particularly the pectorals and caudal—develop distinctly black margins (Wedemeyer *et al.*, 1980; Gorbman *et al.*, 1982). Details of the smolt transformation vary in different salmonid species; indeed, the salmonids (salmon, trout, char) form a spectrum extending from pink salmon *O. gorbuscha*, which are already silvery as emerging fry and able to enter saltwater when they come out of the gravel, to some species of *Salvelinus* (*alpinus, fontinalis, malma*) that migrate only short distances into the sea for a few months in the summer and the lake char *Salvelinus namaycush*, which is not known to smolt or to enter the sea at all; most species of *Oncorhynchus* and *Salmo* are intermediate between these extremes and spend 1–3 or more years in fresh water before smolting.

Many years ago, Rounsefell (1958) arranged the North American salmonid species in a decreasing order of anadromy (Fig. 1). His histogram is still instructive and appropriate for coastal regions, even though it is now recognized that there are no obligatory ocean migrants and that all species can complete their life cycles in lakes and streams and may become well adapted to a totally freshwater environment (Andrews, 1963; Berg, 1979; Collins, 1975; Peden and Edwards, 1976). The smolting changes associated with anadromy are characteristic of most species of *Oncorhynchus* and *Salmo*, but in the anadromous brook trout *Salvelinus fontinalis*, several of the most typical of them (elevated plasma T4, elevated gill ATPase, and increased osmoregulatory ability in sea water) are absent and smolting in this anadromous species has been questioned (McCormick *et al.*, 1985). Further, it is of interest that some populations of non-anadromous Atlantic salmon do not appear to smolt (Birt and Green, 1986).

The spectrum of anadromous salmonids has been discussed elsewhere with speculation concerning the evolution of the anadromous habit and the likely phylogenetic relationships of the different species (Hoar, 1976). This chapter focuses on the physiology of the typical smolt transformation, its control and modulation by the environment, and the implications of smolting physiology in salmon culture—an important and growing area of aquaculture, where high priority must be given to the production of smolts at as early an age as possible. A related topic considered here is the tendency of male salmonids to

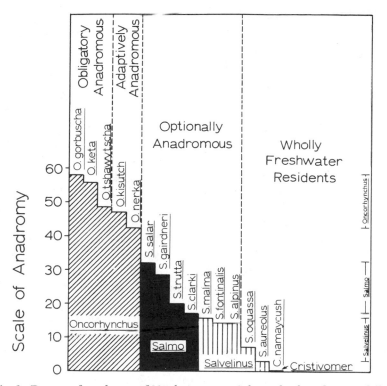

Fig. 1. Degree of anadromy of North American Salmonidae based on six different criteria. [Based on Rounsefell (1958).]

mature at an earlier age than the females. This interesting feature of salmonid development has important implications in salmon culture and, in some spieces, seems to have relevance to the age of smolting. Again the prototype is the Atlantic salmon in its early freshwater stages. Sexually mature and reproductively active male parr (aged 1+ years or older) were carefully investigated by Jones (1959) and others (King *et al.*, 1939) a quarter of a century ago. In several recent papers, Lundqvist (1983) has discussed the physiology of sexual maturation of male parr in relation to the smolt transformation.

The literature on the physiology of smolting is now extensive. There are several excellent reviews and key references may be readily located in their bibliographies (see Folmar and Dickhoff, 1980; Wedemeyer *et al.*, 1980) and in three recent symposia (Bern and Mahnken, 1982; Thorpe *et al.*, 1985; McCormick and Saunders, 1987).

II. THE PHYSIOLOGY OF THE SALMON SMOLT

A. Body Form and Coloration

Several morphological as well as physiological changes occur during smolting (Gorbman et al., 1982; Winans, 1984; Winans and Nishioka, 1987). Superficially, however, both scientist and layman recognize a salmon smolt by its silvery appearance and its relatively slim, streamlined body. Smolt contrast sharply with parr or presmolts, which weigh more per unit length and display brightly colored pigment spots and bars formed by a variety of chromatophores in which the distribution of pigment granules can change cryptically. During smolting, body lipids not only decrease quantitatively but also change qualitatively (Section II,C).

Silvering is due to the synthesis of two purines, guanine and hypoxanthine. The needle-like crystals of these substances are deposited in two distinct skin layers: one directly beneath the scales and the other deep in the dermis adjacent to the muscles. Both the scale layer and the skin layer of purines are present in parr but become thicker in smolts; further, both layers contain guanine with lesser amounts of hypoxanthine, and there is a sharp increase in the ratio of guanine to hypoxanthine during the course of smolting (Markert and Vanstone, 1966; Vanstone and Markert, 1968; Eales, 1969; Hayashi, 1970). The formation of these reflecting layers occurs more rapidly in larger fish and at higher temperatures, but neither temperature nor alteration in photoperiod is required to induce purine deposition; this radical change in appearance seems to be controlled endogenously—a part of the smolt transformation (Johnston and Eales, 1970).

Clearly, the smolt transformation involves changes in purine nitrogen metabolism with an end result (a silvery fish) that is adaptive to the survival of the pelagic postsmolts in the marine habitat. However, the physiological basis and biochemical significance of the phenomenon are not at all clear. Silvering in juvenile salmon has most often been discussed in relation to the increased secretion of thyroid hormones, known to occur in many smolting salmonids (Section II,E). Experimentally, the feeding of thyroid gland tissue or the injection of thyroid stimulating hormone (TSH) or treatment with thyroid hormones will induce purine deposition in salmonids as well as in several other species of fish (Chua and Eales, 1971; Primdas and Eales, 1976). Thyroid treatment can also alter melanophore density, resulting in a lighter-colored fish. Eales (1979) reviews the experimental work on which these conclusions are based. Although a silvery body is advantageous in a pelagic life in deep waters, the physiological basis of purine deposition and the biochemical significance of the altered metabolism remain open to debate.

Guanine and hypoxanthine are insoluble by-products of nitrogen metabolism. They are disposed of in many different ways in various animal groups; their excretion has most often been examined in relation to the quantity of water available for their removal. In some animals with limited supplies of water, these purines and others such as uric acid are permanently stored in the tissues; in other animals they are metabolized to more soluble products. The smolt transformation prepares the juvenile salmon for life in a hyperosmotic environment where excretion of water (and hence substances dissolved in it) must be sharply reduced. These matters are discussed in textbooks of comparative physiology (for example, Hoar, 1983). The suggestion may be made that, in the first instance, purine deposition in salmon skin is metabolically more economic than the several enzymatic oxidations that would be required to turn the purines into more soluble substances such as allantoic acid or urea; in the end, the deposition of this reflecting material may have served as a preadaptation to a successful life in the sea. A number of tantalizing biological questions remain.

B. Growth and Size Relations

Several species of juvenile salmon, trout, and char have been shown to become progressively more tolerant of saltwater as they grow older and larger (Houston, 1961; Conte and Wagner, 1965; Parry, 1966; Wagner et al., 1969; McCormick and Naiman, 1984). However, in salmonids that undergo a distinct smolt transformation, size alone is not the determining factor for a successful life in ocean waters. The transition from a freshwater to a marine life requires a smolt transformation—the coordinated physiological, biochemical and behavioral processes that occur during a very limited time span (Wedemeyer et al., 1980). Nevertheless, the importance of a minimum size, age, or state of growth is emphasized since this appears necessary before smolting can occur.

In an oft-quoted paper, Elson (1957) argued that juvenile Atlantic salmon must attain a minimum length before smolting will occur. After examining data from both sides of the Atlantic, he concluded that parr that reach or exceed 10 cm in length at the end of the growing season are likely to become smolts the next season of smolt descent; otherwise, they remain an additional year or longer in the parr state. Although this rule of minimum length is useful, it has been shown that age is also important; Evans et al. (1984) reported that older fish tend to smolt at a relatively smaller size. Further, rate of growth, rather than length or age, appears to be a significant factor in some species of salmonids (Wagner et al., 1969; Thorpe, 1977, 1986).

Thorpe (1977) and his associates described a bimodality in the growth of Atlantic salmon from Scottish rivers. In populations of hatchery-reared fish, individuals grow at similar rates until the late summer or autumn, when a distinct bimodality develops, which becomes progressively more marked as the season advances (Thorpe et al., 1980). The more rapidly growing fish become 1+year smolts, while the more slowly growing ones remain in fresh water to become 2+year or older smolts. This bimodality is not related to sex or precocious male maturity (Thorpe et al., 1982; Evans et al., 1985; Villarreal and Thorpe, 1985). Genetic factors appear to be responsible for much of the variation in growth rate of Atlantic salmon (Gunnes and Gjedrem, 1978; Refstie and Steine, 1978; Thorpe and Morgan, 1978; Thorpe et al., 1980). Growth bimodality is also the normal pattern in laboratory stocks of Atlantic salmon in Eastern Canada (Bailey et al., 1980; Kristinsson et al., 1985), where the growing conditions are similar to those in Scotland. However, it has not been observed in some more slowly developing laboratory populations of Salmo salar in the Baltic region (Eriksson et al., 1979). Bimodality is difficult to demonstrate in nature, probably because of overlapping year classes and local environmental effects.

Several metabolic differences have been associated with the two modal groups of salmon. In the upper-mode fish, Higgins (1985) recorded faster rates of growth and metabolism characteristic of smolts, while Kristinsson et al. (1985) noted that, in December, the faster-growing, upper-mode fish had plasma thyroxine levels that were three times greater than those in the lower mode fish, indicating basic changes in the physiology at this stage. Effects of thyroid hormones on growth will be noted in a later section (Section II,E). Kristinsson et al. (1985) argue that entrance into the faster-growing phase marks a new developmental stage, which young salmon must reach in the autumn in order to smolt the following spring. This implies that the "decision" to smolt is made during the autumn under a decreasing photoperiod; thus, the short autumn period of rapid growth would be the first of a series of events that culminate in a smolt.

C. Metabolic and Biochemical Changes

Biochemically, the fully transformed smolt is quite a different fish from the parr that gave rise to it. Metabolism and body composition are altered in many ways: the rate of oxygen consumption increases with heightened catabolism of carbohydrate, fat, and protein. There are qualitative as well as quantitative changes in the lipids and blood proteins; gill enzyme systems concerned with ion regulation adjust adaptively. A markedly altered pattern of endocrinology accompanies these many biochemical changes (Section II,E).

Baraduc and Fontaine (1955) reported that the oxidative metabolism of the Atlantic salmon smolt was about 30% higher than that of the parr, even though smolts were in general larger than parr. These authors speculated concerning the possible involvement of thyroid hormone in the altered metabolism. This elevation in metabolism during smolting appears to be characteristic of typical *Salmo* smolts (Power, 1959; Malikova, 1957; Withey and Saunders, 1973; Higgins, 1985). Associated with the changes in oxidative metabolism, well-documented alterations in body composition indicate an elevated catabolism of the body reserves. There are predictable changes in plasma glucose, amino acid nitrogen, and free fatty acids; glycogen and lipid reserves are depleted and there is an elevation in moisture content. Atlantic salmon, steelhead trout (*Salmo gairdneri*), and coho salmon (*Oncorhynchus kisutch*) have been investigated by a number of workers; the literature can be readily traced through several papers and the reviews already cited (Fontaine and Hatey, 1950; Wendt and Saunders, 1973; Farmer *et al.*, 1978; Saunders and Henderson, 1978; Woo *et al.*, 1978; Sheridan *et al.*, 1985a; Sweeting *et al.*, 1985; Sheridan, 1986). Sheridan *et al.* (1985b) showed that the decline in tissue glycogen and fat is due not only to their increased breakdown but also to a greatly reduced synthesis. In spite of the marked loss of fat reserves, Atlantic salmon smolts are more buoyant than parr—probably an important factor in their rapid migration to sea (Saunders, 1965).

Changes in fatty acid composition of smolting Atlantic salmon were noted by Lovern (1934) more than half a century ago. Fatty acids as well as other lipid components have been investigated by many others since that time. Most recently, Sheridan *et al.* (1983, 1985a) have carefully studied the lipids of smolting steelhead trout and the effects of several hormones on lipid metabolism related to smolting in coho salmon (Sheridan, 1986). These studies confirm earlier reports of smolting salmon developing relatively high amounts of long-chain, polyunsaturated fatty acids and relatively low amounts of linoleic acid—values characteristic of typical marine fish, in contrast to freshwater species. Sheridan (1986) emphasizes that the lipid changes of smolting salmon anticipate life in the ocean—a preadaptation for the change in environment—and that growth hormone, prolactin, thyroid hormones, and cortisol are involved (Section II,E). Presumably, these changes in tissue lipids are significant in adaptation to the marine habitat. There are several physiological possibilities: degree of fatty acid saturation, cholesterol/phospholipid ratios, and fluidity of fats have been found important in the control of cell permeability and compensation for temperature change; mechanisms concerned with both permeability and temperature compensation are involved in the marine and adult life of the salmon. The physiology of lipids in relation to cell permeability and temperature compensation are discussed in many places (see, for example, Hoar, 1983; Isaia, 1984).

Several workers have investigated the blood proteins of smolting salmon—especially the multiple hemoglobins. In general, there is an increase in the complexity of the hemoglobin system with additional components added prior to migration (Vanstone et al., 1964; Wilkins, 1968; Giles and Vanstone, 1976). Giles and Vanstone (1976) emphasize that in coho salmon the increased complexity occurs during smolting, with the appearance of two new anodic and four new cathodic components; this new pattern is retained throughout the remainder of life and cannot be induced in parr by manipulation of environmental oxygen, salinity, or temperature. Sullivan et al. (1985) confirm the increase in complexity of the hemoglobins at the time of coho smolting and show that triiodothyronine (T3) treatment accelerates the change at certain environmental salinities while thiourea has the opposite effect. A change in the hemoglobin system also occurs in smolting Atlantic salmon; Koch (1982) likewise argues that the thyroid hormones are probably involved in its expression.

Giles and Randall (1980) investigated the oxygen equilibria of the polymorphic hemoglobins of coho fry and adults, comparing oxygen affinity, Bohr shift, heat of oxygenation, and the influence of adenosine triphosphate. Adult hemoglobins preadapt emigrating smolt to the ocean environment, where several factors create lower oxygen tensions than those encountered by fry and parr (see also Vanstone et al., 1964); in later life, the adult types of hemoglobin may be important in the spawning migration when salmon are liable to experience rapid changes in temperature and pH and may require sudden bursts of swimming activity (Giles and Randall, 1980). The functional significance of the polymorphs found in coho fry may relate to the unloading of oxygen in the tissues of a relatively small animal living in a well-oxygenated environment.

Electrophoretic comparisons of the hemoglobins of several species of salmon in the freshwater and marine stages show that there are consistent changes that probably preadapt the fish for changes in habitat (Vanstone et al., 1964; Hashimoto and Matsuura, 1960; Bradley and Rourke, 1984). In chum salmon (Oncorhynchus gorbuscha), a species that migrates to saltwater during the fry stage without a typical parr–smolt transformation, there is a considerable increase in the proportion of hemoglobin with high oxygen affinity as the fish grow longer and heavier (Hashimoto and Matsuura, 1960). Again, the adaptive nature of these changes in gas transport is indicated. The interesting point emerging from the work on hemoglobins is that smolting salmon of both Salmo and Oncorhynchus species experience an adaptive change in their gas transport proteins at the time of the parr–smolt transformation and, further, that these changes usually occur well in advance of the actual change in habitat (Bradley and Rourke, 1984).

A third important group of biochemical changes relate to the gill enzyme systems and problems of ionic balance. In early life, salmon live in a hypoosmotic environment and are subject to osmotic flooding with water that leaches out essential ions during its removal by the kidneys. In later life, the marine fish lives in a hyperosmotic environment, is subject to loss of water by osmosis, and must make good the deficit by drinking seawater and excreting the salts. In the first case, gill cells actively absorb salt from the fresh waters; in the second instance, gill cells excrete ions, particularly the monovalent ones, sodium and chloride. These matters are considered more fully in the next section; here, the discussion focuses on the enzyme systems responsible for ion balance.

Zaugg and McLain (1970) were the first to demonstrate an increase in gill Na^+,K^+-ATPase activity during the parr–smolt transformation of coho salmon. Subsequently, comparable changes have been reported in Atlantic salmon, steelhead trout, and chinook salmon (Zaugg and McLain, 1972; Zaugg and Wagner, 1973; Giles and Vanstone, 1976; McCartney, 1976; Johnson et al., 1977; Saunders and Henderson, 1978; Ewing and Birks, 1982; Boeuf et al., 1985, and reviews cited). The smolting change in gill ATPases begins well in advance of migration both in Salmo and Oncorhynchus; it peaks during the migratory phase and entry into seawater. In seawater, it rises somewhat after 4–5 days and stabilizes at the higher level, but if smolts remain in fresh water beyond the normal time of migration, the gill enzyme activity declines to the freshwater level (Section IV,A). Nonsmolting and nonmigratory strains of brown trout (S. *trutta*) and rainbow trout (S. *gairneri*) do not show a seasonal increase in the gill Na^+,K^+-ATPases (Boeuf and Harache, 1982); moreover, neither the anadromous nor the nonanadromous forms of the brook trout (*Salvelinus fontinalis*) shows a seasonal change in gill ATPase (McCormick *et al.*, 1985). The contrasting picture of ATPase activity in freshwater parr and migrating salmon smolts is similar to that seen in comparisons of typical freshwater and marine teleosts. It is another of the major adaptations found in the typical parr–smolt transformation.

The physiology of ion transport and the role of the gill ATPases will not be detailed here. These topics were reviewed in Volume X,B of this series (Hoar and Randall, 1984); see in particular the chapters by Isaia (1984) and de Renzis and Bornancin (1084). Hormones involved are also considered in Volume X,B (Rankin and Bolis, 1984) and will be noted in Section II,E of this chapter.

The oxidative metabolism of gill tissues is high and may amount to as much as 7% of the fish's total oxygen consumption (Mommsen, 1984). Thus, several enzyme systems other than the Na^+,K^+-ATPases may be expected to reflect the changes in metabolic demands when freshwater salmon migrate into the ocean. The higher levels of succinic dehydrogenase (SDH) and cytochrome *c* oxidase found in smolting juvenile salmon reflect these altered metabolic demands (Chernitsky, 1980, 1986; Blake *et al.*, 1984; Langdon and Thorpe, 1984, 1985).

Carbonic anhydrase is another enzyme of importance in gill physiology. Its role in CO_2, H^+, HCO_3^-/Cl^- exchange mechanisms, and ammonia movement across fish gills is reviewed by Randall and Daxboeck (1984). The higher values of this ion reported in smolts adapted to seawater are probably related to the problems of gas exchange, ion, and acid–base regulation in waters of higher salinity (Milne and Randall, 1976; Dimberg *et al.*, 1981; Zbanyszek and Smith, 1984).

D. Osmotic and Ionic Regulation

Although the early developmental stages of all salmonids take place in fresh water, most species spend longer or shorter periods of their actively growing life in the sea. In the pink (*Oncorhynchus gorbuscha*) and chum salmon (*O. keta*), the capacity to hypoosmoregulate (excrete salts in the hyperosmotic marine environment and maintain the plasma electrolytes at about one-third seawater concentration) develops in the alevin stages (Weisbart, 1968; Kashiwagi and Sato,

1969); in some populations of chinook (*O. tshawytscha*) it occurs in fry or fingerlings (Clarke and Shelbourne, 1985). In other species of *Oncorhynchus* and in the *Salmo* and *Salvelinus* species, full hypoosmoregulatory ability is attained after a variable period of freshwater residency (usually 1 year or longer), and in *Salmo* and *Oncorhynchus* it requires the parr-smolt transformation. Studies of salinity relations of juvenile freshwater and downstream migrant salmon are now voluminous (Folmar and Dickhoff, 1980; Wedemeyer *et al.*, 1980; McCormick and Saunders, 1987).

The general physiological problems of an anadromous fish were stated in the previous section. The hypoosmotic freshwater habitat requires the excretion of large amounts of water and the acquisition of salts; the hyperosmotic marine environment demands the rigid conservation of water, the drinking of seawater, and the excretion of salt to provide fresh water for the tissues. Gills, opercular epithelia, kidneys, urinary bladder, and intestinal epithelia play active roles in this regulation. The glomerular filtration rate is altered during smolting [see Section II,D,2 and compare Holmes and Stainer (1966) with Eddy and Talbot (1985)]; the rectal and hindgut fluids of the marine salmon are strongly hypertonic and are responsible for the removal of the divalent ions acquired through body surfaces and by drinking sea water; transport of monovalent ions across the gut wall is increased to effect a concentration gradient that will move water from the gut into the body tissues; the gills excrete the excess monovalent ions (sodium and chloride) acquired from the gut.

Plasma electrolytes remain relatively constant throughout the freshwater life of juvenile salmonids. Folmar and Dickhoff (1980) review the literature and note values of 133–155 meq/l for Na^+, 3–6 meq/l for K^+, and 111–135 meq/l for Cl^-. Although some workers have found a decline in the plasma and tissue Cl^- during the presmolt stages (Fontaine, 1951; Kubo, 1955; Houston and Threadgold, 1963), more recent studies emphasize a relative constancy throughout the entire juvenile freshwater life (Parry, 1966; Conte *et al.*, 1966; Miles and Smith, 1968; Saunders and Henderson, 1970).

Even though plasma and tissue electrolytes remain constant throughout the parr stages, experiments have shown that salmonids differ in their capacities to deal with electrolytes when transferred to saline waters (Fig. 2). Although all species can tolerate mild changes in ambient salinity, the Oncorhynchids are in general more resistant to saltwater than species of *Salmo*, while the genus *Salvelinus* is the least resistant of the three genera. This is in line with Rounsefell's degrees of anadromy (Fig. 1), but the order of the species shown by

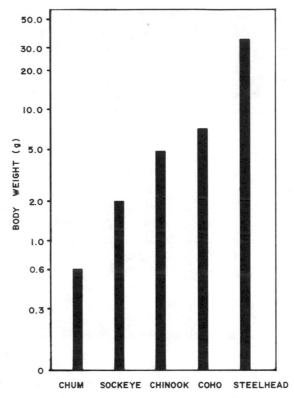

Fig. 2. Mean body weight at which samples of five salmonid species have demonstrated optimal hypoosmoregulatory capacity typical of smolts. [From Clarke (1982).]

Rounsefell (1958) is not in line with present-day studies (Hoar, 1976; Boeuf and Harache, 1982, 1984; Clarke, 1982). The important points to emerge from the many studies of salinity resistance of juvenile freshwater salmonids are:

1. All species can tolerate mild changes in ambient salinity.
2. Larger individuals have a greater tolerance than smaller members of a population (Houston, 1961; Conte and Wagner, 1965; Clarke and Blackburn, 1978; Jackson, 1981).
3. Salinity tolerance changes seasonally and is greater during the spring and early summer than in the autumn and winter (Conte and Wagner, 1965; Boeuf et al., 1978; Lasserre et al., 1978).
4. When given a choice between fresh water and saltwater, juvenile salmonids may show a distinct preference for one or the other. Saltwater preference is strongest at the peak of the smolt

stage when the juveniles migrate, but like salinity tolerance, preference varies with the species, season, and status of smolt development (Baggerman, 1960a,b; McInerney, 1964; Otto and McInerney, 1970; Iwata *et al.*, 1985, 1986).

5. Transfer of young salmon to more saline waters, whether in the parr or smolt stage, is followed by transient changes in tissue electrolytes, after which the plasma values stabilize at or near the previous levels (*Oncornhychus* and *Salmo*), although the distribution of ions in the tissues may be different in the two environments (Eddy and Bath, 1979); in *Salvelinus alpinus* the stable values in sea water are about 25% higher than in fresh water (Gordon, 1957). In chum salmon, Black (1951) found that the adjustment period lasted 12 h; Miles and Smith (1968) reported 36 h for coho salmon; Houston (1959) gave 8–170 h for the adjustment phase in steelhead trout; Prunet and Boeuf (1985) found that Atlantic salmon smolts adjust to seawater in 4 days, while nonsmolting rainbow trout require 2 weeks; and Leray *et al.* (1981) and Jackson (1981) found a crisis period of 30–40 h in rainbow trout, with a new steady state established in 4–5 days. It is evident that this adjustment period varies with species, size, stage of smolting, temperature, and photo-period. Literature reviewed in later sections shows that the adjustment period is marked by changes in the functional mor-phology of the osmoregulatory organs and in secretion of hor-mones. Folmar and Dickhoff (1980) and Wedemeyer *et al.* (1980) summarize the extensive literature and provide bibliog-raphies.

6. A capacity to acclimate is present in all species and stages of freshwater salmonids, and a gradual transfer to more saline waters is more successful than a sudden exposure even in spe-cies such as the chum salmon that are able to tolerate seawater as fry (Iwata and Komatsu, 1984).

7. Further, a smolt transformation is critical to successful growth in full seawater and, even though parr of some species (Section V) can be acclimated to seawater during favorable periods, they fail to grow normally and become "stunted" or die (Otto, 1971; Clarke and Nagahama, 1977; Bern, 1978; Folmar *et al.*, 1982a).

8. The smolt stage itself is brief and smolts that do not enter seawater during this narrow "window" of time (Boeuf and Harache, 1982) revert to the parr condition (Houston, 1961; Koch, 1982; Conte and Wagner, 1965, and reviews cited). Sev-

eral workers have noted that this "narrow window" corresponds to the period of maximum gill enzyme activity.

9. Environmental factors, particularly temperature and photoperiod, modulate the time of smolting, desmolting, and the salinity preference and tolerance changes (Section IV).

The studies of the physiology of osmoregulation in marine fishes focus on the oral ingestion of seawater and the ionic extrusion mechanisms. Research on euryhaline species such as *Fundulus*, on the catadromous eels *Anguilla*, and on the anadromous salmonids have shown that transfer from fresh water to seawater activates these two processes. In rainbow trout, there is a rapid response of the drinking reflex, which reaches a peak within a few hours and then declines to a lower constant level, while the salt extrusion mechanisms are activated more slowly over a period of several days (Shehadeh and Gordon, 1969; Potts *et al.*, 1970; Bath and Eddy, 1979; Fig. 4 in Evans, 1984). The organs and tissues involved in these adjustments are the gills, opercular epithelia, kidneys, urinary bladder, intestine, and possibly the integument (Parry, 1966; Shehadeh and Gordon, 1969; Loretz *et al.*, 1982).

Studies of the euryhaline teleosts have shown the nature of the physiological changes required during the parr–smolt transformation. However, the research is spotty in respect to the temporal development of the changes during smolting. Present findings suggest that these changes occur well in advance of the smolt migration and preadapt the young salmon for life in a hyperosmotic habitat. In summarizing the physiology, the desirability of more studies of the actual development of the hyposmoregulatory mechanisms during smolting is noted.

1. GILLS

The water permeability of gills has been measured in several euryhaline fishes (Platichthys, Fundulus, Anguilla, Salmo). In general, diffusional permeability of water is sharply reduced in the marine habitat (Isaia *et al.*, 1979; Evans, 1984; Isaia, 1984), but there do not seem to be permeability measurements that might demonstrate gradual changes during the parr–smolt transformation and preadaptation to the marine habitat.

The chloride cells of the gills and the opercular epithelia have been of major interest to salmonid physiologists. After many years of controversy, biologists now recognize the central role of these mitochondria-rich cells in processes of salt extrusion (Conte, 1969; Foskett and Scheffey, 1982; Zadunaisky, 1984). In general, these cells are more numerous and larger in the marine environment (Evans, 1984). In smolting salmonids, their proliferation commences well in advance of entrance into seawater. In studies of Atlantic salmon, Langdon and Thorpe (1984) describe seasonal changes in abundance and size of the gill chloride cells coinciding with variations in gill enzyme activity; a springtime peak occurred in both parr and smolts but was much more marked in the smolts. Loretz *et al.* (1982) studied samples of opercular epithelia of coho salmon at 2-week intervals from February to May. Only scattered chloride cells were found until late May when a three-fold increase in cells was noted. These studies suggest chloride-cell proliferation in the opercular epithelium during smolting. Several other studies record increased density and size of chloride cells in smolts (Threadgold and Houston, 1961; Chernitsky, 1980; Burton and Idler, 1984), but studies of the time course of their development from parr to smolt stages are few (Richman *et al.*, 1987).

2. KIDNEY

There are few studies of kidney function during the parr–smolt transformation. The changes that must occur are largely inferred from comparison of freshwater and marine teleosts. In general, the freshwater teleosts produce a copious, dilute (hypotonic) urine, while the marine teleosts excrete a scant volume of isotonic or slightly hypotonic urine (see Tables VII, IX, and X in Hickman and Trump, 1969). Comparative studies of euryhaline species in fresh water and saltwater show that they make the expected adjustments in physiological mechanisms when moved from one habitat to another (see Table XII in Hickman and Trump, 1969). It must be noted, however, that urine output in euryhaline teleosts is a resultant of adjustments in two basic renal mechanisms: glomerular filtration and tubular reabsorption (Fig. 47b in Hickman and Trump, 1969). In fresh water, glomerular filtration is high and tubular reabsorption is low, while in seawater the filtration rate is greatly reduced and the reabsorption accelerated.

Although salmon in fresh water and saltwater behave like other euryhaline fishes, the two basic mechanisms have not been carefully assessed during the parr–smolt transformation. In one of the few relevant studies, Holmes and Stainer (1966) found a reduction of almost 50% in the urine flow of *S. gairdneri* smolts in comparison with pre- and postsmolts (all stages studied in fresh water). Values for presmolts and postsmolts were essentially the same (near 4.5 ml kg^{-1} h^{-1}). At the same time, inulin clearance techniques gave a glomerular filtration rate (GFR) that was about 50% lower in the smolts than in the pre- and postsmolts. These results suggest a basic change in kidney function (GFR) during the parr–smolt transformation before migration; this could be preadaptive for a successful life in saltwater, where urine output is sharply reduced (Potts et al., 1970).

In another relevant paper, Eddy and Talbot (1985) obtained quite different results with S. salar. These investigators reported a sharp increase in urine production coinciding with silvering indicative of the smolt stage; values for parr and presmolts were about 50% lower than those of smolts; again, all values were measured in fresh water ($1–1.5$ mg kg^{-1} h^{-1} in presmolts and $2.5–3$ ml kg^{-1} h^{-1} in silvery "smolts"). Eddy and Talbot (1985) note that their Atlantic salmon smolts were much smaller (25–60 g) than the trout used by Holmes and Stainer (1966), which weighed 150–200 g and that the smaller fish were more difficult to catheterize. They also suggest that "handling diuresis" may have affected the steelhead data and note temperature differences between the two sets of experiments and the problems of accurately assessing the smolt stage. Clearly, more studies will be required to provide an accurate picture of changes in renal physiology during smolting. Structural as well as physiological differences have been described in comparisons of *Salmo gairdneri* adapted to fresh water and to seawater (Henderson et al., 1978; Colville et al., 1983), and further investigations of renal physiology during smolting are likely to prove interesting.

3. URINARY BLADDER

In many teleosts, the urinary bladder is an organ of significance in electrolyte balance (Lahlou and Fossat, 1971; Hirano et al., 1973). In some species the organ is concerned with both water and salt transfers, but in the salmonids it may play only a minor role. In S. gairdneri, Hirano et al. (1973) found that the urinary bladder was osmotically impermeable in both fresh water and saltwater, although there seemed to be an active uptake of Na and Cl ions. In a study of fresh- and saltwater-adapted yearling coho salmon, Loretz et al. (1982) reported reduced electrolyte absorption in seawater. It is suggested that Na and Cl absorption may be necessary in freshwater salmonids to balance ion losses, but in seawater reabsorption is not required and may even be detrimental. Although it appears that the urinary bladder plays a relatively minor role in the water and ion balance of salmonids, its importance has not really been assessed in the smolt transformation.

4. INTESTINE

The intestine of the euryhaline fish is also an organ of water and ion regulation, greatly increasing the fluid absorption in the osmotically desiccating marine habitat (Shehadeh and Gordon, 1969; Lahlou et al., 1975; Morley et al., 1981). Collie and Bern (1982) used in vitro intestinal sac preparations to study fluid absorption. Fluid absorption was increased in smolting coho salmon prior to entrance into seawater; the higher level prevailed for several months before returning to the lower level. Thus, increased fluid absorption from the intestine seems to preadapt the young salmon for life in the ocean and occurs in concert with alterations in renal and branchial osmoregulatory mechanisms. The timing suggests a relationship with the thyroxin surge (Section II,E; Collie and Bern, 1982; Loretz et al., 1982). The nutrient-absorbing role of the intestine also changes during smolting. An increased proline influx has been recorded and suggests that the higher nutritional demands of rapid growth initiated during smolting are met in part by an increased absorption efficiency (Collie, 1985). Experimentally, cortisol and growth hormone have been shown to increase intestinal proline absorption in coho and may regulate the process during smolting (Collie, 1985; Collie and Stevens, 1985).

E. Hormones and Smolting

The salmonids, like many other temperate-zone animals, have a seasonally changing physiology that is manifest in cycles of growth, precisely timed migrations, and seaons of reproduction geared to the most favorable seasons for the birth and development of the young. A working hypothesis developed in an earlier paper stated (Hoar, 1965):

. . . That several species of salmonids undergo physiological and behavioural cycles which, each springtime, preadapt them for life in the ocean; if they do not reach the ocean, the cyle is reversed and the physiology appropriate to life in fresh water again appears. Under natural conditions, changing photoperiods trigger the cycle at the appropriate season, but the cycle is endogenous and does not disappear under constant conditions of illumination. The theory is that this is a general phenomenon in the salmonids, and some evidence for it will probably be found in all species at all stages in their development. . . . This cyclical physiology of the salmon . . . has been very susceptible to modificaton through genetical processes. The smolt transformation, which is an obligatory part of the life cycle of Atlantic salmon, steelhead trout, and coho, is suppressed or lost in species such as the pinks and chums.

The hypothesis is still relevant. In most salmonids, the smolt transformation is tightly timed by photoperiod, and by lunar cycles (*Oncorhynchus*) or flooding streams (*Salmo*), to occur in the springtime, but the amago salmon (*O. rhodurus*) and some populations of chinook salmon (*O. tshawytscha*) undergo a typical smolt transformation in the autumn (Ewing *et al.*, 1979; Nagahama *et al.*, 1982; Nagahama, 1985).

Seasonal changes in vertebrate physiology are timed and regulated by a neuroendocrine system. Environmental cues act through the peripheral sense organs and brain to trigger secretory activity in the hypothalamus; hypothalamic releasing hormones regulate secretion of the hormones of the anterior pituitary gland, which in turn controls endocrine organs such as the thyroid and the interrenal glands through its trophic hormones or secretes hormones that act directly on target tissues (prolactin, growth hormone).

The parr—smolt transformation occurs in association with a general surge in endocrine activity that can be detected in most, if not all, of the endocrine organs. Hormones most thoroughly studied and considered most likely to be involved in the transformation are the thyroid hormones, prolactin, growth hormone, and the corticosteroids, but changes have also been studied in other endocrine factors, particularly the gonadal steroids (Hunt and Eales, 1979; Sower *et al.*, 1984; Patiño and Schreck, 1986; Ikuta *et al.*, 1987), and the secretions of the Stannius corpuscles and the urophysis (Bern, 1978; Aida *et al.*, 1980; Nishioka *et al.*, 1982) (Fig. 3).

1. THYROID HORMONES

A dramatic increase in thyroid activity is generally recognized in smolting salmon. This increase was first reported half a century ago in Atlantic salmon (Hoar, 1939). The histophysiological evidence presented at that time was later confirmed for *S. salar* (reviews by Fontaine, 1954, 1975) and extended to other smolting salmonids by Robertson (1948), Hoar and Bell (1950) and others. Radioiodine techniques gave confirmatory results (Eales, 1963, 1965). The advent of radioimmunoassay (RIA) techniques has permitted measurements of the time course of the smolting surge in plasma thyroxine (T4) and correlated these changes with gill enzyme activity in both *Salmo* and *Oncorhynchus* (Folmar and Dickhoff, 1980, 1981; Boeuf and Prunet, 1985; Dickhoff *et al.*, 1985) (Fig. 4); with lunar cycles in *Oncorhynchus* and *S. gairdneri* (Grau *et al.*, 1981) (Fig. 5 and Section IVD); with stream flow-rate in *Salmo* (Youngson and Simpson, 1984; Lin *et al.*, 1985a; Youngson *et al.*, 1986); and with the migration time and subsequent yield (Fig. 6). Although T4 levels change dramatically, the triiodothyronine (T3) levels are quite stable (Fig. 4). The literature is now extensive for the different species of smolting salmonids, and

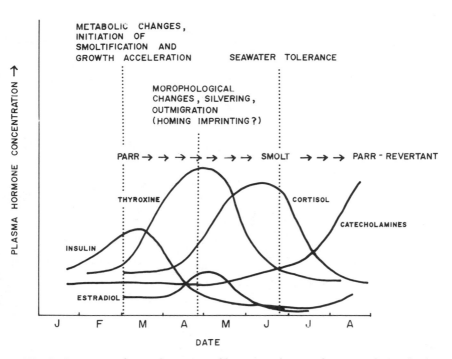

Fig. 3. Summary of typical patterns of hormone changes during smolting of coho salmon. (Graph presented at Tenth International Symposium of Comparative Endocrinology, Copper Mountain, Colorado, 1985. Courtesy of W. W. Dickhoff.)

only selected references are noted here (Nishikawa *et al.*, 1979; Eales, 1979; Folmar and Dickhoff, 1980; Dickhoff *et al.*, 1982; Specker and Richman, 1984; Specker *et al.*, 1984; Grau *et al.*, 1985c; Lin *et al.*, 1985a,b; Dickhoff and Sullivan, 1987).

Thus, there is ample evidence that thyroid activity is usually elevated at the time of the typical parr–smolt transformation. However, the evidence for a triggering role of the thyroid hormones in the onset of smolting is absent. Rather, it appears that thyroid hormones intensify several physiological and behavioral changes of smolting but that these changes are triggered by other hormones or occur endogenously (Hoar, 1976). Two groups of studies are relevant: those concerned with the general sensitivity of the thyroid to many different stimuli (what may be termed its "lability"), and the experimental attempts to alter the time of the parr–smolt transformation with thyroid hormones, thyroid-stimulating hormone (TSH), or antithyroid compounds such as thiourea.

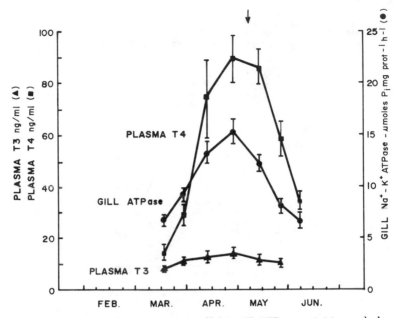

Fig. 4. Relationships between mean gill (Na+K)-ATPase activities and plasma T4 and T3 concentrations of coho salmon in fresh water at the Sandy Hatchery. Bars indicate ± SEM, n = 10 samples of three fish; arrow indicates hatchery release data. [From Folmar and Dickhoff (1981).]

First then, many studies have related changes in thyroid activity to altered environmental conditions: season, temperature, and light (Swift, 1959; Eales, 1965, 1979; White and Henderson, 1977; Osborne *et al.*, 1978; Donaldson *et al.*, 1979; Eales *et al.*, 1982; Grau *et al.*, 1982; Leatherland, 1982; Specker *et al.*, 1984). Thyroid activity is also affected by the level of iodine in the ambient water (Black and Simpson, 1974; Sonstegard and Leatherland, 1976), by toxic substances (Moccia *et al.*, 1977; Leatherland and Sonstegard, 1978, 1981; Leatherland, 1982), and by salinity (Specker and Richman, 1984; Nishioka *et al.*, 1985a; Specker and Kobuke, 1987). Under certain conditions, changes in water flow (Youngson and Simpson, 1984; Youngson *et al.*, 1986) and the transfer from one freshwater environment to another freshwater environment may alter thyroid activity (Grau *et al.*, 1985b; Lin *et al.*, 1985b; Nishioka *et al.*, 1985a; Virtanen and Soivio, 1985). A relationship between nutritional status and thyroid activity has been noted (Leatherland *et al.*, 1977; Donaldson *et al.*, 1979; Eales, 1979; Eales and Shostak, 1985; Milne *et al.*, 1979) as

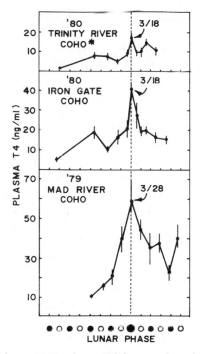

Fig. 5. Patterns of plasma T4 in three California coho salmon stocks plotted on a lunar calendar: ●, new moon; ○, full moon; asterisk, Trinity stock raised at Mad River Hatchery. [From Grau *et al.* (1981). Copyright 1981 by Am. Assoc. Adv. Sci.]

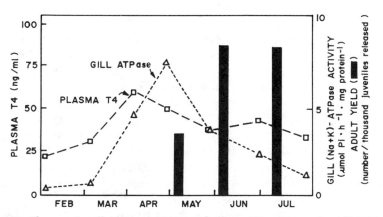

Fig. 6. Changes in gill ATPase activity and plasma T4 concentration of 1979-brood coho salmon reared at the Corvallis Fish Research Laboratory in relation to yield (bars) from groups of juveniles released from Big Creek Hatchery in May, June, and July. [From Ewing *et al.* (1985).]

well as changes in relation to the reproductive conditions of the fish (Sage, 1973; White and Henderson, 1977; Eales, 1979; Hunt and Eales, 1979; Sower *et al.*, 1984) and rearing density (Leatherland and Cho, 1985). Obviously, caution is required in the evaluation of thyroid activity in relation to the physiology of smolting.

Second, experiments designed to test the hypothesis of a causal relation between thyroid activity and the initiation of the parr–smolt transformation have provided scant evidence for this theory. Although body levels of thyroid hormone can be altered by injections of TSH, by feeding, injecting, or immersing fish in thyroid hormones, or by the use of some of the goitrogens such as thiourea (Chan and Eales, 1976; Milne and Leatherland, 1978, 1980; Eales, 1979, 1981; Eales and Shostak, 1985; Miwa and Inui, 1983; Specker and Schreck, 1984; Lin *et al.*, 1985a; Nishioka *et al.*, 1987), the development of typical smolts at seasons when smolting does not normally occur has not been achieved. Nonetheless, some components of the behavior of a typical smolt and some of the physiological changes characteristic of smolting can be induced. The most relevant of the studies seem to be those associated with salinity preference, general activity, growth, silvering, and some of the effects on metabolism.

Baggerman (1960a,b, 1963) induced changes in salinity preferences of juvenile Pacific salmon by the use of TSH, thyroid hormones, and thiourea. These changes were consistent with the theory that thyroid hormone is concerned with the change from a freshwater preference in the parr to a saltwater preference in a smolt. Increased swimming activity and other behavioral changes associated with downstream migration have been recorded in juvenile salmonids following thyroid treatment (Godin *et al.*, 1974), but it should be noted that these changes are not specific to the salmonids and have been similarly induced in some other teleosts—the stickleback *Gasterosteus aculeatus*, for example (Baggerman, 1962; Woodhead, 1970; Katz and Katz, 1978). Not all of the findings are consistent with Baggerman's hypothesis; Birks *et al.* (1985) injected T4 and/or thiourea into steelhead trout near the time of downstream migration and concluded that thyroid stimulation *reduced* the migration tendency. These workers argue that high thyroid activity antagonizes seaward migration and note that downstream migration of juvenile salmonids occurs at the time of falling thyroid levels.

Donaldson *et al.* (1979) review many experiments showing that dietary supplements of thyroid or thyroid hormones may stimulate body growth. Supplements of T3 are particularly active; these effects of thyroid treatment are much more marked when combined with pituitary growth hormone (Barrington *et al.*, 1961; Massey and Smith, 1968; Donaldson *et al.*, 1979; Eales, 1979; McBride *et al.*, 1982; Miwa and Inui, 1985). Donaldson *et al.* (1979), in their critical review, conclude that although thyroid hormone is probably necessary for normal growth, this hormone alone does not seem to stimulate growth but exerts its effect in combination with growth hormone or favorable nutrition.

Increased purine deposition with body silvering frequently follows thyroid treatment. This feature is also more marked when the thyroid hormone is combined with growth hormone; it too is not specific to nonsmolting salmonids. Moreover, the dark pigmentation of the dorsal and caudal (*Salmo*) or pelvic (*Oncorhynchus*) fins seen in typical smolts has not been induced in parr by thyroid treatment (Robertson, 1949; Chua and Eales, 1971; Ikuta *et al.*, 1985; Miwa and Inui, 1983, 1985). Fin darkening is probably caused by a pituitary factor. Komourdjian *et al.* (1976a) report darkening of fin margins and yellowing of the operculae and fin surfaces of Atlantic salmon parr injected with porcine growth hormone (GH); Langdon *et al.* (1984) noted fin darkening in juvenile Atlantic salmon treated with adrenocorticotropic hormone (ACTH). However, it should be remembered that the pituitary preparations used may have contained small amounts of different pituitary hormones. Moreover, ACTH and the melanophore-stimulating hormone (MSH) are biochemically related, having a common core of seven amino acids. Obviously, further studies of the endocrinology of fin pigmentation of smolts is required.

Thyroid hormone is involved in several aspects of intermediary metabolism. Eales (1979) reviews the literature concerning its role in fish metabolism; Donaldson *et al.* (1979) discuss its relationship to growth; Folmar and Dickhoff (1980) comment on the possible regulatory action of the thyroid hormones in smolting. Effects of growth (protein nitrogen metabolism) and purine nitrogen metabolism have been noted in previous sections of this chapter. Carbohydrate metabolism is also altered during the parr–smolt transformation (Section II,C). Lower glycogen and blood glucose appear to be characteristic of smolts, but parr treated with thyroid preparations have failed to show consistent changes in this regard. In a recent paper, Miwa and Inui (1983) treated amago salmon (*O. rhodurus*) with T4 and/or thiourea; the changes in carbohydrate metabolism characteristic of smolting salmon did not occur. The earlier literature is summarized in the reviews already cited.

Changes in the lipid metabolism of juvenile Atlantic salmon during the smolt transformation were reported by Lovern (1934) more than half a century ago. Subsequent research has confirmed Lovern's findings and described several other changes in the lipid metabolism of both smolting *Salmo* and *Oncorhynchus* (Sheridan *et al.*, 1983, 1985a,b; Sheridan, 1985). Treatment with thyroid hormone has been shown to decrease stored lipid in *Salvelinus* as well as *Salmo* and *Oncorhynchus* (Narayansingh and Eales, 1975; Sheridan, 1986), but evidently not in all teleost species (Eales, 1979). In a recent paper, Sheridan (1986) reports that treatment of coho salmon parr with T4 mobilized lipids in both liver and mesenteric fat with an accompanying increase in lipase activity; the plasma nonesterified fatty acids increased; lipase activity also increased in dark muscle. Smolts treated in a similar manner with T4 did not show these changes.

Finally, a very critical question concerns the possible role of thyroid hormones in salinity tolerance. Do thyroid hormones improve the resistance of salmon parr to saline waters and enable them to thrive in the sea? Most experiments have given negative answers, although there is suggestive evidence of a greater requirement for thyroid hormones in saltwater. Eales (1979) reviewed the experimental work and found negative or conflicting evidence of a role for the thyroid in osmotic and ionic regulation; experiments are difficult to interpret because of different levels of iodine in the ambient waters. More recent research has not altered these conclusions (Folmar and Dickhoff, 1979, 1980; Miwa and Inui, 1983; Specker and Richman, 1984; Specker and Kobuke, 1987). Miwa and Inui (1985) report that T4 alone does not improve sea water tolerance, but growth hormone alone or in combination with T4 increases the tolerance and induces a significant elevation in the gill Na^+, K^+-ATPase in amago salmon. Specker and Richman (1984) found that bovine TSH induced a thyroidal response during smolting (as measured by plasma T4); this response was greater during the early period of smolting, while the increase in T4 titers appeared sooner and was of greater duration in fish transferred to sea water. These results suggest an effect of salinity on the kinetics of T4 entry into and exiting from the bloodstream.

In summary, it is now clear that thyroid hormones do not trigger the parr–smolt transformation. At best, treatment of salmon parr with thyroid preparations creates "pseudo-smolts" (Eales, 1979). The thyroid seems to play an important role in enhancing smolting characteristics that are regulated endogenously or by other hormonal factors.

2. GROWTH HORMONE AND PROLACTIN

Growth hormone (GH), also called somatotropin (STH), is growth-promoting in teleost fishes, while prolactin (PRL), which is closely related to GH biochemically, serves several different functions. In the present context, the most relevant functions of prolactin are the conservation of sodium and the decrease of gill water permeability of euryhaline fishes in fresh water (Clarke and Bern, 1980; Loretz and Bern, 1982; Hirano, 1986). These recognized functions of GH and PRL suggest that they may both be important in the physiology of smolting. That the salmonids differ somewhat from euryhaline fish such as *Fundulus heteroclitus* was suggested many years ago by Smith (1956), who reported increased salinity tolerance in *Salmo trutta fario* injected with mammalian anterior pituitary powder and attributed this to growth hormone. A few years later, J. E. McInerney found that injections of purified STH (Nordic) induced a salinity preference change in young coho salmon (review by Hoar, 1966). Since that time, it has been established that in several salmonids, GH not only promotes growth but *increases survival in seawater*. There are several lines of evidence for the actions of these two closely related pituitary factors in smolting salmon: the effects of hypophysectomy, studies of the cytology and ultrastructure of the pituitary, hormone assays, and injections of GH or PRL.

Most hypophysectomized euryhaline fishes die if retained in fresh water. This is not true of several salmonids tested, although there is a decline in the sodium levels in fresh water (Komourdjian and Idler, 1977; Nishioka *et al.*, 1985b).

There have been several important cytological studies of the pituitary gland in relation to smolting. From these it is evident that both the growth-hormone-secreting and the prolactin-secreting cells are activated during the parr–smolt transformation. Among the more recent studies, Leatherland *et al.* (1974) reported greater activity of both GH and PRL pituitary cells in kokanee salmon smolts (*O. keta*). Likewise, Nishioka *et al.* (1982), in an ultrastructural study of various endocrine glands, noted that the PRL cells were active in coho salmon during the parr and smolt stages but substantially more active in freshwater smolts than seawater smolts; this argues for an involvement of PRL in the salmon smolt in fresh water. Nishioka *et al.* (1982) also found that GH cells were active in both parr and smolts but more active in the smolts *both in fresh water and in seawater*. Further cytological evidence (light and electron microscopy) of GH function in the marine environment comes from Nagahama *et al.* (1977), who noted greater GH activity in yearling coho parr when transferred to seawater. In contrast, prolactin cells of the parr were markedly more active in fresh water than in the seawater fish; when parr were transferred from saltwater to fresh water, however, a stimulation of the PRL cells was indicated—findings that again argue for a role of PRL in the freshwater coho salmon. Another paper by these workers provides confirmatory evidence (Clarke and Nagahama, 1977). Nagahama (1985) also finds that prolactin is involved in the adaptation of the amago salmon to fresh water.

Data on plasma PRL levels (RIA analyses) support the cytological findings. Prunet and Boeuf (1985) compared nonsmolting rainbow trout with smolting Atlantic salmon. Plasma prolactin levels declined significantly after transfer of the trout from fresh water to seawater, although the plasma osmotic pressure increased. In contrast, smolted Atlantic salmon adapted quickly to seawater and had similar prolactin levels in both environments. Prunet et al. (1985) measured plasma and pituitary prolactin levels in rainbow trout adapted to different salinities. Transfer from fresh water to sea water decreased plasma PRL; although a transient rise in pituitary PRL followed the transfer, the values after three weeks were lower than in the freshwater controls. The reverse transfer (SW to FW) induced, within one day, a rise in the plasma PRL. These findings indicate an important role for PRL in the freshwater adaptation of sedentary rainbow trout. Hirano's (1986) data for chum salmon are confirmatory.

Finally, there are many reports of the effects of GH and PRL injections into salmonids. Growth stimulation has been reported to follow GH injections into both intact (Higgs et al., 1976; Komourdjian et al., 1976a; Markert et al., 1977) and hypophysectomized salmonids (Komourdjian et al., 1978). Increased seawater tolerance was also reported in Komourdjian's experiments, which were carried out on rainbow trout and Atlantic salmon. Clarke et al. (1977) found that GH from either teleost or mammalian sources lowered plasma sodium in underyearling sockeye salmon in both fresh water and seawater; survival was high in both environments. Miwa and Inui (1985) reported that ovine growth hormone increased significantly the survival of amago salmon parr transferred to 27‰ seawater. Bolton et al. (1987a) studied the effects of ovine or chum salmon GH on the plasma sodium, calcium, and magnesium levels of S. gairdneri transferred for 24 h to 80% sea water (3 i.p. injections of hormone at 3 day intervals). They conclude that the seawater-adapting actions of GH are specific to the hormone and are not consequent to an increase in size.

In summary, the evidence from several lines of study is consistent in showing increased activity of both GH and PRL pituitary cells at the time of smolting. Further, prolactin plays a special role in the osmoregulatory physiology of the smolt in fresh water, while GH is the important factor in hydromineral regulation in the sea (review by Hirano, 1986).

3. THE INTERRENAL GLAND

Many years ago, M. Fontaine and his co-workers presented histophysiological evidence of increased interrenal activity in smolting
Atlantic salmon (Fontaine and Olivereau, 1957, 1959; Olivereau,
1962, 1975). Their findings were subsequently confirmed in several
species of *Salmo* and in *Oncorhynchus* by further light microscopy
(review by Specker, 1982), by ultrastructure (Nishioka *et al.*, 1982),
and by biochemical studies of the plasma corticosteroids (Fontaine
and Hatey, 1954; Specker and Schreck, 1982). The adrenocorticotropic hormone (ACTH) producing cells of the pituitary are also more
active in the smolt that in the parr (Olivereau, 1975; Nishioka *et al.*,
1982).

Corticosteriods, especially cortisol, are essential to the life of a
teleost fish. They function both in water-electrolyte homeostasis and
in intermediary metabolism (Chester-Jones *et al.*, 1969; Chester-
Jones and Henderson, 1980). Their importance in the salmonids has
been amply demonstrated, but the details of their actions throughout
the life cycle are still not clear (review by Specker, 1982). The action
of the corticosteroids appears to differ somewhat among various species of euryhaline teleosts and may not be consistent among all the
salmonids (see Nichols *et al.*, 1985). Investigations of interrenal physiology are complicated by "stress handling," which is known to alter
corticosteriod dynamics (Wedemeyer, 1972; Donaldson, 1981; Redding *et al.*, 1984; Leatherland and Cho, 1985).

Elevation of the corticosteroids during the smolting episode of the
coho salmon is shown in Fig. 7. The rise is correlated with the increase in gill ATPase activity noted in Section II,C (see also Patiño *et
al.*, 1985), but a cause-and-effect relationship has not been established
(compare Langhorne and Simpson, 1986 and Richman and Zaugg,
1987). However, the data are suggestive of an important function of
the interrenal in improving the salinity tolerance of the smolt by stimulating gill Na^+, K^+-ATPase. Young (1986), using an in vitro system for
the incubation of interrenal tissue of coho salmon, found a marked
increase in the sensitivity of the tissues to ACTH during April, and
this was correlated with peak plasma T4 titers and enhanced osmoregulatory capacities. Langdon *et al.* (1984) reported that ACTH, but not
cortisol, increases the gill ATPase activity in juvenile Atlantic salmon,
while both ACTH and cortisol increase the succinic dehydrogenase in
suspensions of gill cell homogenates; neither treatment affects size or
abundance of chloride cells in intact gills. Studies of several salmonids transferred from fresh to salt water support the view that cortico-

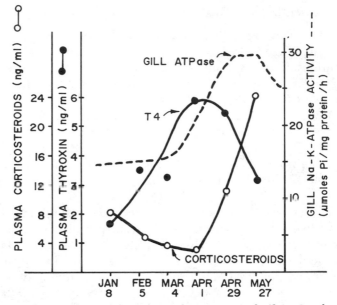

Fig. 7. Plasma concentrations of thyroxine and corticosteroids (from Specker and Schreck, 1982) and gill (Na+K)-ATPase activity (from Zaugg and McLain, 1970, 1976) in laboratory-reared, constant 10°C, coho salmon during smolting. [From Specker (1982).]

steroids improve salinity tolerance (Nichols and Weisbart, 1985; Nichols *et al.*, 1985), but substantially more research will be required to clarify the details of interrenal physiology in the preadaptation of smolting salmon to marine life.

Sheridan (1986) provided recent evidence of an action of cortisol in regulating smolt-associated changes in intermediary metabolism. Cortisol implants into coho parr caused a significant reduction in total lipid and triacylglycerol content of liver and dark muscle—changes characteristic of the parr-smolt transformation. These changes were accompanied by increased lipase activity. The changes did not occur in coho smolt similarly treated. The data suggest that the rise in cortisol is responsible for some of the changes in lipid metabolism observed at the time of the smolt transformation. Cortisol may also play a special part in stress-related physiology and the functioning of the immune system during this phase of the salmon life cycle (Specker, 1982; Barton *et al.*, 1985), but substantially more research on the interrenal hormones is required.

III. SEXUAL MATURATION: AN ALTERNATE STRATEGY IN DEVELOPING MALE PARR

Variable numbers of male parr may become sexually mature. This interesting phenomenon has been examined most thoroughly in the Atlantic salmon, where it was noted more than a century ago (see Orton et al., 1938). It is a true sexual maturation, resulting in functional males that participate in spawning with large adult females and produce viable offspring (Orton et al., 1938; King et al., 1939; Jones, 1959). The percentage of males that mature as parr varies in different populations. It may be as low as 5–10% (Evropeizeva, 1959; Bailey et al., 1980) but is often in excess of 50% (Evropeizeva, 1959, 1960; Österdahl, 1969; Dodd et al., 1978; Lundqvist, 1983; Myers, 1984). In an extreme situation, 100% of male Atlantic salmon in the Black Sea population are said to mature as parr, and only the females are anadromous (Leyzerovich, 1973). It would be interesting to have more information on the biology of these Black Sea populations; Österdahl (1969) suggested that even though parr may contribute substantially to the genetics of a population, large adult males are necessary in the spawning behavior, where they interact with the adult females in nesting, spawning reactions, and defense of the redd. However, Myers and Hutchings (1985) have more recently shown that large males are not necessary for natural spawning; mature male parr mated successfully with Atlantic salmon grilse.

Precocious maturation is not confined to Atlantic salmon but has been described also in S. gairdneri (Schmidt and House, 1979; Skarphedinsson et al., 1985), in S. trutta (Jonsson and Sandlund, 1979), in several species of Oncorhynchus (Robertson, 1957; Gebhards, 1960; MacKinnon and Donaldson, 1976; Hard et al., 1985), and in Salvelinus (Leyzerovich, 1973). It appears to be a viable and biologically important alternate tactic in the life history of many salmonids (Saunders and Schom, 1985).

The age of precocious maturation is variable. In nature, male Atlantic salmon parr may first mature at 1+ years of age or later. They may mature again in subsequent years if they remain in fresh water, or they may become smolts but remain in fresh water, revert to the parr condition, and mature; further, postsmolts (i.e., smolts during their first year in sea water) may occasionally mature (Saunders and Henderson, 1965; Sutterlin et al., 1978; Lundqvist and Fridberg, 1982); a year later they may return as grilse. In hatcheries with very favorable

growing conditions, both Atlantic and coho salmon parr may mature at
0+ years of age (Saunders *et al.*, 1982); maturation of parr at 0+ has
also been recorded in natural habitats but only in the lower reaches of
the River Scorff (France), where growing conditions are very favorable (Baglinière and Maisse, 1985). Again, as in other salmonid characteristics, there is a spectrum in the biology of early maturation
among different species and stocks of salmonids. The present discussion focuses on the relationship of early maturation to the smolting
process.

There are several implications of early maturation in fish culture.
The precocious males are likely to suffer increased mortality and/or
delay in reaching the rapidly growing marine phase. Sex ratios of
smolts and adults are affected (Leyzerovich, 1973; Mitans, 1973; Dalley *et al.*, 1983; Myers, 1984); further, there appear to be genetic
factors determining precocious maturity, and this too has implications
in salmon breeding programs (Saunders and Sreedharan, 1977; Glebe
et al., 1978; Naevdal *et al.*, 1978; J. E. Thorpe, 1987; Thorpe *et al.*,
1983). It is therefore important to understand, and if possible regulate,
the reproductive physiology of the male parr. Research interest has
focused on the endocrinology, the genetics (Section V), the relative
rates of growth of precocious parr and smolting juveniles, and the
relationships between precocious sexual maturation and the parr–
smolt transformation.

Studies of the reproductive physiology of maturing male parr have
shown that changes in the pituitary gonadotrops (Lindahl, 1980) and
in the pituitary and plasma gonadotropins parallel those of maturing
adult fish (Crim and Evans, 1978). The levels of testosterone and 11-
ketotestosterone rise in a comparable manner in the precocious parr
and the maturing adults, although the actual values are lower in the
parr (Stuart-Kregor *et al.*, 1981). In Atlantic salmon, at any rate, the
endocrinological regulation of precocious maturation is evidently the
same as that of the maturing adult males (Dodd *et al.*, 1978).

Is rate of growth a determining factor in the onset of precocious
sexual maturation? Eriksson *et al.* (1979) summarized the conflicting
literature and concluded that genetic factors rather than rates of
growth were determinants of early maturity. In a study of hatchery-
reared Baltic *Salmo salar*, all individuals were found to grow at comparable rates until the end of the second summer, when the fish were
1+ years old (unimodal population as discussed in Section II). At that
time, about 50% of the males matured. In short, maturation of 50% of
the males took place at a time when immature males, maturing males,

and all females were similar in size. Rates of growth did not seem to be the determining factor in the onset of maturity. The following season, when the fish were 2+ years old, another group of males matured—again at a time when the immatures, the maturing parr, and the females formed a unimodal group. Following maturation (at either 1+ or 2+ years), growth rate declines while the gonads are ripe, but the interesting point is that growth rate does not appear to determine the onset of maturation in these populations. In contrast, data on size dependence have been presented by several workers who claim that the larger, more rapidly growing individuals mature (Evropeizeva, 1960; Dalley et al., 1983; Myers et al., 1986 and citations in Eriksson et al., 1979), while Sutterlin et al. (1978), in a study of postsmolts in sea water, found that it was the smallest individuals that matured. In support of Eriksson et al. (1979), Naevdal et al. (1978), in a study of a hatchery-reared population of Atlantic salmon, found indirect evidence supporting the concept that sexual maturation is randomly distributed and not related to size or rate of growth. These workers conclude that precocious maturation retards the growth of the male parr but that growth before maturation is uniform, thus supporting the idea that genetic factors are involved.

Thorpe and Morgan (1980) discuss a "two-threshold hypothesis" for critical sizes for smolting and precocious maturation. In a population of Scottish salmon where parr may smolt and emigrate at 1+ year (if they have attained the critical size), it is the more slowly developing individuals that remain in fresh water and become sexually mature at 1+ years. These workers discuss the "two-threshold hypothesis" and suggest that if the critical size for maturation is greater than the critical size for smolting, then populations with a high proportion of *small* 1-year smolts will not tend to mature in their first autumn; populations with a high proportion of *large* 1-year smolts may be expected to show a high incidence of precocious parr in the first autumn.

What is the effect of precocious maturation on the subsequent tendency to smolt and migrate? Thorpe, (1986, 1987) has examined the interrelationships among these three relevant factors: growth rate, sexual maturation, and time of smolting. Although he finds a strong positive correlation between growth rate and maturation rate, the "decision" to smolt is also critical to the onset of sexual maturity. In sibling populations of Atlantic salmon, relatively large individuals that smolted at 1 year did not mature until at least 2.5 years, while in the same population some smaller fish that did not smolt at 1 year matured 6 months later without emigrating to sea. Moreover, at age 2 years, the fish that matured at 1.5 years failed to undergo smolting as completely as their immature siblings (conclusion based on gill enzyme studies). Thorpe (1986, 1987) argues that the processes of smolting and reproduction are mutually incompatible. Smolting is a commitment to life in saline waters; reproduction demands a freshwater habitat. This concept of incompatibility is supported by several investigators. Nagahama (1985) compared plasma levels of testosterone and 11-ketotestosterone in amago salmon at three stages: in precociously maturing males, in smolting and in desmolting fish. The values were highest in early maturing male fish and lowest in the smolts; they increased during desmolting. Findings of Miwa and Inui (1986) also support this concept of incompatibility. These workers fed sex steriods to sterilized amago salmon and noted reduced gill Na^+, K^+-ATPase activity, reduced numbers of chloride cells, and a thicker skin and gill epithelia than in the smolting controls; the steroids prevented epidermal silvering and an increase in seawater tolerance (see also Aida et al., 1984; Ikuta et al., 1985, 1987; McCormick and Naiman, 1985).

Several other investigations are also relevant to the effects of parr maturation on the subsequent tendency of salmon to smolt and migrate. In the first place, Thorpe and Morgan (1980) and Saunders et al. (1982) found that male parr that matured in autumn may migrate the following spring while their body fats are depleted and their energy reserves at a low level as a consequence of spawning (Thorpe and Morgan, 1980; Saunders et al., 1982). This situation seems to srgue against some theories that suggest a role for changing energy reserves in triggering the smolt transformation (see Saunders et al., 1982), but further investigation on this point is indicated. In the second place, some workers have noted a strong tendency of precocious parr to remain in fresh water and to mature the following autumn (Leyzerovich, 1973; Eriksson et al., 1979). This situation plus the increased mortality associated with sexual maturation reduce the proportion of males with respect to females in the migrating smolt population (Mitans, 1973; Dalley et al., 1983; Myers, 1984). The increased mortality of the males may be related to their spawning activity, the harsh winter conditions that postspawners face while energy reserves are depleted, or predation; it occurs both in nature and under laboratory conditions (Clarke et al., 1985).

In summary, the age of sexual maturation is very flexible in male salmon. Although maturation in the parr stages in best known in S. salar, it has also been recorded in other salmonids. There are several consequences of parr maturation: (a) sex ratios of migrant smolts and returning adults are biased toward the female, since sexual maturation reduces the relative numbers of males—at least in natural populations; (b) maturation seems to reduce the tendency to smolt and migrate the following spring (Österdahl, 1969)—in this connection, it may be relevant that precocious male parr have relatively low capacities to adapt to seawater (Aida et al., 1984; Lundqvist et al., 1986); (c) male smolts may adopt the alternative tactic of remaining in fresh water and spawning as parr that have readapted to fresh water (Lundqvist and Fridberg, 1982; Thorpe, 1986, 1987); and (d) variability in the life history of salmon may be an important safeguard against loss of small stocks through several successive years of reproductive failure (Saunders and Schom, 1985). Thus, several factors related to parr maturation may affect the smolt populations of Atlantic salmon and the productivity of a river system.

IV. ENVIRONMENTAL MODULATION OF THE SMOLT TRANSFORMATION

The season of the parr–smolt transformation is highly predictable. Like the dates of migration, reproduction, and the emergence of small fry from the gravel, parr become smolts in relation to the seasonal changes in their surroundings—especially the cyclical variations in day length and temperature. The environmental regulation of smolting has now been critically studied for more than two decades (see reviews by Hoar, 1965, 1976). The early experiments demonstrated a strong effect of photoperiod but indicated an underlying endogenous rhythmicity; such seasonal changes as skin silvering, growth, salinity preference, and tolerance could be advanced by accelerated photoperiods or delayed by retarded photoperiods, but the changes could not be entirely suppressed. An underlying circannual rhythm was suspected.

A circannual rhythm has been defined as one that persists under constant environmental conditions but deviates by a fixed amount from the annual cycle of 365.26 days. Circannual rhythms were first charted in the golden hamster (Citellus lateralis) by Pengelley and Asmundson (1969), but are now recognized in many animals. The pioneer work on the teleost fishes is summarized by Baggerman (1980), Eriksson and Lundqvist (1982), and Lam (1983).

A. The Circannual Rhythm of Smolting

Eriksson and Lundqvist (1982) kept individually marked Baltic S. *salar* under constant conditions of light [light–dark (LD) 12 : 12] and water temperature (11.0 ± 0.5°C) for 14 months. The fish were given a surplus of food, and their condition factor (weight per unit length) and skin coloration were recorded at regular intervals. Under these constant conditions, a period of high condition was followed by a period of lower condition factor, and this in turn by another period of high condition. Likewise, parr-like appearance was followed by silver smolt, and this in turn by reverted parr. The mean time between two smoltings was about 10 months. In the absence of environmental cues, the cycle of growth and skin coloration runs at its own frequency, and this differs substantially from 1 year. The endogenous rhythmicity noted by earlier workers is truly circannual, as defined by workers in this field. Factors such as photoperiod and temperature synchronize or entrain the rhythm to the annual cycle; they serve as *zeitgebers* in the biological clock terminology. Eriksson and Lundqvist's (1982) findings are in line with Brown's (1945) experimental work on the growth of brown trout held under constant conditions (11.5°C and LD 12 : 12); her fish showed seasonal changes in growth even though the conditions in the experimental tanks remained constant. Likewise, Wagner (1974a), in a study of juvenile steelhead trout raised at constant temperature in total darkness, found that some fish that reached a certain critical size developed smolt characteristics indicating an endogenous rhythm of smolting. Atlantic salmon, coho and sockeye salmon, and also steelhead trout in many investigations of photoperiod effects show evidence of endogenous rhythms of physiology that are synchronized by daylength (Section IV,B).

B. Modulation of the Rhythm by Photoperiod

Among salmonids, photoperiod is the most usual synchronizer of seasonally changing physiological processes of sexual maturation, spawning, growth, smolting, and migration (reviews by Poston, 1978; Wedemeyer *et al.*, 1980; Lundqvist, 1983). Many biologists have now studied the role of photoperiod (PP) in salmonids—especially the Atlantic salmon, steelhead trout, and sockeye and coho salmon. The most critical of the studies on *S. salar* were done by R. L. Saunders and his associates in Eastern Canada; at about the same time, H. H. Wagner carried out the first comprehensive analyses of smolting steelhead trout, and these studies have been followed by several noteworthy investigations of *Oncorhynchus*—particularly those of W. C. Clarke and associates on coho and sockeye salmon. Poston (1978) tabulates several photoperiod effects noted in other species of salmonids.

Saunders and Henderson (1970) compared Atlantic salmon held under (a) constant day lengths of 13 h light; (b) simulated natural PP (increasing daylength in springtime); and (c) the reciprocal of natural PP (decreasing daylength from early March and increasing daylength from late June). Growth rates, degree of silvering, thyroid activity (as shown by histology), excitability, and plasma osmotic and chloride levels were examined. The young salmon under constant and natural PP showed no particular departure from the normal conditions of smolting; however, reciprocal PP affected both growth and the excitability of the fish, indicating a disturbed endocrine physiology. In contrast to the natural sequence in smolting, the fish under reciprocal PP developed a high condition factor (weight in relation to length) in fresh water, while in the sea they grew more slowly, ate less, and had lower efficiencies of food conversion. The reciprocal-PP fish were also less sensitive to external stimuli than those under natural or constant PP. No differences between natural and reciprocal PPs were noted in the degree of silvering, thyroid histology, and plasma osmotic and chloride levels. Only growth processes and excitability were altered in these tests.

In a subsequent study with reciprocal photoperiods, the experimental fish were found to have lower metabolic rates (standard rate of O_2 consumption) compared with natural-PP fish when the measurements were made in total darkness in seawater (Withey and Saunders, 1973). In a third study, Saunders and Henderson (1978) compared gill ATPase activity, body lipids, and moisture in fish held under different photoperiods. Under both natural and reciprocal PPs, gill ATPases increased markedly during the late winter and spring while the levels of total body lipids declined and moisture increased. The point of interest, however, is that these changes occurred earlier and were more marked in fish held under reciprocal PP, indicating that the long nights of winter trigger preadaptations of Atlantic salmon for the smolt phase. Salinity resistance (measured at 40‰ salinity) increased in a comparable manner under all photoperiod regimes. The lipid and moisture data are in line with the findings of Komourdjian *et al.* (1976b), who attributed the changes to the stimulation of the pituitary secretion of growth hormone and ACTH by the longer day lengths. Saunders's investigations of Atlantic salmon in Eastern Canada have been followed by studies of photoperiod effects on *S. salar* of the

Baltic region (Eriksson and Lundqvist, 1982; Clarke *et al.*, 1985); findings are similar with the two different populations of salmon.

Wagner's (1974a,b) investigation of the effects of daylength on steelhead trout differed from those of Saunders in that accelerated and decelerated as well as reversed or reciprocal PPs were tested. In the accelerated or decelerated regimes, the simulated seasonal change was advanced or retarded by 6 min per week. In this way, the anticipated season of smolting was advanced or retarded by 6 to 8 weeks. The coefficient of condition and migratory behavior were assessed in the various groups by periodic weight/length measurements and by releases of fish into a natural stream with subsequent trapping and enumeration downstream. Wagner's (1974a) experiments demonstrated the importance of both photoperiod and the rate of change of PP in regulating the time of smolting. Increasing day length, rather than day length or total exposure to light, was considered the prime stimulator of smolting. Reports by Zaugg and Wagner (1973) and Zaugg (1981) confirm photoperiod as the synchronizer of an endogenous rhythm of smolting and show that advanced PP will accelerate gill Na^+,K^+-ATPase changes characteristic of smolting in the steelhead by 1 month, although seawater survival does not seem to be closely associated with PP and smolting (Wagner, 1974b).

Clarke *et al.* (1978, 1981) studied osmoregulatory performance (24-h seawater challenge) and body lipid and liver glycogen levels in coho and sockeye salmon subject to constant or increasing or decreasing daylengths at different seasons and at different temperatures. Like the two species of *Salmo* considered above, these two oncorhynchids are physiologically responsive to photoperiod and temperature; advanced photoperiods accelerate growth and improve osmoregulatory ability, while temperature controls the rate of the response. Sensitivity of fry to photoperiod varies seasonally.

More recent studies have emphasized that the outcome of daylength manipulations may depend on the age/size of the fish *and* the period of exposure to short days before the initiation of the advanced PP regime. In studies of juvenile coho salmon, Brauer (1982) fixed daylength at 12.27 h for 1 month before starting to increase the day in late March; the delay (phase adjustment) resulted in improved growth, better food conversion, and seawater adaptability. Clarke and Shelbourn (1986) commenced their experiment at an earlier stage with free-swimming coho fry in February. Daylength was fixed at 9.75 h for 1 or for 2 months before starting to increase the light period at the natural rate of increase. Thus, three groups of coho were compared: fish under natural daylength, fish with 1 month delay and fish with 2 months delay. Delayed PPs produced fish of more uniform size (absence of bimodality seen under natural PP), with greater capacities to hypoosmoregulate following the 24-h seawater challenge, and improved growth in seawater. The sustained exposure to short days synchronized smolting in these coho.

Finally, photoperiod studies of chinook juveniles emphasize the species variations in responsiveness to day-length manipulation. Chinook salmon, unlike the four species discussed above, migrate to sea over an extended period and develop an early tolerance to seawater. There are two varieties of chinook salmon (Healey, 1983): the spring chinook, which spend a year or more in fresh water as juveniles and tend to produce yearling smolts (stream type), and the fall chinook, which produce underyearling migrants (ocean type). Ewing *et al.* (1979) studied a population of spring chinooks that showed peaks in gill ATPase activity in October of their first year, and the following May and October of their second year. Fish were reared under controlled photoperiods for 2 years; only the October peak of the first year was modified and found to be suppressed when photoperiods were advanced by 3 months. Clarke *et al.* (1981), in their study of fall chinooks, found no evidence of photoperiod regulation of growth or osmoregulation; these chinooks seemed insensitive to photoperiod manipulation; growth/size was considered the important factor in determining their entrance into the sea.

In summary, there is a circannual rhythm of physiological changes associated with the parr–smolt transformation. After juvenile salmon reach a certain critical size they become smolts with the capacity to hypoosmoregulate and grow in seawater; behavioral changes result in a downstream migration. Many of the smolting changes are reversible, and fish that cannot reach the sea revert to the parr condition. These cycles of physiology occur at intervals of about 1 year under constant light conditions; under natural light conditions they occur at predictable seasons according to the geographic location. If the seasonal changes in day length are advanced (simulating early spring conditions), the smolt transformation can be accelerated, while the prolongation of winter light conditions (short daylength) will delay the changes. The most important components of light in the manipulation of daylength (photoperiod) are *direction* and *rate of change*. Further, the period of darkness (short days of winter) that precedes the lengthening days is important in synchronizing the changes; prolongation of the season of short days accelerates the changes that occur in the subsequent period of lengthening photoperiod. Of the several changes associated with smolting, growth, hypoosmoregulatory capacity, and migratory tendency seem most sensitive to photoperiod manipulation, while body silvering seems independent. However, generalizations must be made cautiously, since the response varies not only with the photoperiod regime but also with the age/size of the fish, the season when light control is initiated, and with the temperature (Section IV,C).

C. Temperature Effects

The smolt migration was correlated with rising springtime temperatures by some of the very early fisheries scientists. Foerster (1937) correlated the commencement of sockeye smolt migration from Cultus Lake, British Columbia, with the vernal rise in lake temperature; the winter minimal water temperatures approximated 2.5°C and the threshold migration temperature seemed to be about 4.4°C. Cessation of migration appeared to be related to the warming of the lake surface waters, which formed a "temperature blanket." White (1939) reported that peaks of Atlantic salmon migration from Forest Glen Brook, Cape Breton Island, occurred at low light intensities (night) when the water temperature rose sharply above 10–12°C. In this section of the review, experimental work on the effects of temperature on the physiological changes of smolting is considered—particularly in relation to the photoperiod regulation of the parr–smolt transformation.

Temperature effects were studied in many of the photoperiod investigations reviewed in the previous section. Some experiments were performed at a single constant temperature (Saunders and Henderson, 1970, 1978; Komourdjian et al., 1976b; Ewing et al., 1980a; Brauer, 1982; Clarke et al., 1985); in others, the ambient temperature was used (Lundqvist, 1980); in still others, comparisons were made at a lower and at a higher temperature (Zaugg et al., 1972; Ewing et al., 1979; Pereira and Adelman, 1985), and some experiments have been designed to test temperature specifically with three or more temperatures maintained at a constant level (Knutsson and Grav, 1976; Clarke et al., 1978, 1981; Clarke and Shelbourn, 1980, 1985, 1986). Wagner's (1974a) studies of steelhead differed in the use of changing temperatures rather than constant temperatures; he compared fish subjected to a simulated normal stream temperature with fish subjected to a seasonally advanced temperature, an accelerated temperature, and a decelerated temperature regime.

In general, it is concluded that temperature controls the rate of the physiological response to photoperiod. Smolting occurs sooner at

higher temperatures, and further, changing temperatures seem to be
more stimulating than a constant temperature (Wagner, 1974a; Jonsson and Rudd-Hansen, 1985). However, these generalizations cannot
be applied directly to the manipulation of the time of smolting without due regard to (a) species variations, (b) the inhibitory effects of
high as well as low temperatures, and (c) the fact that high temperatures accelerate the reversal of the smolt to parr condition as well as
the parr–smolt transformation (review by Wedemeyer *et al.*, 1980).

Although Atlantic salmon (*S. salar*) smolt at temperatures as high
as 15°C (Saunders and Henderson, 1970; Komourdjian *et al.*, 1976b;
Johnston and Saunders, 1981), the normal smolting increase in gill
ATPase is suppressed or declines in steelhead trout (*S. gairdneri*) at
temperatures in excess of 13°C (Zaugg *et al.*, 1972; Zaugg and Wagner,
1973; Zaugg, 1981). It should be noted, however, that the smolt runs
of Atlantic salmon peak at about 10–12°C and are normally over before
water temperatures of 15°C are reached (White, 1939). Further, it
seems relevant that Knutsson and Grav (1976), in studies of *S. salar*,
reported optimal growth under long photoperiods at 15°C but more
pronounced effects of photoperiod on seawater adaptation at 11°C.

The experiments of Zaugg and associates (see also Adams *et al.*,
1973, 1975) suggest an optimum temperature for the parr–smolt transformation. Higher temperatures accelerate smolting up to about
10–12°C, but above these temperatures at least some of the smolting
changes (gill enzyme activity and migratory behavior) are inhibited or
occur only briefly (Zaugg and McLain, 1976) (Fig. 8).

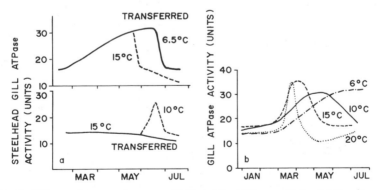

Fig. 8. Effects of temperature on gill Na+, K+–ATPase activity. (a) Yearling steelhead trout held at 6.5°C and transferred to 15°C (upper panel); held at 15°C and transferred to 10°C (lower panel). [From Wedemeyer *et al.* (1980) after Zaugg *et al.* (1972).] (b)
Summary of changes during development of coho salmon. [From Wedemeyer *et al.*
(1980) after Zaugg and McLain (1976). Courtesy *Marine Fisheries Review.*]

Finally, the significance of temperature in the smolt-to-parr reversal is emphasized. Zaugg and McLain (1976) noted that the time span
of elevated gill ATPase activity was much longer in steelhead maintained at 6–10°C than at temperatures in excess of 13°C; only a very
brief period of enzyme activity occurred at 20°C (Fig. 8). Higher temperatures accelerate the smolting changes, but unless the young
salmon can enter saltwater within a very short time span, the physiology will revert to that of the parr and the young fish will be trapped in
fresh water for another year (review by Wedemeyer *et al.*, 1980).

D. Other Modulating Environmental Factors

Although photoperiod and temperature are the most reliable cues
for synchronizing the smolting season, three other environmental
variables may be important in timing the transformation: the lunar
cycle, the runoff associated with flooding rivers in springtime, and the
nature of the dissolved solids, which may change naturally with
stream flow or because of industrial activities.

1. LUNAR CYCLE

In juvenile coho salmon, chinook salmon, and steelhead trout,
Grau (1982) and Grau *et al.* (1981, 1982) observed a strong correlation
between the new moon and the peak surge in plasma thyroxine. Figure 5 shows a consistent peaking of plasma T4 at the time of the new
moon when values for juvenile coho are plotted on a lunar calendar.
In a subsequent study of coho, raised in Hawaii, where photoperiod
has no significant effect on the timing or magnitude of changes in
thyroid hormone, Grau *et al.* (1985a) found three peaks in plasma
thyroid hormones; two of these occurred at the time of the new moon,
while the third was associated with the full moon—suggesting that
the periodicity is semilunar.

Juvenile chinook salmon (*O. tshawytscha*) were also examined in
some detail (Grau, 1982; Grau *et al.*, 1982). Chinook juveniles do not
show a single mass migration but move seaward in groups at several
different times throughout the spring and summer. The hatchery fish
examined showed four peaks in plasma T4, and these peaks occurred
in the samples taken closest to the new moon, again supporting the
theory that the lunar-phased peak in thyroid hormone is concerned
with the initiation of migration. To test the practical implications of
these findings, the return of coho to the Trinity River Hatchery in
California was studied in relation to the time of juvenile release; re-

coveries from groups released on the new moon closest to the expected peak of plasma T4 were approximately twofold greater than the previous releases that were not lunar-based (Nishioka et al., 1983). These arguments find some support in field studies of downstream migrating juvenile coho salmon—particularly Mason's (1975) study where the downstream movements of coho fry peaked with the new moon while seaward migration of smolts peaked with the full moon. Movements of fry and smolt showed no obvious relation to either water temperature or stream flow; lunar rhythmicity was the important factor in timing movements of coho in Lyman Creek, Vancouver Island. Mason (1975) suggests that in streams subject to greater variations in flow and temperature, these factors, rather than lunar periodicity, might trigger migration.

Yamauchi et al. (1984, 1985) examined the concept of lunar phasing of T4 surges and migration in masu salmon. Their fish were best able to osmoregulate in seawater at the peak of smolting (as judged by external characters). T4 was high during smolting and peaked at the time of the new moon in April, when the onset of migration occurred. The largest migrations occurred immediately after a rainfall around the time of the new moon, thus showing the importance of rainfall as well as the lunar cycle.

Although studies of juvenile oncorhynchids have usually supported a concept of lunar periodicity in thyroid function and downstream migration, data are less consistent in studies of Salmo. Boeuf and Prunet (1985) found suggestive evidence that T4 peaked at the new moon, but Youngson et al. (1983, 1985, 1986) emphasized the effects of stream flow on downstream migration; they considered stream flow rather than lunar phasing important in Atlantic salmon migrations and noted that elevated stream flow occurs around the time of the new moon and thus facilitates migration. In a study of S. salar smolts tagged internally with ultrasonic telemetry transmitters, movement through the estuary of the Penobscot River, Maine, was dependent on water currents and seemingly on no other environmental variable, although the rising springtime water temperature (above 5°C) initiated the migration (Fried et al., 1978). The first reports of elevated plasma T4 in steelhead trout at the time of the new moon (Grau et al., 1981, 1982) were not confirmed in a later investigation (Lin et al., 1985a).

In summary, present evidence of a lunar or semilunar synchronization of some physiological changes during smolting (especailly thyroid secretion) is reasonably persuasive for several species of Oncorhynchus but questionable for Salmo. However, more research is

required to confirm this suggested distinction between the two salmonid genera, since exceptions have also been found in coho salmon.
The importance of further research is apparent. If the lunar cycle
provides a *zeitgeber* in addition to photoperiod, then hatchery opera-
tions are likely to be improved by relating juvenile releases to the
lunar calendar.

2. DISSOLVED SOLIDS AND pH

Composition of dissolved solids may change during seasons of
heavy runoff with seepage from the land. It is conceivable that particular substances present during the spring runoff may affect the physiology of juvenile salmon. Although naturally occurring dissolved
solids are considered significant in regulating reproduction of some
tropical and subtropical fish (Lam, 1983), salmon smolting is not
known to be affected by them. There is, however, ample evidence that
industrial contaminants and pesticides affect the physiology of young
salmon. Effects of such substances on thyroid activity were noted in
Section II,E,1. There is also evidence that salinity tolerance and migratory tendencies are altered by contaminants. Lorz and McPherson
(1976) found that chronic copper exposure during smolting partially or
completely inactivated the gill ATPase system of coho salmon and
that the normal migratory behavior of the fish was suppressed. Damage from copper was not apparent until the fish were moved into
saltwater, when there was a high mortality. Davis and Shand (1978)
obtained similar results with young sockeye salmon. Several other
heavy metals and a number of organic pollutants have been tested and
the damaging effects evaluated (Lorz and McPherson, 1976; review
by Wedemeyer *et al.*, 1980). The damage varies, but the fact remains
that smolting physiology is not normal; this stage of salmon development as well as the earlier stages is adversely affected by heavy metals
and pesticides (Folmar *et al.*, 1982b; Nichols *et al.*, 1984).

Another topic of current concern is the acidification of fresh waters
by acid precipitation ("acid rain") and its effect on fish populations.
Saunders *et al.* (1983) reared Atlantic salmon in waters at about pH 6.5
and about pH 4.5. Smolting, as judged by salinity tolerance and gill
ATPase, was impaired at the lower pH. These workers conclude that
smolting does not proceed normally in *S. salar* subjected to low pH.
This is not surprising, but the knowledge is important in culturing
salmon and in predicting the effects of industrial activity on the future
of the wild stocks.

V. SOME PRACTICAL PROBLEMS IN SMOLT PRODUCTION

The history of salmonid culture can be traced to a European monk, Dom Pinchon, in the fifteenth century. His simple methods of hatching eggs and culturing fish were the basis of more sophisticated techniques developed in the eighteenth century, leading in France to the first attempts (1842) to exploit these ideas practically for the enhancement of salmonid stocks (notes from Day, 1887). By the latter half of the nineteenth century, salmonid hatcheries were widely established in Asia, Europe, and North America (Bowen, 1970; Thorpe, 1980). The objective of these early efforts was to release large numbers of juvenile salmon and trout into natural environments; the assumption was that the larger the number of fish released, the greater would be the return to the fishermen. Hatcheries usually produced fry and fingerlings, but in some cases (sockeye salmon, for example) the fish were cultured to the smolt stage (Foerster, 1968). Realization that returns of mature fish may not be directly related to the numbers of eggs hatched and juveniles released has resulted in many changes in objectives and orientation. In the past quarter century, the emphasis has been to enhance salmon stocks through stream improvement, the production and release of juveniles under more natural conditions, and the protection of growing fish in ponds and sea pens (Thorpe, 1980).

Whether salmon and trout are cultured to improve commercial and sport fishing or to grow fish to marketable size in sea pens, an understanding of the physiology of smolting is basic to success. It is obvious that, in both cases, *early smolting* is economically advantageous in minimizing culture costs and losses through predation and disease in fresh water. To obtain maximum advantage from the more favorable growing conditions of the marine environment, the fish should migrate or be transferred to the sea at the earliest age consistent with normal growth and ability to survive in seawater.

Much of the recent research on the physiology of smolting has been directed to practical problems associated with salmonid ranching (Thorpe, 1980) and the culture of salmon in sea pens. There have been many workshops and discussions and several important symposia devoted to these topics. Two recent symposia have been particularly valuable: one held at La Jolla, California, in 1981 focused on Pacific salmon, especially the coho (*Aquaculture*, Vol. 28, pp. 1–270, 1982); the other at the University of Stirling in 1984 was directed primarily to the Atlantic salmon (*Aquaculture*, Vol. 45, pp. 1–404, 1985). The organizers of these symposia identified several problems as central to the production of quality smolts for release in streams or culture in sea pens.

In very broad terms, there are three major thrusts toward production of smolts of good quality: (a) achievement of rapid growth to smolting size; (b) successful osmotic and ionic regulation in the marine environment; and (c) a high growth rate in the sea with a successful return to fresh water. The recent research has been concentrated in three main areas: (1) the environmental regulation of growth, early smolting, and seawater adaptability; (2) the endocrinological basis of smolting with the objective of manipulating hormones in the regulation of smolting physiology; and (3) improvement of the stock through genetics. Space permits only brief reference to the recent literature with an emphasis on the importance of further studies. The hazard of general comments is recognized. There are many species of salmonid fishes, with smolting biology as variable as their morphology, physiology and behavior. Further, the smolt transformation is not a single event but involves many changes that occur at different rates over an extended time period. The generalizations that follow are made with some reservations.

A. Minimizing the Juvenile Freshwater Phase

For practical reasons, the abbreviation of the parr stage remains a major objective of salmon farmers. In nature the environment is dominant in regulating the age of smolting. For example, Atlantic salmon smolts in the southern part of their range may migrate seaward after one summer in the rivers (1+ year smolts), but at Ungava Bay, the most northern extent of their distribution, the average age of smolt migration is 5+ years and some juveniles are 8+ years at migration (Power, 1969). Manipulation of temperature and photoperiod is the most economical approach to the production of 1+ year smolts (Wedemeyer *et al.*, 1980) or even 0+ year smolts that have been produced under hatchery conditions (Brannon *et al.*, 1982; Ísaksson, 1985; Zaugg *et al.*, 1986.

Current research has shown that manipulation of temperature and day length must be based on a careful study of the species involved. There is an optimum temperature, which may vary with the species. Low temperatures delay growth and smolting; high temperatures increase growth rate (Donaldson and Brannon, 1975; Brannon *et al.*,

1982; Piggins and Mills, 1985) but may suppress some important changes of smolting (Fig. 8) and also accelerate the reverse change of smolt to parr (Wedemeyer *et al.*, 1980). Photoperiod control must also be based on studies of the particular species. Day length per se is less important than the direction and rate of change in day length; short days preceding long days are involved, and exposure to continuous light or days of constant length may result in satisfactory growth but a failure to smolt (Section IV,B). Saunders *et al.* (1985a) report that juvenile Atlantic salmon under continuous light grow faster than salmon under simulated natural photoperiods but fail to smolt and do not grow like natural smolts when transferred to sea cages.

Several rearing conditions, in addition to temperature and day length, are crucial to growth and early smolting. Adequate diet, in terms of energy-providing and growth-promoting foods, is the most obvious of these (Cowey, 1982). In addition, experiments with supplements of inorganic salts have shown their potential for elevating gill ATPase and increasing survival in seawater; tests have been carried out with Atlantic salmon parr (Basulto, 1976) and with coho and chinook salmon (Zaugg, 1982; Nishioka *et al.*, 1985a). Dietary supplements of thyroid hormones and anabolic steroids have been carefully tested for growth-promoting and smolt-inducing effects. Thyroid hormones, especially T3, increase growth in Atlantic salmon parr but do not induce early smolting (Saunders *et al.*, 1985b). Improved seawater tolerance as well as growth enhancement have been shown in several tests with T3 supplements to the diets of coho salmon, chinook salmon, and steelhead trout (Higgs *et al.*, 1979, 1982; Fagerlund *et al.*, 1980; McBride *et al.*, 1982). The anabolic steroids (androgens) also have demonstrable growth-promoting effects when added to the diet of young Pacific salmon and steelhead trout, but there is no evidence of improved seawater tolerance (Fagerlund *et al.*, 1980; Higgs *et al.*, 1982; McBride *et al.*, 1982). Related to these dietary tests are several demonstrations of the growth-promoting effects of injected growth hormone, which also improves saltwater tolerance (Komourdjian *et al.*, 1976a,b; Higgs *et al.*, 1977). However, from an economic angle, injection procedures have considerably less potential for use in mass production of young salmon.

Rearing density is also important in smolt production. High density in rearing ponds depresses growth rate through competition for food and increases the hazards of infection, with resulting delays in smolting and reduced survival of juveniles (Refstie, 1977; Schreck *et al.*, 1985). Crowding also causes stress responses in the adrenocortical system, with elevated plasma corticosteroids and detectable histological changes in the interrenal gland (Noakes and Leatherland, 1977; Specker and Schreck, 1980; Schreck, 1982).

Finally, the importance of genetic factors is now widely recognized in attempts to improve growth and achieve early smolting (Wilkins and Gosling, 1983). Several different studies of Atlantic salmon have demonstrated growth variation and differences in the time of smolting of juveniles cultured from gametes obtained in different geographic regions (Refstie *et al.*, 1977; Refstie and Steine, 1978; Thorpe and Morgan, 1978; Riddell and Leggett, 1981; Riddell *et al.*, 1981). The potential of breeding programs in the production of early age smolts is recognized for both *Salmo* (Refstie and Steine, 1978; Bailey *et al.*, 1980) and *Oncorhynchus* (Saxton *et al.*, 1984). The hazards of reduced genetic variability as a result of selection programs has also been noted (Allendorf and Utter, 1979).

B. Successful Transfer to the Marine Habitat

Smolting does not commit a salmon or a trout to life in saltwater. In sea-going populations, reversion to the parr condition in fresh water is a viable option. The mechanisms triggering or regulating smolt–parr reversion have not been well investigated; studies of seawater adaptability should be investigated throughout the year in relation to temperature and photoperiod. However, this option of smolt–parr reversion in salmonids means that the time of smolt release or transfer from hatchery to sea pens is critical to successful salmon management. It is now felt that poor adult returns from cultured smolts are sometimes related to inappropriate times of release or transfer; losses of as many as 70% from premature transfer of smolts have been noted (Folmar and Dickhoff, 1981; see also Wedemeyer *et al.*, 1980; Bilton *et al.*, 1982). For both Atlantic and Pacific salmon, the physiological condition of the smolt-like freshwater fish is recognized as important to growth and survival in the marine habitat. The unsatisfactory saltwater growth of Atlantic salmon "smolts" cultured under continuous or unchanging light conditions has been noted (Saunders *et al.*, 1985a). Another example for Atlantic salmon has been reported by Gudjónsson (1972), who reared *S. salar* in constant light and obtained good growth and healthy juveniles using heated (geothermal) waters in Iceland; however, there were no returns of sea-run fish—juveniles of smolt size did not migrate when released but remained in the streams. In contrast, salmon raised in similar heated waters under outdoor conditions of changing day length gave good adult returns (see also Wedemeyer *et al.*, 1980).

Marine survival of young coho salmon has been intensively studied in relation to time of transfer from fresh water to seawater (Folmar and Dickhoff, 1980, 1981). This species seems to be extremely sensitive to time of transfer, and fish moved to sea pens either too early or too late fail to realize their full potential and may become "stunted" or revert to the parr condition (Fig. 9). Morphological and physiological characteristics of "stunts" and "parr-revertants" have been tabulated and the literature summarized in some detail by Folmar et al. (1982a). Many of the studies have focused on the endocrine system, which appears to be hypofunctional (Clarke and Nagahama, 1977; Bern, 1978; Fryer and Bern, 1979); "stunts" are said to be "hypoendocrine" (Nishioka et al., 1982; Grau et al., 1985a). However, recent studies of plasma GH levels in normal and stunted yearling coho salmon show that the plasma levels of GH are consistently higher in the stunts, suggesting that deficiencies in the receptor and/or the mediating system may be the cause of stunting (Bolton et al., 1987b). Several organs other than the endocrine organs are also different in the stunts from the normal fish; liver, muscle, and the organs concerned with osmotic and ion regulation show describable differences (see Folmar et al., 1982a).

Various smolting characteristics have been considered in predicting the best time for smolt release or transfer from hatchery to sea pens. A minimum size is essential (Section II,B), and, if the fish are in the smolt stage, the larger the smolts, the better seem to be the returns (Mahnken et al., 1982). Superficial features such as silvering and low condition factor are not reliable predictive indices, since

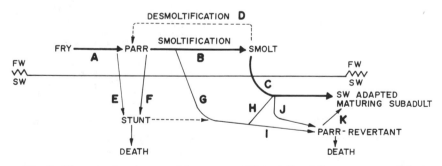

Fig. 9. Diagram showing possible sources of "stunts" and "parr-revertants" during the smolting and postsmolting stages in the life of the coho salmon. (A, B, C) Normal sequence in anadromous salmon. (D) Smolts that do not enter seawater revert to freshwater parr condition. (E, F) Premature transfer to seawater during the parr stage, resulting in "stunting." (G, I) Transfer to seawater before the completion of smolting or (H, J) failure of smolts to make adequate growth may result in death of undersize fish. Further details in original papers. [From Folmar et al. (1982a).]

these features may persist for a time after the readiness to migrate has passed and while the smolt is reverting to a parr (Wedemeyer *et al.*, 1980).

Physiological testing of osmoregulatory competence (seawater challenge test) can be used to follow smolting changes (Clarke, 1982; Saunders *et al.*, 1982; Hogstrand and Haux, 1985), but this, or the assessment of chloride-cell density and ion transport, may not always be practical in large-scale production programs. Also among the more technically demanding prediction indices are the well-marked changes in the endocrine system (especially the thyroid and interrenal glands) and in the gill ATPases. A surge in thyroid hormone occurs reliably in smolting salmonids (Grau *et al.*, 1985a). Tests indicate that the best time for coho release follows the T4 surge when the cortisol peaks. Of many smolting characteristics examined, Folmar and Dickhoff (1981) reported that only one was statistically related to the percent survival after 6 months in seawater; this statistic was the percent of the area beneath the T4 curve prior to seawater transfer. Comparisons of plasma T4 in smolts leaving fishways voluntarily with levels in fish that remain behind show that the former are consistently higher (Grau *et al.*, 1985a). Thus, the thyroxine surge seems to be a reliable index for coho, and this may be used to advantage with a lunar calendar. Loretz *et al.* (1982) suggested that the appropriate time for seawater entry of hatchery-reared coho salmon may be several weeks after the new moon–related thyroxine peak.

R. D. Ewing and associates have assessed gill ATPases in relation to the most favorable time for release of young chinooks (Ewing *et al.*, 1980a,b; Ewing and Birks, 1983). For this salmonid, gill ATPase activity seems to be a reliable index of smolting but one that is not always a necessary prerequisite to migration. Ewing *et al.* (1984) also found that the T4 surge is a poor index for predicting release time of chinooks.

The recognition of a *migratory readiness* as distinct from *physiological development* has been emphasized (Solomon, 1978). Physiological changes occur over an extended time span and are environmentally regulated (particularly by photoperiod and temperature); when the fish are in a proper state of migratory readiness, a proximal stimulus (lunar phase, stream flooding) initiates migration. Thus, behavioral and environmental factors as well as physiological changes must often be considered. Marked species variations are recognized, and, at this stage in fisheries science, there appears to be no simple and precise index of smolting climax and the most favorable time for release or transfer. The search for more reliable tests in predicting releasc times is likely to remain an important objective of salmonid biologists for some time.

ACKNOWLEDGMENTS

The manuscript was read critically by H. A. Bern, W. Craig Clarke, W. W. Dickhoff, Hans Lundqvist, and R. L. Saunders. I am grateful for their many helpful suggestions.

REFERENCES

Adams, B. L., Zaugg, W. S., and McLain, L. R. (1973). Temperature effect on parr–smolt transformation in steelhead trout (*Salmo gairdneri*) as measured by gill sodium–potassium stimulated adenosine triphosphatase. *Comp. Biochem. Physiol.* **44**, 1333–1339.

Adams, B. L., Zaugg, W. S., and McLain, L. R. (1975). Inhibition of saltwater survival and Na$^+$-K$^+$-ATPase elevation in steelhead trout (*Salmo gairdneri*) by moderate water temperatures. *Trans. Am. Fish. Soc.* **104**, 766–769.

Aida, K., Nishioka, R. S., and Bern, H. A. (1980). Changes in the corpuscles of Stannius of coho salmon (*Oncorhynchus kisutch*) during smoltification and seawater adaptation. *Gen. Comp. Endocrinol.* **41**, 296–304.

Aida, K., Kato, T., and Awaji, M. (1984). Effects of castration on the smoltification of precocious male masu salmon (*Oncorhynchus masou*). *Bull. Jpn. Soc. Sci. Fish.* **50**, 565–571.

Allendorf, F. W., and Utter, F. M. (1979). Population genetics. *In* "Fish Physiology" (W. S. Hoar, D. J. Randall, and J. R. Brett, eds.), Vol. 8, pp. 407–454. Academic Press, New York.

Andrews, J. (1963). Healthy salmon at U.B.C. show expert he's wrong. *Vancouver Sun*, Vancouver, British Columbia, Jan. 29.

Baggerman, B. (1960a). Salinity preference, thyroid activity and the seaward migration of four species of Pacific salmon (*Oncorhynchus*). *J. Fish. Res. Board Can.* **17**, 295–322.

Baggerman, B. (1960b). Factors in the diadromous migrations of fish. *Symp. Zool. Soc. London* **1**, 33–60.

Baggerman, B. (1962). Some endocrine aspects of fish migration. *Gen. Comp. Endocrinol., Suppl.* **1**, 188–205.

Baggerman, B. (1963). The effect of TSH and antithyroid substances on salinity preference and thyroid activity in juvenile Pacific salmon. *Can. J. Zool.* **41**, 307–319.

Baggerman, B. (1980). Photoperiodic and endogenous control of the annual reproductive cycle in teleost fishes. *In* "Environmental Physiology of Fishes" (M. A. Ali, ed.), pp. 533–567. Plenum, New York.

Baglinière, J. L., and Maisse, G. (1985). Precocious maturation and smoltification in wild Atlantic salmon in the Armorican Massif, France. *Aquaculture* **45**, 249–263.

Bailey, J. K., Saunders, R. L., and Buzeta, M. I. (1980). Influence of parental smolt age and sea age on growth and smolting of hatchery-reared Atlantic salmon (*Salmo salar*). *Can. J. Fish. Aquat. Sci.* **37**, 1379–1386.

Baraduc, M. M., and Fontaine, M. (1955). Étude comparée du métabolismé respiratoire du jeune Saumon sédentaire (parr) et migrateur (smolt). *C. R. Seances Soc. Biol. Ses. Fil.* **149**, 1327–1329.

Barrington, E. J. W., Barron, N., and Piggins, D. J. (1961). The influence of thyroid powder and thyroxine upon the growth of rainbow trout (*Salmo gairdneri*). *Gen. Comp. Endocrinol.* **1**, 170–178.

Barton, B. A., Schreck, C. B., Ewing, R. D., Hemmingsen, A. R., and Patiño, R. (1985). Changes in plasma cortisol during stress and smoltification of coho salmon, *Oncorhynchus kisutch*. *Gen. Comp. Endocrinol.* **59**, 468–471.

Basulto, S. (1976). Induced saltwater tolerance in connection with inorganic salts in the feeding of Atlantic salmon (*Salmo salar* L.). *Aquaculture* **8**, 45–55.

Bath, R. N., and Eddy, F. B. (1979). Salt and water balance in rainbow trout (*Salmo gairdneri*) rapidly transferred from fresh water to sea water. *J. Exp. Biol.* **83**, 193–202.

Berg, R. E. (1979). External morphology of the pink salmon, *Oncorhynchus gorbuscha*, introduced into Lake Superior. *J. Fish. Res. Board Can.* **36**, 1283–1287.

Bern, H. A. (1978). Endocrinological studies on normal and abnormal salmon smoltification. *In* "Comparative Endocrinology" (P. J. Gaillard and H. H. Boer, eds.), pp. 97–100. Elsevier/North-Holland, Amsterdam.

Bern, H. A., and Mahnken, C. V. W., eds. (1982). Salmonid smoltification. Proceedings of a symposium sponsored by the Pacific sea grant advisory program and the California Sea Grant College Program. *Aquaculture* **28**, 1–270.

Bilton, H. T., Alderdice, D. F., and Schnute, J. T. (1982). Influence of time and size at release of juvenile coho salmon (*Oncorhynchus kisutch*) on returns at maturity. *Can. J. Fish. Aquat. Sci.* **39**, 426–447.

Birks, E. K., Ewing, R. D., and Hemmingsen, A. R. (1985). Migration tendency in juvenile steelhead trout, *Salmo gairdneri* Richardson, injected with thyroxine and thiourea. *J. Fish Biol.* **26**, 291–300.

Birt, T. P., and Green, J. M. (1986). Parr–smolt transformation in sexually mature male anadromous and nonanadromous Atlantic salmon, *Salmo salar*. *Can. J. Fish. Aquat. Sci.* **43**, 680–686.

Black, J. J., and Simpson, C. L. (1974). Thyroid enlargement in Lake Erie coho salmon. *J. Natl. Cancer Inst.* (*U.S.*) **53**, 725–730.

Black, V. S. (1951). Changes in body chloride, density, and water content of chum (*Oncorhynchus keta*) and coho (*O. kisutch*) salmon fry when transferred from fresh water to sea water. *J. Fish. Res. Board Can.* **8**, 164–177.

Blake, R. L., Roberts, F. L., and Saunders, R. L. (1984). Parr–smolt transformation of Atlantic salmon (*Salmo salar*): Activities of two respiratory enzymes and concentrations of mitochondria in the liver. *Can. J. Fish. Aquat. Sci.* **41**, 199–203.

Boeuf, G., and Harache, Y. (1982). Criteria for adaptation of salmonids to high salinity seawater in France. *Aquaculture* **28**, 163–176.

Boeuf, G., and Harache, Y. (1984). Adaptation osmotique à l'eau de mer de differentes espèces (*Salmo trutta*, *Salmo gairdneri*, *Salvelinus fontinalis*) et hydride (*Salmo trutta* ♀ × *Salvelinus fontinalis* ♂) de salmonides. *Aquaculture* **40**, 343–358.

Boeuf, G., and Prunet, P. (1985). Measurements of gill (Na$^+$-K$^+$)-ATPase activity and plasma thyroid hormones during smoltification in Atlantic salmon (*Salmo salar* L.). *Aquaculture* **45**, 111–119.

Boeuf, G., Lasserre, P., and Harache, Y. (1978). Osmotic adaptation of *Oncorhynchus kisutch* Walbaum. II. Plasma osmotic and ionic variations, and gill (Na$^+$-K$^+$)-ATPase activity of yearling coho salmon transferred to sea water. *Aquaculture* **15**, 35–52.

Boeuf, G., Le Roux, A., Gaignon, J. L., and Harache, Y. (1985). Gill (Na$^+$-K$^+$)-ATPase activity and smolting in Atlantic salmon (*Salmo salar* L.) in France. *Aquaculture* **45**, 73–81.

Bolton, J. P., Collie, N. L., Kawauchi, H., and Hirano, T. (1987a). Osmoregulatory actions of growth hormone in rainbow trout (Salmo gairdneri). J. Endocrinol. 112, 63–68.

Bolton, J. P., Young, G., Nishioka, R. S., Hirano, T., and Bern, H. A. (1987b). Plasma growth hormone levels in normal and stunted yearling coho salmon (Oncorhynchus kisutch). J. Exp. Zool. 242, 379–382.

Bowen, J. T. (1970). A history of fish culture as related to the development of fishery programs. In "A Century of Fisheries in North America" (N. G. Benson, ed.), Spec. Publ. No. 7. Am. Fish. Soc., Washington, D.C.

Bradley, T. M., and Rourke, A. W. (1984). An electrophoretic analysis of plasma proteins from juvenile Oncorhynchus tshawytscha (Walbaum). J. Fish Biol. 24, 703–709.

Brannon, E., Feldmann, C., and Donaldson, L. (1982). University of Washington zero-age coho smolt production. Aquaculture 28, 195–200.

Brauer, E. P. (1982). The photoperiod control of coho salmon smoltification. Aquaculture 28, 105–111.

Brown, M. E. (1945). The growth of brown trout (Salmo trutta L.). II. The growth of two-year-old trout at a constant temperature of 11.5°C. J. Exp. Biol. 22, 130–144.

Burton, M. P., and Idler, D. R. (1984). Can Newfoundland landlocked salmon, Salmo salar L., adapt to sea water? J. Fish Biol. 24, 59–64.

Chan, H. H., and Eales, J. G. (1976). Influence of bovine TSH on plasma thyroxine levels and thyroid function in brook trout, Salvelinus fontinalis (Mitchill). Gen. Comp. Endocrinol. 28, 461–472.

Chernitsky, A. G. (1980). Functional state of chloride cells of Baltic salmon (Salmo salar L.) at different stages of its life cycle. Comp. Biochem. Physiol. A 67A, 519–522.

Chernitsky, A. G. (1986). Quantitative evaluation of the degree of parr-smolt transformation in wild smolts and hatchery juveniles of Atlantic salmon (Salmo salar L.) by SDH activity of chloride cells. Aquaculture 59, 287–297.

Chester-Jones, I., and Henderson, I. W. (1980). "General, Comparative and Clinical Endocrinology," Vol. 3, pp. 395–523. Academic Press, London.

Chester-Jones, I., Chan, D. K. O., Henderson, I. W., and Ball, J. N. (1969). The adrenocortical steroids, adrenocorticotropin and the Corpuscles of Stannius. In "Fish Physiology" (W. S. Hoar and D. J. Randall, eds.), Vol. 2, pp. 321–376. Academic Press, New York.

Chua, D., and Eales, J. G. (1971). Thyroid function and dermal purines in brook trout, Salvelinus fontinalis (Mitchill). Can. J. Zool. 49, 1557–1561.

Clarke, W. C. (1982). Evaluation of the seawater challenge test as an index of marine survival. Aquacultures 28, 177–183.

Clarke, W. C., and Bern, H. A. (1980). Comparative endocrinology of prolactin. In "Hormonal Proteins and Peptides" (C. H. Li, ed.), Vol. 8, pp. 105–197. Academic Press, New York.

Clarke, W. C., and Blackburn, J. (1978). Seawater challenge test performed on hatchery stocks of chinook and coho salmon in 1977. Tech. Rep.—Fish. Mar. Serv. (Can.) 761, 1–19.

Clarke, W. C., and Nagahama, Y. (1977). Effect of premature transfer to sea water on growth and morphology of the pituitary, thyroid, pancreas, and interrenal in juvenile coho salmon (Oncorhynchus kisutch). Can. J. Zool. 55, 1620–1630.

Clarke, W. C., and Shelbourn, J. E. (1980). Growth and smolting of under-yearling coho salmon in relation to photoperiod and temperature. Proc. N. Pac. Aquacult. Symp., 1980, pp. 209–216.

Clarke, W. C., and Shelbourn, J. E. (1985). Growth and development of seawater adaptability by juvenile fall chinook salmon (*Oncorhynchus tshawytscha*) in relation to temperature. *Aquaculture* **45**, 21–31.

Clarke, W. C., and Shelbourn, J. E. (1986). Delayed photoperiod produces more uniform growth and greater seawater adaptability in underyearling coho salmon (*Oncorhynchus kisutch*). *Aquaculture* **56**, 287–299.

Clarke, W. C., Farmer, S. W., and Hartwell, K. M. (1977). Effect of teleost pituitary growth hormone on growth of *Tilapia mossambica* and on growth and seawater adaptation of sockeye salmon *Oncorhynchus nerka*. *Gen. Comp. Endocrinol.* **33**, 174–178.

Clarke, W. C., Shelbourn, J. E., and Brett, J. R. (1978). Growth and adaptation to sea water in 'underyearling' sockeye (*Oncorhynchus nerka*) and coho (*O. kisutch*) salmon subjected to regimes of constant or changing temperature and day length. *Can. J. Zool.* **56**, 2413–2421.

Clarke, W. C., Shelbourn, J. E., and Brett, J. R. (1981). Effect of artificial photoperiod cycles, temperature, and salinity on growth and smolting in underyearling coho (*Oncorhynchus kisutch*), chinook (*O. tshawytscha*) and sockeye (*O. nerka*) salmon. *Aquaculture* **22**, 105–116.

Clarke, W. C., Lundqvist, H., and Eriksson, L.-O. (1985). Accelerated photoperiod advances seasonal cycle of seawater adaptation in juvenile Baltic salmon, *Salmo salar* L. *J. Fish Biol.* **26**, 29–35.

Collie, N. L. (1985). Intestinal nutrient transport in coho salmon (*Oncorhynchus kisutch*) and the effects of development, starvation, and seawater adaptation. *J. Comp. Physiol.* **B156**, 163–174.

Collie, N. L., and Bern, H. A. (1982). Changes in intestinal fluid transport associated with smoltification and seawater adaptation in coho salmon, *Oncorhynchus kisutch* (Walbaum). *J. Fish Biol.* **21**, 337–348.

Collie, N. L., and Stevens, J. J. (1985). Hormonal effects on L-proline transport in coho salmon (*Oncorhynchus kisutch*) intestine. *Gen. Comp. Endocrinol.* **59**, 399–409.

Collins, J. J. (1975). Occurrence of pink salmon (*Oncorhynchus gorbuscha*) in Lake Huron. *J. Fish. Res. Board Can.* **32**, 402–404.

Colville, T. P., Richards, R. H., and Dobbie, J. W. (1983). Variations in renal corpuscular morphology with adaptation to sea water in the rainbow trout, *Salmo gairdneri* Richardson. *J. Fish Biol.* **23**, 451–456.

Conte, F. P. (1969). Salt secretion. *In* "Fish Physiology" (W. S. Hoar and D. J. Randall, eds.), Vol. 1, pp. 241–292. Academic Press, New York.

Conte, F. P., and Wagner, H. H. (1965). Development of osmotic and ionic regulation in juvenile steelhead trout *Salmo gairdneri*. *Comp. Biochem. Physiol.* **14**, 603–620.

Conte, F. P., Wagner, H. H., Fessler, J., and Gnose, C. (1966). Development of osmotic and ionic regulation in juvenile coho salmon *Oncorhynchus kisutch*. *Comp. Biochem. Physiol.* **18**, 1–15.

Cowey, C. B. (1982). Special issue on fish biochemistry. *Comp. Biochem. Physiol. B* **73B**, 1–180.

Crim, L. W., and Evans, D. M. (1978). Seasonal levels of pituitary and plasma gonadotropin in male and female Atlantic salmon parr. *Can. J. Zool.* **56**, 1550–1555.

Dalley, E. L., Andrews, C. W., and Green, J. M. (1983). Precocious male Atlantic salmon parr (*Salmo salar*) in insular Newfoundland. *Can. J. Fish. Aquat. Sci.* **40**, 647–652.

Davis, J. C., and Shand, I. G. (1978). Acute and sublethal copper sensitivity, growth and saltwater survival in young Babine Lake sockeye salmon. *Tech. Rep.—Fish. Mar. Serv. (Can.)* **847**, 1–55.

Day, F. (1887). "British and Irish Salmonidae." Williams & Norgate, London.

de Renzis, G., and Bornancin, M. (1984). Ion transport and gill ATPases. In "Fish Physiology" (W. S. Hoar and D. J. Randall, eds.), Vol. 10B, pp. 65–104. Academic Press, New York.

Dickhoff, W. W., and Sullivan, C. V. (1987). The thyroid gland in smoltification. Am. Fish Soc. Symp. 1, 197–210.

Dickhoff, W. W., Darling, D. S., and Gorbman, A. (1982). Thyroid function during smoltification of salmonid fishes. Gunma Symp. Endocrinol. 19, 45–61.

Dickhoff, W. W., Sullivan, C. V., and Mahnken, C. V. W. (1985). Thyroid hormones and gill ATPase during smoltification of Atlantic salmon (Salmo salar). Aquaculture 45, 376.

Dimberg, K., Höglund, L. B., Knutsson, P. G., and Riddersträle, Y. (1981). Histochemical localization of carbonic anhydrase in gill lamellae from young salmon (Salmo salar L.) adapted to fresh water and salt water. Acta Physiol. Scand. 112, 218–220.

Dodd, J. M., Stuart-Kregor, P. A. C., Sumpter, J. P., Crim, L. W., and Peter, R. E. (1978). Premature sexual maturation in the Atlantic salmon (Salmo salar L.). In "Comparative Endocrinology" (P. J. Gaillard and H. H. Boer, eds.), pp. 101–104. Elsevier/ North-Holland, Amsterdam.

Donaldson, E. M. (1981). The pituitary–interrenal axis as an indicator of stress in fish. In "Stress and Fish" (A. D. Pickering, ed.), pp. 11–48. Academic Press, New York.

Donaldson, E. M., Fagerlund, U. H. M., Higgs, D. A., and McBride, J. R. (1979). Hormonal enhancement of growth. In "Fish Physiology" (W. S. Hoar, D. J. Randall, and J. R. Brett, eds.), Vol. 8, pp. 455–597. Academic Press, New York.

Donaldson, L. R., and Brannon, E. L. (1975). The use of warmed water to accelerate the production of coho salmon. Fisheries (Am. Fish. Soc., Bethesda) 1(4), 12–15.

Eales, J. G. (1963). A comparative study of thyroid function in migrant juvenile salmon. Can. J. Zool. 41, 811–824.

Eales, J. G. (1965). Factors influencing seasonal changes in thyroid activity in juvenile steelhead trout, Salmo gairdneri. Can. J. Zool. 43, 719–729.

Eales, J. G. (1969). A comparative study of purines responsible for silvering in several freshwater fishes. J. Fish. Res. Board Can 26, 1927 1031.

Eales, J. G. (1979). Thyroid in cyclostomes and fishes. In "Hormones and Evolution" (E. J. W. Barrington, ed.), Vol. 1, pp. 341–436. Academic Press, London.

Eales, J. G. (1981). Extrathyroidal effects of low concentrations of thiourea on rainbow trout, Salmo gairdneri. Can. J. Fish. Aquat. Sci. 38, 1283–1285.

Eales, J. G., and Shostak, S. (1985). Correlation between food ration, somatic growth parameters and thyroid function in Arctic charr, Salvelinus alpinus L. Comp Biochem. Physiol. A 80A, 553–558.

Eales, J. G., Chang, J. P., Van der Kraak, G., Omeljaniuk, R. J., and Uin, L. (1982). Effects of temperature on plasma thyroxine and iodine kinetics in rainbow trout, Salmo gairdneri. Gen. Comp. Endocrinol. 47, 295–307.

Eddy, F. B., and Bath, R. N. (1979). Ionic regulation in rainbow trout (Salmo gairdneri) adapted to fresh water and dilute sea water. J. Exp. Biol. 83, 181–192.

Eddy, F. B., and Talbot, C. (1985). Urine production in smolting Atlantic salmon, Salmo salar L. Aquaculture 45, 67–72.

Elson, P. F. (1957). The importance of size in the change from parr to smolt in Atlantic salmon. Can. Fish. Cult. 21, 1–6.

Eriksson, L.-O., and Lundqvist, H. (1982). Circannual rhythms and photoperiod regulation of growth and smolting in Baltic salmon (Salmo salar L.). Aquaculture 28, 113–121.

Eriksson, L.-O., Lundqvist, H., and Johansson, H. (1979). On the normality of size distribution and the precocious maturation in a Baltic salmon, *Salmo salar* L., parr population. *Aquilo, Ser. Zool.* **19**, 81–86.

Evans, D. H. (1984). The roles of gill permeability and transport mechanisms in euryhalinity. In "Fish Physiology" (W. S. Hoar and D. J. Randall, eds.), Vol. 10B, pp. 239–283. Academic Press, New York.

Evans, G. T., Rice, J. C., and Chadwick, E. M. P. (1984). Patterns of growth and smolting of Atlantic salmon (*Salmo salar*) parr. *Can. J. Fish. Aquat. Sci.* **41**, 783–797.

Evans, G. T., Rice, J. C., and Chadwick, E. M. P. (1985). Patterns of growth and smolting of Atlantic salmon (*Salmo salar*) parr in a southwestern Newfoundland river. *Can. J. Fish. Aquat. Sci.* **42**, 539–543.

Evropeizeva, N. V. (1959). Experimental analysis of the young salmon (*Salmo salar* L.) in the stage of transition to the sea. *Rapp P.-V. Réun, Cons. Int. Explor. Mer.* **148**, 29–39.

Evropeizeva, N, V, (1960). Correlation between the process of early gonad ripening, and transformation to the seaward-migrating stage, among male Baltic salmon (*Salmo salar* L.) held in ponds (in Russian). *Zool. Zh.* **39**, 777–779; *Fish. Res. Board Can., Transl. Ser.* **430**.

Ewing, R. D., and Birks, E. K. (1982). Criteria for parr–smolt transformation in juvenile chinook salmon (*Oncorhynchus tshawytscha*). *Aquaculture* **28**, 185–194.

Ewing, R. D., Johnson, S. L., Pribble, H. J., and Lichatowich, J. A. (1979). Temperature and photoperiod effects on gill (Na+K)-ATPase activity in chinook salmon (*Oncorhynchus tshawytscha*). *J. Fish. Res. Board Can.* **36**, 1347–1353.

Ewing, R. D., Pribble, H. J., Johnson, S. L., Fustish, C. A., Diamond, J., and Lichatowich, J. A. (1980a). Influence of size, growth rate, and photoperiod on cyclic changes in gill (Na+K)-ATPase activity in chinook salmon (*Oncorhynchus tshawytscha*). *Can. J. Fish. Aquat. Sci.* **37**, 600–605.

Ewing, R. D., Fustish, C. A., Johnson, S. L., and Pribble, H. J. (1980b). Seaward migration of juvenile chinook salmon without elevated gill (Na+K)-ATPase activities. *Trans. Am. Fish. Soc.* **109**, 349–356.

Ewing, R. D., Evenson, M. D., Birks, E. K., and Hemmingsen, A. R. (1984). Indices of parr–smolt transformation in juvenile steelhead trout (*Salmo gairdneri*) undergoing volitional release at Cole River Hatchery, Oregon. *Aquaculture* **40**, 209–221.

Ewing, R. D., Hemmingsen, A. R., Evenson, M. D., and Lindsay, R. L. (1985). Gill (Na+K)-ATPase activity and plasma thyroxine concentrations do not predict time of release of hatchery coho (*Oncorhynchus kisutch*) and chinook salmon (*Oncorhynchus tshawytscha*) for maximum adult returns. *Aquaculture* **45**, 359–373.

Fagerlund, U. H. M., Higgs, D. A., McBride, J. R., Plotnikoff, M. D., and Dosanjh, B. S. (1980). The potential for using the anabolic hormones 17α-methyltestosterone and (or) 3,5,3'-triiodo-L-thyronine in fresh water rearing of coho salmon (*Oncorhynchus kisutch*) and the effects on subsequent seawater performance. *Can. J. Zool.* **58**, 1424–1432.

Farmer, G. J., Ritter, J. A., and Ashfield, D. (1978). Seawater adaptation and parr–smolt transformation of juvenile Atlantic salmon, *Salmo salar. J. Fish. Res. Board Can.* **35**, 93–100.

Foerster, R. E. (1937). The relation of temperature to the seaward migration of young sockeye salmon (*Oncorhynchus nerka*). *J. Biol. Board Can.* **3**, 421–438.

Foerster, R. E. (1968). The sockeye salmon, *Oncorhynchus nerka*. *Bull. Fish. Res. Board Can.* **162**, 1–422.

Folmar, L. C., and Dickhoff, W. W. (1979). Plasma thyroxine and gill Na+-K+ ATPase changes during seawater acclimation of coho salmon, *Oncorhynchus kisutch*. *Comp. Biochem. Physiol. A* **63A**, 329–332.

Folmar, L. C., and Dickhoff, W. W. (1980). The parr–smolt transformation (smoltification) and seawater adaptation in salmonids: A review of selected literature. *Aquaculture* **21**, 1–37.

Folmar, L. C., and Dickhoff, W. W. (1981). Evaluation of some physiological parameters as predictive indices of smoltification. *Aquaculture* **23**, 309–324.

Folmar, L. C., Dickhoff, W. W., Mahnken, C. V. W., and Waknitz, F. W. (1982a). Stunting and parr-reversion during smoltification of coho salmon (*Oncorhynchus kisutch*). *Aquaculture* **28**, 91–104.

Folmar, L. C., Dickhoff, W. W., Zaugg, W. S., and Hodgins, H. O. (1982b). The effects of aroclor 1254 and No. 2 fuel oil on smoltification and seawater adaptation of coho salmon (*Oncorhynchus kisutch*). *Aquat. Toxicol.* **2**, 291–299.

Fontaine, M. (1951). Sur la diminution de la teneur en chlore du muscle des jeunes saumons (smolts) lors de la migration d'avalaison. *C. R. Hebd. Seances Acad. Sci.* **232**, 2477–2479.

Fontaine, M. (1954). Du déterminisme physiologique des migrations. *Biol. Rev. Cambridge Philos. Soc.* **29**, 390–418.

Fontaine, M. (1975). Physiological mechanisms in the migration of marine and amphihaline fish. *Adv. Mar. Biol.* **13**, 241–355.

Fontaine, M., and Hatey, J. (1950). Variations de la teneur du foie en glycogène chez le jeune saumon (*Salmo salar* L.) au cours de la smoltification. *C. R. Seances Soc. Biol. Ses Fil.* **144**, 953–955.

Fontaine, M., and Hatey, J. (1954). Sur la teneur en 17-hydroxycorticostéroides du plasma de saumon (*Salmo salar* L.). *C. R. Hebd. Seances Acad. Sci.* **239**, 319–321.

Fontaine, M., and Olivereau, M. (1957). Interrénal antérieur et smoltification chez *Salmo salar* (L.). *J. Physiol. (Paris)* **49**, 174–176.

Fontaine, M., and Olivereau, M. (1959). Interrénal antérieur et smoltification chez *Salmo salar* L. *Bull. Soc. Zool. Fr.* **84**, 161–162.

Foskett, J. K., and Scheffey, C. (1982). The chloride cell: Definitive identification as the salt-secreting cell in teleosts. *Science* **215**, 164–166.

Fried, S. M., McCleave, J. D., and LaBar, G. W. (1978). Seaward migration of hatchery-reared Atlantic salmon, *Salmo salar*, smolts in the Penobscot River estuary, Maine: Riverine movements. *J. Fish Res. Board Can.* **35**, 76–87.

Fryer, J. N., and Bern, H. A. (1979). Growth hormone binding to tissues of normal and stunted juvenile coho salmon, *Oncorhynchus kisutch*. *J. Fish Biol.* **15**, 527–533.

Gebhards, S. V. (1960). Biological notes on precocious male chinook salmon parr in the Salmon River drainage, Idaho. *Prog. Fish-Cult.* **22**(3), 121–123.

Giles, M. A., and Randall, D. J. (1980). Oxygenation characteristics of the polymorphic hemoglobins of coho salmon (*Oncorhynchus kisutch*) at different developmental stages. *Comp. Biochem. Physiol. A* **65A**, 265–271.

Giles, M. A., and Vanstone, W. E. (1976). Ontogenetic variation in the multiple hemoglobins of coho salmon (*Oncorhynchus kisutch*) and the effect of environmental factors on their expression. *J. Fish. Res. Board Can.* **33**, 1144–1149.

Glebe, B. D., Saunders, R. L., and Sreedharan, A. (1978). Genetic and environmental influence in expression of precocious sexual maturity of hatchery-reared Atlantic salmon (*Salmo salar*) parr. *Can. J. Genet. Cytol.* **20**, 444.

Godin, J.-G., Dill, P. A., and Drury, D. E. (1974). Effects of thyroid hormones on behaviour of yearling Atlantic salmon (*Salmo salar*). *J. Fish. Res. Board Can.* **31**, 1787–1790.

Gorbman, A., Dickhoff, W. W., Mighell, J. L., Prentice, E. F., and Waknitz, F. W. (1982). Morphological indices of developmental progress in the parr–smolt coho salmon, *Oncorhynchus kisutch*. *Aquaculture* **28**, 1–19.

Gordon, M. S. (1957). Observations on osmoregulation in the Arctic char (*Salvelinus alpinus* L.). *Biol. Bull. (Woods Hole, Mass.)* **112**, 28–32.

Grau, E. G. (1982). Is the lunar cycle a factor timing the onset of salmon migration? *In* "Salmon and Trout Migratory Behavior Symposium" (E. L. Brannon and E. O. Salo, eds.), pp. 184–189. Univ. of Washington Press, Seattle.

Grau, E. G., Dickhoff, W. W., Nishioka, R. S., Bern, H. A., and Folmar, L. C. (1981). Lunar phasing of the thyroxine surge preparatory to seaward migration of salmonid fish. *Science* **221**, 607–609.

Grau, E. G., Specker, J. L., Nishioka, R. S., and Bern, H. A. (1982). Factors determining the occurrence of the surge in thyroid activity in salmon during smoltification. *Aquaculture* **28**, 49–57.

Grau, E. G., Fast, A. W., Nishioka, R. S., Bern, H. A., Barclay, D. K., and Katase, S. A. (1985a). Variations in thyroid hormone levels and in performance in the seawater challenge test accompanying development in coho salmon raised in Hawaii. *Aquaculture* **45**, 121–132.

Grau, E. G., Fast, A. W., Nishioka, R. S., and Bern, H. A. (1985b). The effect of transfer to novel conditions on blood thyroxine in coho salmon. *Aquaculture* **45**, 377.

Grau, E. G., Nishioka, R. S., Specker, J. L., and Bern, H. A. (1985c). Endocrine involvement in the smoltification of salmon with special reference to the role of the thyroid gland. *In* "Current Trends in Comparative Endocrinology" (B. Lofts and W. N. Holmes, eds.), pp. 491–493. Hong Kong Univ. Press, Hong Kong.

Gudjónsson, T. (1972). Smolt rearing techniques, stocking and tagged adult salmon recaptures in Iceland. *Int. Atl. Salmon Found., Spec. Publ.* **4**, 227–235.

Gunnes, K., and Gjedrem, T. (1978). Selection experiments with salmon. IV. Growth of Atlantic salmon during two years in the sea. *Aquaculture* **15**, 19–33.

Hard, J. J., Wertheimer, A. C., Heard, W. R., and Martin, R. M. (1985). Early male maturity in two stocks of chinook salmon (*Oncorhynchus tshawytscha*) transplanted to an experimental hatchery in southeastern Alaska. *Aquaculture* **48**, 351–359.

Hashimoto, K., and Matsuura, F. (1960). Comparative studies on two hemoglobins of salmon. V. Change in proportion of two hemoglobins with growth. *Bull. Jpn. Soc. Sci. Fish.* **26**, 931–937.

Hayashi, S. (1970). Biochemical studies on the skin of fish. II. Seasonal changes of purine content of masu salmon from parr to smolt. *Bull. Jpn. Soc. Sci. Fish.* **37**, 508–512.

Healey, M. C. (1983). Coastwide distribution and ocean migration patterns of stream- and ocean-type chinook salmon, *Oncorhynchus tshawytscha*. *Can. Field Nat.* **97**, 427–433.

Henderson, I. W., Brown, J. A., Oliver, J. A., and Haywood, G. P. (1978). Hormones and single nephron function in fishes. *In* "Comparative Endocrinology" (P. J. Gaillard and H. H. Boer, eds.), pp. 217–222. Elsevier/North-Holland, Amsterdam.

Hickman, C. P., Jr., and Trump, B. F. (1969). The kidney. *In* "Fish Physiology" (W. S. Hoar and D. J. Randall, eds.), Vol. 1, pp. 91–239. Academic Press, New York.

Higgins, P. J. (1985). Metabolic differences between Atlantic salmon (*Salmo salar*) parr and smolts. *Aquaculture* **45**, 33–53.

Higgs, D. A., Donaldson, E. M., Dye, H. M., and McBride, J. R. (1976). Influence of bovine growth hormone and L-thyroxine on growth, muscle composition, and histological structure of the gonads, thyroid, pacreas, and pituitary of coho salmon (*Oncorhynchus kisutch*). *J. Fish. Res. Board Can.* **33**, 1585–1603.

Higgs, D. A., Fagerlund, U. H. M., McBride, J. R., Dye, H. M., and Donaldson, S. M.

(1977). Influence of combinations of bovine growth hormone, 17 α-methyltestosterone, and L-thyroxine on growth of yearling coho salmon (*Oncorhynchus kisutch*). *Can. J. Zool.* **55**, 1048–1056.

Higgs, D. A., Fagerlund, U. H. M., McBride, J. R., and Eales, J. G. (1979). Influence of orally administered L-thyroxine or 3,5,3'-triiodo-L-thyronine on growth, food consumption, and food conversion of underyearling coho salmon (*Oncorhynchus kisutch*). *Can. J. Zool.* **57**, 1974–1979.

Higgs, D. A., Fagerlund, U. H. M., Eales, J. G., and McBride, J. R. (1982). Application of thyroid and steriod hormones as anabolic agents in fish culture. *Comp. Biochem. Physiol. B* **73B**, 143–176.

Hirano, T. (1986). The spectrum of prolactin action in teleosts. *Prog. Clin. Biol. Res.* **205**, 53–74.

Hirano, T., Johnson, D. W., Bern, H. A., and Utida, S. (1973). Studies on water and ion movements in the isolated urinary bladder of selected freshwater, marine and euryhaline teleosts. *Comp. Biochem. Physiol. A* **45A**, 529–540.

Hoar, W. S. (1939). The thyroid gland of the Atlantic salmon. *J. Morphol.* **65**, 257–295.

Hoar, W. S. (1965). The endocrine system as a chemical link between the organism and its environment. *Trans. R. Soc. Can., Ser. IV*, **3**, 175–200.

Hoar, W. S. (1966). Hormonal activities of the pars distalis in cyclostomes, fish and amphibia. *In* "The Pituitary Gland" (G. W. Harris and B. T. Donovan, eds.), Vol. 1, pp. 242–294. Butterworth, London.

Hoar, W. S. (1976). Smolt transformation: Evolution, behavior, and physiology. *J. Fish. Res. Board Can.* **33**, 1233–1252.

Hoar, W. S. (1983). "General and Comparative Physiology," 3rd ed. Prentice-Hall, Englewood Cliffs, New Jersey.

Hoar, W. S., and Bell, G. M. (1950). The thyroid gland in relation to the seaward migration of Pacific salmon. *Can. J. Res., Sect. D* **28**, 126–136.

Hoar, W. S., and Randall, D. J., eds. (1984). "Fish Physiology," Vols. 10A and 10B. Academic Press, New York.

Hogstrand, C., and Haux, C. (1985). Evaluation of the sea-water challenge test on sea trout, *Salmo trutta*. *Comp. Biochem. Physiol., A* **82A**, 261–266.

Holmes, W. N., and Stainer, I. M. (1966). Studies on the renal excretion of electrolytes by the trout (*Salmo gairdneri*). *J. Exp. Biol.* **44**, 33–46.

Houston, A. H. (1959). Osmoregulatory adaptation of steelhead trout (*Salmo gairdneri* Richardson) to sea water. *Can. J. Zool.* **37**, 729–748.

Houston, A. H. (1961). Influence of size upon the adaptation of steelhead trout (*Salmo gairdneri*) and chum salmon (*Oncorhynchus keta*) to sea water. *J. Fish. Res. Board Can.* **18**, 401–415.

Houston, A. H., and Threadgold, L. T. (1963). Body fluid regulation in smolting Atlantic salmon. *J. Fish. Res. Board Can.* **20**, 1355–1369.

Hunt, D. W. C., and Eales, J. G. (1979). The influence of testosterone propionate on thyroid function of immature rainbow trout, *Salmo gairdneri* Richardson. *Gen. Comp. Endocrinol.* **37**, 115–121.

Ikuta, K., Aida, K., Okumoto, N., and Hanyu, I. (1985). Effects of thyroxine and methyltestosterone on smoltification of masu salmon (*Oncorhynchus masou*). *Aquaculture* **45**, 289–303.

Ikuta, K., Aida, K., Okumoto, N., and Hanyu, I. (1987). Effects of sex steriods on smoltification of masu salmon, *Oncorhynchus masou*. *Gen. Comp. Endocrinol.* **65**, 99–110.

Isaia, J. (1984). Water and nonelectrolyte permeation. *In* "Fish Physiology" (W. S. Hoar and D. J. Randall, eds.), Vol. 10B, pp. 1–38. Academic Press, New York.

Isaia, J., Payan, P., and Girard, J.-P. (1979). A study of the water permeability of the gills of freshwater- and seawater-adapted trout (*Salmo gairdneri*): Mode of action of epinephrine. *Physiol. Zool.* **52**, 269–279.

Ísaksson, Á. (1985). The production of one-year smolts and prospects of producing zero-smolts of Atlantic salmon in Iceland using geothermal resources. *Aquaculture* **45**, 305–319.

Iwata, M., and Komatsu, S. (1984). Importance of estuarine residence for adaptation of chum salmon (*Oncorhynchus keta*) fry to seawater. *Can. J. Fish. Aquat. Sci.* **41**, 744–749.

Iwata, M., Ogura, H., Komatsu, S., Suzuki, K., Nishioka, R. S., and Bern, H. A. (1985). Changes in salinity preference of chum and coho salmon during development. *Aquaculture* **45**, 380–381.

Iwata, M., Ogura, H., Komatsu, S., and Suzuki, K. (1986). Loss of seawater preference in chum salmon (*Oncorhynchus keta*) fry retained in fresh water after migration. *J. Exp. Zool.* **240**, 369–376.

Jackson, A. J. (1981). Osmotic regulation in rainbow trout (*Salmo gairdneri*) following transfer to sea water. *Aquaculture* **24**, 143–151.

Johnson, S. L., Ewing, R. D., and Lichatowick, J. A. (1977). Characterization of gill (Na+K)-ATPase activated adenosine triphosphatase from chinook salmon, *Oncorhynchus tshawytscha*. *J. Exp. Zool.* **199**, 345–354.

Johnston, C. E., and Eales, J. G. (1970). Influence of body size on silvering of Atlantic salmon (*Salmo salar*) at parr-smolt transformation. *J. Fish. Res. Board Can.* **27**, 983–987.

Johnston, C. E., and Saunders, R. L. (1981). Parr–smolt transformation of yearling Atlantic salmon (*Salmo salar*) at several rearing temperatures. *Can. J. Fish. Aquat. Sci.* **38**, 1189–1198.

Jones, J.W. (1959). "The Salmon," Chapter 7, pp. 116–129. Collins, London.

Jonsson, B., and Ruud-Hansen, J. (1985). Water temperature as primary influence on timing of seaward migration of Atlantic salmon (*Salmo salar*) smolts. *Can. J. Fish. Aquat. Sci.* **42**, 593–595.

Jonsson, B., and Sandlund, O. T. (1979). Environmental factors and life histories of isolated river stocks of brown trout (*Salmo trutta m. fario*) in Søre river system, Norway. *Environ. Biol. Fishes* **4**, 43–54.

Kashiwagi, M., and Sato, R. (1969). Studies on the osmoregulation of chum salmon, *Oncorhynchus keta* (Walbaum). 1. The tolerance of eyed period eggs, alevins and fry of chum salmon to sea water. *Tohoku J. Agric. Res.* **20**, 41–47.

Katz, A. H., and Katz, H. M. (1978). Effects of DL-thyroxine on swimming speed in the pearl danio, *Brachydanio albolineatus* (Blyth). *J. Fish Biol.* **12**, 527–530.

King, G. M., Jones, J. W., and Orton, J. H. (1939). Behaviour of mature male salmon parr, *Salmo salar* juv. L. *Nature (London)* **143**, 162–163.

Knutsson, S., and Grav, T. (1976). Seawater adaptation in Atlantic salmon (*Salmo salar* L.) at different experimental temperatures and photoperiods. *Aquaculture* **8**, 169–187.

Koch, H. J. A. (1982). Hemoglobin changes with size in the Atlantic salmon (*Salmo salar* L.). *Aquaculture* **28**, 231–240.

Komourdjian, M. P., and Idler, D. R. (1977). Hypophysectomy of rainbow trout, *Salmo gairdneri*, and its effect on plasmatic sodium regulation. *Gen. Comp. Endocrinol.* **32**, 536–542.

Komourdjian, M. P., Saunders, R. L., and Fenwick, J. C. (1976a). The effect of porcine somatotropin on growth, and survivial in seawater of Atlantic salmon (*Salmo salar*) parr. *Can. J. Zool.* **54**, 531–535.

Komourdjian, M. P., Saunders, R. L., and Fenwick, J. C. (1976b). Evidence for the role of growth hormone as a part of a 'light-pituitary axis' in growth and smoltification of Atlantic salmon (Salmo salar). Can. J. Zool. 54, 544–551.

Komourdjian, M. P., Burton, M. P., and Idler, D. R. (1978). Growth of rainbow trout, Salmo gairdneri, after hypophysectomy and somatotropin therapy. Gen. Comp. Endocrinol. 34, 158–162.

Kristinsson, J. B., Saunders, R. L., and Wiggs, A. J. (1985). Growth dynamics during the development of bimodal length–frequency distribution in juvenile Atlantic salmon (Salmo salar L.). Aquaculture 45, 1–20.

Kubo, T. (1955). Changes in some characteristics of blood of smolts of Oncorhynchus masou during seaward migration. Bull. Fac. Fish., Hokkaido Univ. 6, 201–207.

Lahlou, B., and Fossat, B. (1971). Méchanisme du transport de l'eau et du sel à travers le vessie urinaire d'un poisson téléostéen en eau douce, la truite arc-en-ciel. C. R. Hebd. Seances Acad. Sci. 273, 2108–2110.

Lahlou, B., Crenesse, D., Bensahla-Talet, A., and Porthe-Nibelle, J. (1975). Adaptation de la truite d'élevage à l'eau de mer. Effets sur les concentrations plasmatiques, les échanges branchiaux et le transport intestinal du sodium. J. Physiol. (Paris) 70, 593–603.

Lam, T. J. (1983). Environmental influences on gonadal activity in fish. In "Fish Physiology" (W. S. Hoar, D. J. Randall, and E. M. Donaldson, eds.), Vol. 9B, pp. 65–116. Academic Press, New York.

Langdon, J. S., and Thorpe, J. E. (1984). Responses of gill Na+-K+ ATPase activity, succinic dehydrogenase activity and chloride cells to seawater adaptation in Atlantic salmon, Salmo salar L., parr and smolt. J. Fish Biol. 24, 323–331.

Langdon, J. S., and Thorpe, J. E. (1985). The ontogeny of smoltification: Developmental patterns of gill Na+/K+-ATPase, SDH, and chloride cells in juvenile Atlantic salmon, Salmo salar L. Aquaculture 45, 83–95.

Langdon, J. S., Thorpe, J. E., and Roberts, R. J. (1984). Effects of cortisol and ACTH on gill Na+/K+-ATPase, SDH and chloride cells in juvenile Atlantic salmon Salmo salar L. Comp. Biochem. Physiol. A 77A, 9–12.

Langhorne, P., and Simpson, T. H. (1986). The interrelationship of cortisol, gill (Na+K) ATPase, and homeostasis during the parr-smolt transformation of Atlantic salmon. (Salmo salar L.). Gen. Comp. Endocrinol. 61, 203–213.

Lasserre, P., Boeuf, G., and Harache, Y. (1978). Osmotic adaptation of Oncorhynchus kisutch Walbaum. I. Seasonal variation of gill Na+–K+ ATPase activity in coho salmon, 0+-age and yearling, reared in fresh water. Aquaculture 14, 365–382.

Leatherland, J. F. (1982). Environmental physiology of the teleostean thyroid gland: A reivew. Environ. Biol. Fishes 7, 83–110.

Leatherland, J. F., and Cho, C. Y. (1985). Effect of rearing density on thyroid and interrenal gland activity and plasma and hepatic metabolite levels in rainbow trout, Salmo gairdneri Richardson. J. Fish Biol. 27, 583–592.

Leatherland, J. F., and Sonstegard, R. A. (1978). Lowering of serum thyroxine and triiodothyronine levels in yearling coho salmon, Oncorhynchus kisutch, by dietary mirex and PCBs. J. Fish. Res. Board Can. 35, 1285–1289.

Leatherland, J. F., and Sonstegard, R. A. (1981). Thyroid funciton, pituitary structure and serum lipids in Great Lakes coho salmon, Oncorhynchus kisutch Walbaum, 'jacks' compared with sexually immature spring salmon. J. Fish Biol. 18, 643–653.

Leatherland, J. F., McKeown, B. A., and John, T. M. (1974). Circadian rhythm of plasma prolactin, growth hormone, glucose and free fatty acid in juvenile kokanee salmon, Oncorhynchus nerka. Comp. Biochem. Physiol. A 47A, 821–828.

Leatherland, J. F., Cho, C. Y., and Slinger, S. J. (1977). Effects of diet, ambient temperature, and holding conditions on plasma thyroxine levels in rainbow trout (*Salmo gairdneri*). *J. Fish Res. Board Can.* **34**, 677–682.

Leray, C., Colin, D. A., and Florentz, A. (1981). Time course of osmotic adaptation and gill energetics of rainbow trout (*Salmo gairdneri*) following abrupt changes in external salinity. *J. Comp. Physiol.* **144**, 175–181.

Leyzerovich, K. A. (1973). Dwarf males in hatchery propagation of the Atlantic salmon [*Salmo salar* (L.)]. *J. Ichthyol.* (*Engl. Transl.*) **13**, 382–391.

Lin, R. J., Rivas, R. J., Nishioka, R. S., Grau, E. G., and Bern, H. A. (1985a). Effects of feeding triiodothyronine (T_3) on thyroxin (T_4) levels in steelhead trout, *Salmo gairdneri*. *Aquaculture* **45**, 133–142.

Lin, R. J., Rivas, R. J., Grau, E. G., Nishioka, R. S., and Bern, H. A. (1985b). Changes in plasma thyroxin following transfer of young coho salmon (*Oncorhynchus kisutch*) from fresh water to fresh water. *Aquaculture* **45**, 381–382.

Lindahl, K. (1980). The gonadotropic cell in parr, precocious parr male and smolt of Atlantic salmon, *Salmo salar*. An immunological, light- and electron microscopical study. *Acta Zool.* (*Stockholm*) **61**, 117–125.

Loretz, C. A., and Bern, H. A. (1982). Prolactin and osmoregulation in vertebrates. An update. *Neuroendocrinology* **35**, 292–304.

Loretz, C. A., Collie, N. L., Richman, N. H., III, and Bern, H. A. (1982). Osmoregulatory changes accompanying smoltification in coho salmon. *Aquaculture* **28**, 67–74.

Lorz, H. W., and McPherson, B. P. (1976). Effects of copper or zinc in fresh water on adaptation to sea water and ATPase activity, and the effects of copper on migratory disposition of coho salmon (*Oncorhynchus kisutch*). *J. Fish. Res. Board Can.* **33**, 2023–2030.

Lovern, J. A. (1934). Fat metabolism in fishes. V. The fat of the salmon in its young freshwater stages. *Biochem. J.* **28**, 1961–1963.

Lundqvist, H. (1980). Influence of photoperiod on growth in Baltic salmon parr (*Salmo salar* L.) with special reference to the effect of precocious sexual maturation. *Can. J. Zool.* **58**, 940–944.

Lundqvist, H. (1983). Precocious sexual maturation and smolting in Baltic salmon (*Salmo salar* L.): Photoperiodic synchronization and adaptive significance of annual biological cycles. Thesis, Dept. of Ecological Zoology, University of Umeå, Umeå, Sweden.

Lundqvist, H., and Fridberg, G. (1982). Sexual maturation versus immaturity: different tactics with adaptive values in Baltic salmon (*Salmo salar* L.) male smolts. *Can. J. Zool.* **60**, 1822–1827.

Lundqvist, H., Clarke, W. C., Eriksson, L.-O., Funegård, P., and Engström, B. (1986). Seawater adaptability in three different river stocks of Baltic salmon (*Salmo salar* L.) during smolting. *Aquaculture* **52**, 219–229.

McBride, J. R., Higgs, D. A., Fagerlund, U. H. M., and Buckley, J. T. (1982). Thyroid and steriod hormones: Potential for control of growth and smoltification of salmonids. *Aquaculture* **28**, 201–209.

McCartney, T. H. (1976). Sodium-potassium dependent adenosine triphosphatase activity in gills and kidneys of Atlantic salmon (*Salmo salar*). *Comp. Biochem. Physiol. A* **53A**, 351–353.

McCormick, S. D., and Naiman, R. J. (1984). Osmoregulation in brook trout, *Salvelinus fontinalis*. II. Effects of size, age and photoperiod on seawater survival and ionic regulation. *Comp. Biochem. Physiol. A* **79A**, 17–28.

McCormick, S. D., and Naiman, R. (1985). Hypoosmoregulation in an anadromous teleost: Influence of sex and maturation. *J. Exp. Zool.* **234**, 193–198.

McCormick, S. D., and Saunders, R. L. (1987). Physiological adaptations to marine life. *Amer Fish Soc. Symp. 1*, 211–229.

McCormick, S. D., Naiman, R. J., and Montgomery, E. T. (1985). Physiological smolt characteristics of anadromous and non-anadromous brook trout (*Salvelinus fontinalis*) and Atlantic salmon (*Salmo salar*). *Can. J. Fish. Aquat. Sci.* **42**, 529–538.

McInerney, J. E. (1964). Salinity preference: An orientation mechanism in salmon migration. *J. Fish. Res. Board Can.* **21**, 995–1018.

MacKinnon, C. N., and Donaldson, E. M. (1976). Environmentally induced precocious sexual development of the male pink salmon (*Oncorhynchus gorbuscha*). *J. Fish. Res. Board Can.* **33**, 2602–2605.

Mahnken, C. V. W., Prentice, E., Waknitz, W., Monan, G., Sims, C., and Williams, J. (1982). The application of recent smoltification research to public hatchery releases: An assessment of size/time requirements for Columbia River Hatchery coho salmon (*Oncorhynchus kisutch*). *Aquaculture* **28**, 251–268.

Malikova, E. M. (1957). Biochemical analysis of young salmon at the time of their transformation to a condition close to the smolt stage, and during retention of smolts in freshwater. (in Russian). *Tr. Latv. Otdel. VNIRO* **2**, 241–255; *Fish. Res. Board Can., Transl. Ser.* **232**, 1–19 (1959).

Markert, J. R., and Vanstone, W. E. (1966). Pigments in the belly skin of coho salmon (*Oncorhynchus kisutch*). *J. Fish. Res. Board Can.* **23**, 1095–1098.

Markert, J. R., Higgs, D. A., Dye, H. M., and MacQuarrie, D. W. (1977). Influence of bovine growth hormone on growth rate, appetite, and food conversion of yearling coho salmon (*Oncorhynchus kisutch*) fed two diets of different composition. *Can. J. Zool.* **55**, 74–83.

Mason, J. C. (1975). Seaward movements of juvenile fishes, including lunar periodicity in the movement of coho salmon (*Oncorhynchus kisutch*) fry. *J. Fish. Res. Board Can.* **32**, 2542–2547.

Massey, B. D., and Smith, C. L. (1968). The action of thyroxine on mitochondrial respiration and phosphorylation in the trout (*Salmo trutta fario* L.). *Comp. Biochem. Physiol.* **25**, 241–255.

Miles, H. M., and Smith, L. S. (1968). Ionic regulation in migrating juvenile coho salmon. *Oncorhynchus kisutch. Comp. Biochem. Physiol.* **26**, 381–398.

Milne, R. S., and Leatherland, J. F. (1978). Effect of ovine TSH, thiourea, ovine prolactin and bovine growth hormone of plasma thyroxine and triiodothyronine levels in rainbow trout, *Salmo gairdneri. J. Comp. Physiol.* **124**, 105–110.

Milne, R. S., and Leatherland, J. F. (1980). Changes in plasma thyroid hormone following administration of exogenous pituitary hormones and steroid hormones to rainbow trout (*Salmo gairdneri*). *Comp. Biochem. Physiol. A* **66A**, 679–686.

Milne, R. S., and Randall, D. J. (1976). Regulation of arterial pH during fresh water to sea water transfer in the rainbow trout *Salmo gairdneri. Comp. Biochem. Physiol. A.* **53A**, 157–160.

Milne, R. S., Leatherland, J. F., and Holub, B. J. (1979). Changes in plasma thyroxine, triiodothyronine and cortisol associated with starvation in rainbow trout (*Salmo gairdneri*). *Environ. Biol. Fishes* **4**, 185–190.

Mitans, A. R. (1973). Dwarf males and the sex structure of a Baltic salmon [*Salmo salar* (L.).] population. *J. Ichthyol (Engl. Transl.)* **13**, 192–197.

Miwa, S., and Inui, Y. (1983). Effects of thyroxine and thiourea on the parr–smolt transformation of amago salmon (*Oncorhynchus rhodurus*). *Bull. Natl. Res. Inst. Aquacult. (Jpn.)* **4**, 41–52.

Miwa, S., and Inui, Y. (1985). Effects of L-thyroxine and ovine growth hormone on smoltification of Amago salmon (*Oncorhynchus rhodurus*). *Gen. Comp. Endocrinol.* **58**, 436–442.

Miwa, S., and Inui, Y. (1986). Inhibitory effects of 17α-methyltestosterone and estradiol-17β on smoltification of sterilized amago salmon (*Oncorhynchus rhodurus*). *Aquaculture* **53**, 21–39.

Moccia, R. D., Leatherland, J. L., and Sonstegard, R. A. (1977). Increasing frequency of thyroid goitres in coho salmon (*Oncorhynchus kisutch*) in the Great Lakes. *Science* **198**, 425–426.

Mommsen, T. P. (1984). Metabolism of the fish gill. *In* "Fish Physiology' (W. S. Hoar and D. J. Randall, eds.), Vol. 10B, pp. 203–238. Academic Press, New York.

Morley, M., Chadwick, A., and El Tounsey, E. M. (1981). The effect of prolactin on the water absorption by the intestine of the trout (*Salmo gairdneri*). *Gen. Comp. Endocrinol.* **44**, 64–68.

Myers, R A (1984). Demographic consequences of precocious maturation of Atlantic salmon (*Salmo salar*). *Can. J. Fish. Aquat. Sci.* **41**, 1349–1353.

Myers, R. A., and Hutchings, J. A. (1985). Mating of anadromous Atlantic salmon, *Salmo salar* L., with mature male parr. *Int. Counc. Explor. Sea, Counc. Meet.* **M:8**, 1–6.

Myers, R. A., Hutchings, J. A., and Gibson, R. J. (1986). Variation in male parr maturation within and among populations of Atlantic salmon, *Salmo salar. Can. J. Fish. Aquat. Sci.* **43**, 1242–1248.

Naevdal, G., Holm, M., Ingebritsen, O., and Møller, D. (1978). Variation in age at first spawning in Atlantic salmon (*Salmo salar*). *J. Fish. Res. Board Can.* **35**, 145–147.

Nagahama, Y. (1985). Involvement of endocrine systems in the amago salmon, *Oncorhynchus rhodurus*. *Aquaculture* **45**, 383–384.

Nagahama, Y., Clarke, W. C., and Hoar, W. S. (1977). Influence of salinity on ultrastructure of the secretory cells of the adenohypophyseal pars distalis in yearling coho salmon *Oncorhynchus kisutch*. *Can. J. Zool.* **55**, 183–198.

Nagahama, Y., Adachi, S., Tashiro, F., and Grau, E. G. (1982). Some endocrine factors affecting the development of seawater tolerance during the parr–smolt transformation of the amago salmon, *Oncorhynchus rhodurus*. *Aquaculture* **28**, 81–90.

Narayansingh, T., and Eales, J. G. (1975). The influence of physiological doses of thyroxine on the lipid reserve of starved and fed brook trout, *Salvelinus fontinalis* (Mitchill). *Comp. Biochem. Physiol. B* **52B**, 407–412.

Nichols, D. J., and Weisbart, M. (1985). Cortisol dynamics during seawater adaptation of Atlantic salmon *Salmo salar. Am. J. Physiol.* **248**, R651–R659.

Nichols, D. J., and Weisbart, M., and Quinn, J. (1985). Cortisol kinetics and fluid distribution in brook trout (*Salvelinus fontinalis*). *J. Endocrinol.* **107**, 57–69.

Nichols, J. W., Wedemeyer, G. A., Mayer, F. L., Dickhoff, W. W., Gregory, S. V., Yasutake, W. T., and Smith, S. D. (1984). Effects of freshwater exposure to arsenic trioxide on the parr–smolt transformation of coho salmon (*Oncorhynchus kisutch*). *Environ. Toxicol. Chem.* **3**, 143–149.

Nishikawa, K., Hirashima, T., Suzuki, S., and Suzuki, M. (1979). Changes in circulating L-thyroxine and L-triiodothyronine of the masu salmon, *Oncorhynchus masou*, accompanying the smoltification, measured by radioimmunoassay. *Endocrinol. Jpn.* **26**, 731–735.

Nishioka, R. S., Bern, H. A., Lai, K. V., Nagahama, Y., and Grau, E. G. (1982). Changes in the endocrine organs of coho salmon during normal and abnormal smoltification— An electron-microscope study. *Aquaculture* **28**, 21–38.

Nishioka, R. S., Young, G., Grau, E. G., and Bern, H. A. (1983). Environmental, behavioral and endocrine bases for lunar-phased hatchery releases of salmon. *Proc. N. Pac. Aquacult. Symp., 2nd, 1983*, pp. 161–172.

Nishioka, R. S., Young, G., Bern, H. A., Jochimsen, W., and Hiser, C. (1985a). Attempts to intensify the thyroxin surge in coho and king salmon by chemical stimulation. *Aquaculture* **45**, 215–225.

Nishioka, R. S., Richman, N. H., Young, G., and Bern, H. A. (1985b). Preliminary studies on the effects of hypophysectomy on coho and king salmon. *Aquaculture* **45**, 385–386.

Nishioka, R. S., Grau, E. G., Lai, K. V., and Bern, H. A. (1987). Effect of thyroid-stimulating hormone on the physiology and morphology of the thyroid gland in coho salmon, *Oncorhynchus kisutch*. *Fish. Physiol. Biochem.* **3**, 63–71.

Noakes, D. L. G., and Leatherland, J. F. (1977). Social dominance and interrenal cell activity in rainbow trout, *Salmo gairdneri* (Pisces, Salmonidae). *Environ. Biol. Fishes* **2**, 131–136.

Olivereau, M. (1962). Modifications de l'interrénal du smolt (*Salmo salar* L.) au cours du passage d'eau douce en eau de mer. *Gen. Comp. Endocrinol.* **2**, 565–573.

Olivereau, M. (1975). Histophysiologie de l'axe hypophysocorticosurrenalien chez le saumon de l'Atlantique (cycle en eau douce, vie thalassique et reproduction). *Gen. Comp. Endocrinol.* **27**, 9–27.

Orton, J. H., Jones, J. W., and King, G. M. (1938). The male sexual stage in salmon parr (*Salmo salar* L. juv.). *Proc. R. Soc. London, Ser. B* **125**, 103–114.

Osborne, R. H., Simpson, T. H., and Youngson, A. F. (1978). Seasonal and diurnal rhythms of thyroidal status in the rainbow trout, *Salmo gairdneri* Richardson. *J. Fish Biol.* **12**, 531–540.

Osterdahl, L. (1969). The smolt run of a small Swedish river. *In* "Symposium on Salmon and Trout in Streams" (T. G. Northcote, ed.), p. 205–215. Inst. Fish., Univ. of British Columbia, Vancouver, Canada.

Otto, R. G. (1971). Effects of salinity on the survival and growth of pre-smolt coho salmon (*Oncorhynchus kisutch*). *J Fish. Res. Board Can.* **28**, 343–349.

Otto, R. G., and McInerney, J. D. (1970). Development of salinity preference in pre-smolt coho salmon, *Oncorhynchus kisutch*. *J. Fish. Res. Board Can.* **27**, 793–800.

Parry, G. (1966). Osmotic adaptation in fishes. *Biol. Rev. Cambridge Philos. Soc.* **41**, 392–444.

Patiño, R., and Schreck, C. B. (1986). Sexual dimorphism of plasma sex steroid levels in juvenile coho salmon, *Oncorhynchus kisutch*, during smoltification. *Gen. Comp. Endocrinol.* **61**, 127–133.

Patiño, R, Schreck, C. B., and Redding, J. M. (1985). Clearance of plasma corticosteroids during smoltification of coho salmon, *Oncorhynchus kisutch*. *Comp. Biochem. Physiol. A* **82A**, 531–535.

Peden, A. E., and Edwards, J. C. (1976). Permanent residence in fresh water of a large chum salmon (*Oncorhynchus keta*). *Syesis* **9**, 363.

Pengelley, E. T., and Asmundson, S. J. (1969). Free-running periods of endogenous circannual rhythms in the golden-mantled ground squirrel, *Citellus lateralis*. *Comp. Biochem. Physiol.* **30**, 177–183.

Pereira, D. L., and Adelman, I. R. (1985). Interactions of temperature, size and photoperiod on growth and smoltification of chinook salmon (*Oncorhynchus tshawytscha*). *Aquaculture* **46**, 185–192.

Piggins, D. J., and Mills, C. P. R. (1985). Comparative aspects of the biology of naturally produced and hatchery-reared Atlantic salmon smolts (*Salmo salar* L.). *Aquaculture* **45**, 321–333.

Poston, H. A. (1978). Neuroendocrine mediation of photoperiod and other environmental influences on physiological responses in salmonids: A review. *Tech. Pap. U.S. Fish. Wild. Serv.* **96**, 1–14.

Potts, W. T. W., Foster, M. A., and Stather, J. W. (1970). Salt and water balance in salmon smolts. *J. Exp. Biol.* **52**, 553–564.

Power, G. (1959). Field measurements of the basal oxygen consumption of Atlantic salmon parr and smolts. *Arctic* **12**, 195–202.

Power, G. (1969). The salmon of Ungava Bay. *Tech. Paper—Arct. Inst. North Am.* **22**, 1–72.

Primdas, F. H., and Eales, J. G. (1976). The influence of TSH and ACTH on purine and pteridine deposition in the skin of rainbow trout (*Salmo gairdneri*). *Can. J. Zool.* **54**, 576–581.

Prunet, P., and Boeuf, G. (1985). Plasma prolactin level during transfer of rainbow trout (*Salmo gairdneri*) and Atlantic salmon (*Salmo salar*) from fresh water to sea water. *Aquaculture* **45**, 167–176.

Prunet, P., Boeuf, G., and Houdebine, L. M. (1985). Plasma and pituitary prolactin levels in rainbow trout during adaptation to different salinities. *J. Exp. Zool.* **235**, 187–196.

Randall, D. J., and Daxboeck, C. (1984). Oxygen and carbon dioxide transfer across fish gills. *In* "Fish Physiology" (W. S. Hoar and D. J. Randall, eds.), Vol. 10A, pp. 263–314. Academic Press, New York.

Rankin, J. C., and Bolis, L. (1984). Hormonal control of water movement across the gills. *In* "Fish Physiology" (W. S. Hoar and D. J. Randall, eds.), Vol. 10B, pp. 177–201. Academic Press, New York.

Redding, J. M., Patiño, R., and Schreck, C. B. (1984). Clearance of corticosteroids in yearling coho salmon, *Oncohynchus kisutch,* in fresh water and seawater and after stress. *Gen. Comp. Endocrinol.* **54**, 433–443.

Refstie, T. (1977). Effect of density on growth and survival of rainbow trout. *Aquaculture* **11**, 329–334.

Refstie, T., and Steine, T. A. (1978). Selection experiments with salmon. III. Genetic and environmental sources of variation in length and weight in the freshwater phase. *Aquaculture* **14**, 221–234.

Refstie, T., Steine, T. A., and Gjedrem, T. (1977). Selection experiments with salmon. II. Proportion of Atlantic salmon smoltifying at 1 year of age. *Aquaculture* **10**, 231–242.

Richman, N. H., III, and Zaugg, W. S. (1987). Effects of cortisol and growth hormone on osmoregulation in pre- and desmoltified coho salmon (*Oncorhynchus kisutch*). *Gen. Comp. Endocrinol.* **65**, 189–198.

Richman, N. H., Tai De Diaz, S. T., Nishioka, R. S., Prunet, P., and Bern, H. A. (1987). Osmoregulatory and endocrine relationships with chloride cell morphology and density during smoltification in coho salmon (*Oncorhynchus kisutch*). *Aquaculture* **60**, 265–285.

Riddell, B. E., and Leggett, W. C. (1981). Evidence of an adaptive basis for geographic variation in body morphology and time of downstream migration in juvenile Atlantic salmon (*Salmo salar*). *Can. J. Fish. Aquat. Sci.* **38**, 308–320.

Riddell, B. E., Leggett, W. C., and Saunders, R. L. (1981). Evidence of adaptive polygenic variation between two populations of Atlantic salmon (*Salmo salar*) native to tributaries of the S. W. Miramichi River, N. B. *Can. J. Fish. Aquat. Sci.* **38**, 321–333.

Robertson, O. H. (1948). The occurrence of increased activity of the thyroid gland in rainbow trout at the time of transformation from parr to silvery smolt. *Physiol. Zool.* **21**, 282–295.

Robertson, O. H. (1949). Production of silvery smolt stage in rainbow trout by intramuscular injection of mammalian thyroid extract and thyrotropic hormone. *J. Exp. Zool.* **110**, 337–355.

Robertson, O. II. (1957). Survival of precociously mature king salmon male parr (Oncorhynchus tshawytscha juv.) after spawning. Calif. Fish Game 43, 119–130.

Rounsefell, G. A. (1958). Anadromy in North American Salmonidae. U.S. Fish. Wildl. Serv. Fish. Bull. 58, 171–185.

Sage, M. (1973). The evolution of thyroidal function in fishes. Am. Zool. 13, 899–905.

Saunders, R. L. (1965). Adjustment of buoyancy in young Atlantic salmon and brook trout by changes in swimbladder volume. J. Fish. Res. Board Can. 22, 335–352.

Saunders, R. L., and Henderson, E. B. (1965). Precocious sexual development in male post-smolt Atlantic salmon reared in the laboratory. J. Fish. Res. Board Can. 22, 1567–1570.

Saunders, R. L., and Henderson, E. B. (1970). Influence of photoperiod on smolt development and growth of Atlantic salmon (Salmo salar). J. Fish. Res. Board Can. 27, 1295–1311.

Saunders, R. L., and Henderson, E. B. (1978). Changes in gill ATPase activity and smolt status of Atlantic salmon (Salmo salar). J. Fish. Res. Board Can. 35, 1542–1546.

Saunders, R. L., and Schom, C. B. (1985). Importance of the variation in life history parameters of Atlantic salmon (Salmo salar). Can. J. Fish. Aquat. Sci. 42, 615–618.

Saunders, R. L., and Sreedharan, A. (1977). The incidence and genetic implications of sexual maturity in male Atlantic salmon parr. Int. Counc. Explor. Sea, Counc. Meet. M:21, 1–8.

Saunders, R. L., Henderson, E. B., and Glebe, B. D. (1982). Precocious sexual maturation and smoltification in male Atlantic salmon (Salmo salar). Aquaculture 28, 211–229.

Saunders, R. L., Henderson, E. B., Harmon, P. R., Johnston, C. E., and Eales, J. G. (1983). Effects of low environmental pH on smolting of Atlantic salmon (Salmo salar). Can. J. Fish. Aquat. Sci. 40, 1203–1211.

Saunders, R. L., Henderson, E. B., and Harmon, P. R. (1985a). Effects of photoperiod on juvenile growth and smolting of Atlantic salmon and subsequent survival and growth in sea cages. Aquaculture 45, 55–66.

Saunders, R. L., McCormick, S. D., Henderson, E. B., Eales, J. G., and Johnston, C. E. (1985b). The effect of orally administered 3,5,3'-triiodo-L-thyronine on growth and salinity tolerance of Atlantic salmon (Salmo salar L.). Aquaculture 45, 143–156.

Saxton, A. M., Hershberger, W. K., and Iwamoto, R. N. (1984). Smoltification in the net-pen culture of coho salmon: Quantitative genetic analysis. Trans. Am. Fish. Soc. 113, 339–347.

Schmidt, S. P., and House, E. W. (1979). Precocious sexual development in hatchery-reared and laboratory-maintained steelhead trout (Salmo gairdneri). J. Fish. Res. Board Can. 36, 90–93.

Schreck, C. B. (1982). Stress and rearing of salmonids. Aquaculture 28, 241–249.

Schreck, C. B., Patiño, R., Pring, C. K., Winton, J. R., and Holway, J. E. (1985). Effect of rearing density on indices of smoltification and performance of coho salmon, Oncorhynchus kisutch. Aquaculture 45, 345–358.

Shehadeh, Z. H., and Gordon, M. S. (1969). The role of the intestine in salinity adaptation of the rainbow trout, Salmo gairdneri. Comp. Biochem. Physiol. 30, 397–418.

Sheridan, M. A. (1985). Changes in the lipid composition of juvenile salmonids associated with smoltification and premature transfer to seawater. Aquaculture 45, 387–388.

Sheridan, M. A. (1986). Effects of thyroxin, cortisol, growth hormone, and prolactin on lipid metabolism of coho salmon, Oncorhynchus kisutch, during smoltification. Gen. Comp. Endocrinol. 64, 220–238.

Sheridan, M. A., Allen, W. V., and Kerstetter, T. H. (1983). Seasonal variations in the lipid composition of the steelhead trout, *Salmo gairdneri* Richardson, associated with the parr–smolt transformation. *J. Fish Biol.* **23**, 125–134.

Sheridan, M. A., Allen, W. V., and Kerstetter, T. H. (1985a). Changes in the fatty acid composition of steelhead trout, *Salmo gairdneri* Richardson, associated with parr–smolt transformation. *Comp. Biochem. Physiol. B* **80B**, 671–676.

Sheridan, M. A., Woo, N. Y. S., and Bern, H. A. (1985b). Changes in the rates of glycogenesis, glycogenolysis, lipogenesis, and lipolysis in selected tissues of the coho salmon associated with parr–smolt transformation. *J. Exp. Zool.* **236**, 35–44.

Skarphedinsson, O., Bye, V. J., and Scott, A. P. (1985). The influence of photoperiod on sexual development in underyearling rainbow trout, *Salmo gairdneri* Richardson. *J. Fish Biol.* **27**, 319–326.

Smith, D. C. W. (1956). The role of the endocrine organs in the salinity tolerance of trout. *Mem. Soc. Endocrinol.* **5**, 83–101.

Solomon, D. J. (1978). Migration of smolts of Atlantic salmon (*Salmo salar* L.) and sea trout (*S. trutta* L.) in a chalkstream. *Environ. Biol. Fishes* **3**, 223–229.

Sonstegard, R. A., and Leatherland, J. F. (1976). The epizootiology and pathogenesis of thyroid hyperplasia in coho salmon (*Oncorhynchus kisutch*) in Lake Ontario. *Cancer Res.* **36**, 4467–4475.

Sower, S. A., Sullivan, C. V., and Gorbman, A. (1984). Changes in plasma estradiol and effects of triiodothyronine on plasma estradiol during smoltification of coho salmon (*Oncorhynchus kisutch*). *Gen. Comp. Endocrinol.* **54**, 486–492.

Specker, J. L. (1982). Interrenal function and smoltification. *Aquaculture* **28**, 59–66.

Specker, J. L., and Kobuke, L. (1987). Seawater acclimation and thyroidal response to thyrotropin in juvenile coho salmon (*Oncorhynchus kisutch*). *J. Exp. Zool.* **241**, 327–332.

Specker, J. L., and Richman, N. H., III (1984). Environmental salinity and the thyroidal response to thyrotropin in juvenile coho salmon (*Oncorhynchus kisutch*). *J. Exp. Zool.* **230**, 329–333.

Specker, J. L., and Schreck, C. B. (1980). Stress responses to transportation and fitness for marine survival in coho salmon (*Oncorhynchus kisutch*) smolts. *Can. J. Fish. Aquat. Sci.* **37**, 765–769.

Specker, J. L., and Schreck, C. B. (1982). Changes in plasma corticosteroids during smoltification of coho salmon, *Oncorhynchus kisutch*. *Gen. Comp. Endocrinol.* **46**, 53–58.

Specker, J. L., and Schreck, C. B. (1984). Thyroidal response to mammalian thyrotropin during smoltification of coho salmon (*Oncorhynchus kisutch*). *Comp. Biochem. Physiol. A* **78A**, 441–444.

Specker, J. L., DiStefano, J. J., III, Grau, E. G., Nishioka, R. S., and Bern, H. A. (1984). Development-associated changes in thyroxine kinetics in juvenile salmon. *Endocrinology* **115**, 399–406.

Stuart-Kregor, P. A. C., Sumpter, J. P., and Dodd, J. M. (1981). The involvement of gonadotrophin and sex steroids in the control of reproduction in the parr and adults of Atlantic salmon, *Salmo salar* L. *J. Fish Biol.* **18**, 59–72.

Sullivan, C. V., Dickhoff, W. W., Mahnken, C. V. W., and Hershberger, W. K. (1985). Changes in the hemoglobin system of the coho salmon *Oncorhynchus kisutch* during smoltification and triiodothyronine and propylthiouracil treatment. *Comp. Biochem. Physiol. A* **81A**, 807–813.

Sutterlin, A. M., Harmon, P., and Young, B. (1978). Precocious sexual maturation in Atlantic salmon (*Salmo salar*) postsmolts reared in a seawater impoundment. *J. Fish Res. Board Can.* **35**, 1269–1271.

Sweeting, R. M., Wagner, G. F., and McKeown, B. A. (1985). Changes in plasma glucose, amino acid nitrogen and growth hormone during smoltification and seawater adaptation in coho salmon, *Oncorhynchus kisutch*. *Aquaculture* **45**, 185–197.

Swift, D. R. (1959). Seasonal variation in the activity of the thyroid gland of yearling brown trout *Salmo trutta* Linn. *J. Exp. Biol.* **36**, 120–125.

Thorpe, J. E. (1977). Bimodal distribution of length of juvenile Atlantic salmon (*Salmo salar* L.) under artificial rearing conditions. *J. Fish Biol.* **11**, 175–184.

Thorpe, J. E., ed. (1980). "Salmon Ranching." Academic Press, London.

Thorpe, J. E. (1986). Age at first maturity in Atlantic salmon, *Salmo salar*: Freshwater period influences and conflicts for smolting. *Can. Spec. Publ. Fish Aquat. Sci.* **89**, 7–14.

Thorpe, J. E. (1987). Smolting versus residency: Developmental conflict in salmonids. *Am. Fish Soc. Symp.* **1**, 244–252.

Thorpe, J. E., and Morgan, R. I. G. (1978). Paternal influence on growth rate, smolting rate and survival in hatchery reared juvenile Atlantic salmon, *Salmo salar*. *J. Fish Biol.* **13**, 549–556.

Thorpe, J. E., and Morgan, R. I. G. (1980). Growth-rate and smolting-rate of progeny of male Atlantic salmon parr, *Salmo salar* L. *J. Fish Biol.* **17**, 451–459.

Thorpe, J. E., Morgan, R. I. G., Ottaway, E. M., and Miles, M. S. (1980). Time of divergence of growth groups between potential 1+ and 2+ smolts among sibling Atlantic salmon. *J. Fish Biol.* **17**, 13–21.

Thorpe, J. E., Talbot, C., and Villarreal, C. (1982). Bimodality of growth and smolting in Atlantic salmon, *Salmo salar* L. *Aquaculture* **28**, 123–132.

Thorpe, J. E., Morgan, R. I. G., Talbot, C., and Miles, M. S. (1983). Inheritance of developmental rates in Atlantic salmon, *Salmo salar* L. *Aquaculture* **33**, 119–128.

Thorpe, J. E., Bern, H. A., Saunders, R. L., and Soivio, A. (1985). Salmon Smoltification II. Proceedings of a workshop sponsored by the Commission of the European Communities. Univ. Stirling, Scotland 3-6 July, 1984. *Aquaculture* **45**, 1–404.

Threadgold, L. T., and Houston, A. H. (1961). An electron microscope study of the 'chloride-secretory cell' of *Salmo salar* L., with reference to plasma-electrolyte regulation. *Nature (London)* **190**, 612–614.

Vanstone, W. E., and Markert, J. R. (1968). Some morphological and biochemical changes in coho salmon, *Oncorhynchus kisutch*, during the parr–smolt transformation. *J. Fish. Res. Board Can.* **25**, 2403–2418.

Vanstone, W. E., Roberts, E., and Tsuyuki, H. (1964). Changes in the multiple hemoglobin patterns of some Pacific salmon, genus *Oncorhynchus*, during the parr–smolt transformation. *Can. J. Physiol. Pharmacol.* **42**, 697–703.

Villarreal, C. A., and Thorpe, J. E. (1985). Gonadal growth and bimodality of length frequency distribution in juvenile Atlantic salmon (*Salmo salar*). *Aquaculture* **45**, 265–288.

Virtanen, E., and Soivio, A. (1985). The patterns of T_3, T_4, cortisol and Na^+-K^+-ATPase during smoltification of hatchery-reared *Salmo salar* and comparison with wild smolts. *Aquaculture* **45**, 97–109.

Wagner, H. H. (1974a). Photoperiod and temperature regulation of smolting in steelhead trout (*Salmo gairdneri*). *Can. J. Zool.* **52**, 219–234.

Wagner, H. H. (1974b). Seawater adaptation independent of photoperiod in steelhead trout (*Salmo gairdneri*). *Can. J. Zool.* **52**, 805–812.

Wagner, H. H., Conte, F. P., and Fessler, J. L. (1969). Development of osmotic and ionic regulation in two races of chinook salmon, *Oncorhynchus tshawytscha*. *Comp. Biochem. Physiol.* **29**, 325–341.

Wedemeyer, G. (1972). Some physiological consequences of handling stress in the juvenile coho salmon (*Oncorhynchus kisutch*) and steelhead trout (*Salmo gairdneri*). *J. Fish. Res. Board Can.* **29**, 1780–1783.

Wedemeyer, G. A., Saunders, R. L., and Clarke, W. C. (1980). Environmental factors affecting smoltification and early marine survival of anadromous salmonids. *Mar. Fish. Rev.* **42**(6), 1–14.

Weisbart, M. (1968). Osmotic and ionic regulation in embryos, alevins, and fry of the five species of Pacific salmon. *Can. J. Zool.* **46**, 385–397.

Wendt, C. A. G., and Saunders, R. L. (1973). Changes in carbohydrate metabolism in young Atlantic salmon in response to various forms of stress. *Int. Atl. Salmon Found., Spec. Publ. Ser.* **4**,(1), 55–82.

White, B. A., and Henderson, N. E. (1977). Annual variations in the circulating levels of thyroid hormones in the brook trout, *Salvelinus fontinalis*, as measured by radioimmunoassay. *Can. J. Zool.* **55**, 475–481.

White, H. C. (1939). Factors influencing descent of Atlantic salmon smolts. *J. Fish. Res. Board Can.* **4**, 323–326.

Wilkins, N. P. (1968). Multiple hemoglobins of the Atlantic salmon (*Salmo salar*). *J. Fish. Res. Board Can.* **25**, 2651–2663.

Wilkins, N. P., and Gosling, E. M., eds. (1983). Genetics in aquaculture. Proceedings of the International Symposium, Galway, Ireland, 1982. *Aquaculture* **33**, 1–426.

Winans, G. A. (1984). Multivariate morphometric variability in Pacific salmon: Technical demonstration. *Can. J. Fish. Aquat. Sci.* **41**, 1150–1159.

Winans, G. A., and Nishioka, R. S. (1987). A multivariate description of changes in body shape of coho salmon (*Oncorhynchus kisutch*) during smoltification. *Aquaculture* **66** (in press).

Withey, K. G., and Saunders, R. L. (1973). Effect of a reciprocal photoperiod regime on standard rate of oxygen consumption of postsmolt Atlantic salmon (*Salmo salar*). *J. Fish. Res. Board Can.* **30**, 1898–1900.

Woo, N. Y. S., Bern, H. A., and Nishioka, R. S. (1978). Changes in body composition associated with smoltification and premature transfer to seawater in coho salmon (*Oncorhynchus kisutch*) and king salmon (*O. tshawytscha*). *J. Fish Biol.* **13**, 421–428.

Woodhead, P. M. J. (1970). The effect of thyroxine upon the swimming of cod. *J. Fish. Res. Board Can.* **27**, 2337–2338.

Yamauchi, K., Koide, K., Adachi, S., and Nagahama, Y. (1984). Changes in seawater adaptability and blood thyroxine concentrations during smoltification of the masu salmon, *Oncorhynchus masou*, and the Amago salmon, *Oncorhynchus rhodurus*. *Aquaculture* **42**, 247–256.

Yamauchi, K., Ban, M., Kasahara, N., Izumi, T., Kojima, H., and Harako, T. (1985). Physiological and behavioral changes occurring during smoltification in the masu salmon, *Oncorhynchus masou*. *Aquaculture* **45**, 227–235.

Young, G. (1986). Cortisol secretion *in vitro* by the interrenal of coho salmon (*Oncorhynchus kisutch*) during smoltification: Relationship with plasma thyroxine and plasma cortisol. *Gen. Comp. Endocrinol.* **63**, 191–200.

Youngson, A. F., and Simpson, T. H. (1984). Changes in serum thyroxine levels during smolting in captive and wild Atlantic salmon, *Salmo salar* L. *J. Fish Biol.* **24**, 29–39.

Youngson, A. F., Buck, R. J. G., Simpson, T. H., and Hay, D. W. (1983). The autumn and spring emigrations of juvenile Atlantic salmon, *Salmo salar* L., from the Girnock Burn, Aberdeenshire, Scotland: Environmental release of migration. *J. Fish Biol.* **23**, 625–639.

Youngson, A. F., Scott, D. C. B., Johnstone, R., and Pretswell, D. (1985). The thyroid system's role in the downstream migration of Atlantic salmon (*Salmo salar* L.) smolts. *Aquaculture* **45**, 392–393.

Youngson, A. F., McLay, H. A., and Olsen, T. C. (1986). The responsiveness of the thyroid system of Atlantic salmon (*Salmo salar* L.) smolts to increased water velocity. *Aquaculture* **56**, 243–255.

Zadunaisky, J. A. (1984). The chloride cell: The active transport of chloride and the paracellular pathways. *In* "Fish Physiology" (W. S. Hoar and D. J. Randall, eds.), Vol. 10B, pp. 129–176. Academic Press, New York.

Zaugg, W. S. (1981). Advanced photoperiod and water temperature effects on gill Na⁺-K⁺ adenosine triphosphatase activity and migration of juvenile steelhead (*Salmo gairdneri*). *Can. J. Fish. Aquat. Sci.* **38**, 758–764.

Zaugg, W. S. (1982). Some changes in smoltification and seawater adaptability of salmonids resulting from environmental and other factors. *Aquaculture* **28**, 143–151.

Zaugg, W. S., and McLain, L. R. (1970). Adenosine triphosphatase activity in gills of salmonids: Seasonal variations and salt water influence in coho salmon, *Oncorhynchus kisutch*. *Comp. Biochem. Physiol.* **35**, 587–596.

Zaugg, W. S., and McLain, L. R. (1972). Changes in gill adenosinetriphosphatase activity associated with parr–smolt transformation in steelhead trout, coho, and spring chinook salmon. *J. Fish. Res. Board Can.* **29**, 167–171.

Zaugg, W. S., and McLain, L. R. (1976). Influence of water temperature on gill sodium, potassium-stimulated ATPase activity in juvenile coho salmon (*Oncorhynchus kisutch*). *Comp. Biochem. Physiol. A* **54A**, 419–421.

Zaugg, W. S., and Wagner, H. H. (1973). Gill ATPase activity related to parr–smolt transformation and migration in steelhead trout (*Salmo gairdneri*): Influence of photoperiod and temperature. *Comp. Biochem. Physiol. B* **45B**, 955–965.

Zaugg, W. S., Adams, B. L., and McLain, L. R. (1972). Steelhead migration: Potential temperature effects as indicated by gill adenosine triphosphatase activities. *Science* **176**, 415–416.

Zaugg, W. S., Bodle, J. E., Manning, J. E., and Wold, E. (1986). Smolt transformation and seaward migration of 0-age progeny of adult spring chinook salmon (*Oncorhynchus tshawytscha*) matured early with photoperiod control. *Can. J. Fish. Aquat. Sci.* **43**, 885–888.

Zbanyszek, R., and Smith, L. S. (1984). Changes in carbonic anhydrase activity in coho salmon smolts resulting from physical training and transfer into seawater. *Comp. Biochem. Physiol. A* **79A**, 229–233.

5

ONTOGENY OF BEHAVIOR AND
CONCURRENT DEVELOPMENTAL CHANGES
IN SENSORY SYSTEMS IN TELEOST FISHES

DAVID L. G. NOAKES

Department of Zoology
College of Biological Science
University of Guelph
Guelph, Ontario, Canada N1G 2W1

JEAN-GUY J. GODIN

Department of Biology
Mount Allison University
Sackville, New Brunswick, Canada E0A 3C0

I. INTRODUCTION

Fishes show considerable inter-individual variation in their behavior (e.g., Dill, 1983; Ringler, 1983; Magurran, 1986). Functional
explanations proposed for such behavioral plasticity include environmental variability, phenotypic differences between individuals that
constrain behavior, and frequency-dependent behavior of competi-
tors (Magurran, 1986). An epigenetic approach to an understanding of
behavior, however, may be more fruitful than a functional approach,

FISH PHYSIOLOGY, VOL. XIB

and should perhaps precede any functional (evolutionary) explanations of behavior (Jamieson, 1986). The epigenetic approach investigates the dynamics of the process of behavioral development within and between generations (Jamieson, 1986). That is, this approach views behavior as an interaction of genetically determined structure, development, and function. Although some individual differences in behavior are genetically based (Noakes, 1986), distinguishing between the effects of genotype, environment, and development is very difficult (Magurran, 1986).

In an earlier volume of this series, Baerends (1971) detailed the ethological analysis of fish behavior. We follow this ethological tradition, and build on that foundation. We too will necessarily stress the causation of behavior, in keeping with the physiological nature of this book. However, ours is a more restricted and specialized consideration, since it deals only with the developmental aspects of behavior. These two aspects of behavior, causation and ontogeny, along with function and phylogeny, are the four major concerns of ethology. All four are to some extent inextricably intertwined in any comprehensive study of behavior, even studies ostensibly directed solely to the ontogeny of behavior.

Our coverage is restricted to teleost fishes, since almost all information is from those species. We have also restricted our consideration to visual, chemosensory, and mechanoreceptor systems, for the same reason. The virtual absence of studies of early behavior of nonteleostean fish species, and on other aspects of behavior, was somewhat surprising to us. If nothing else, the definition of these lacunae may serve a useful purpose in directing attention to needed future research in those areas.

II. DEVELOPMENT: PERIODS AND STAGES

Some General Features of Fish Behavioral Ontogeny

Considerable changes in behavior occur during fish ontogeny. Weak contractions of the developing heart appear to be the first observable movements, occurring about one-third of the way through embryonic life (Huntingford, 1986). Thereafter until hatching, weak and irregular twitching of the trunk musculature (initially of myogenic origin) gradually become stronger, more coordinated, and more regular with time, and movements of the jaws, opercula, and pectoral fins appear (Huntingford, 1986). About half-way through embryogenesis, these movements become neurogenic and responsive to external stimuli (Abu-Gideiri, 1966, 1969; Eaton and Nissanov, 1985; Eaton and DiDomenico, 1986; Huntingford, 1986). Following hatching, former embryonic motor patterns such as rapid flexions of the body persist as escape (startle) responses (Eaton and Nissanov, 1985; Eaton and DiDomenico, 1986) and coordinated swimming patterns in the young and adult fish, or they may be modified into distinct behavioral acts with varying function (Noakes, 1978a,b, 1981; Huntingford, 1986).

As the young fish ages, new behaviors appear in its expanding repertoire that eventually lead to the complete behavioral repertoire of the adult (Noakes, 1978a,b, 1981; Huntingford, 1986). The various changes in behavior that fish undergo during development coincide with several morphological and physiological changes in their nervous system, among other organismal changes (e.g., hormonal secretions), and appear to reflect species-typical adaptations to the immediate physical and social environments of the developing fish (Noakes, 1978a,b, 1981; Huntingford, 1986). These behavioral changes thus represent more than just the developmental substrates of adult behaviors.

Ontogenetic changes in fish behavior may result from a number of interacting, proximate causal factors. These include developmental changes in the nervous system, nonneural physiological and morphological changes in the fish, changes in external stimuli, and experience (Huntingford, 1986). Of these factors, we only review comprehensively the evidence for the role of developmental changes in the nervous system underlying changes in fish behavior during ontogeny. Evidence in support of the other aforementioned factors influencing fish behavioral ontogeny has been reviewed recently by Huntingford (1986), and consequently only brief reference to them is made herein where appropriate.

III. DEVELOPMENT OF SENSORY SYSTEMS AND BEHAVIORAL ONTOGENY

A. Development of the Visual System and Visually Mediated Behavior

Depending on whether the parental fish hide their eggs in the substrate or under or in submerged structures (brood hiders), carry them internally (bearers) or release them in the water column (broadcast spawners), and whether they are brood guarders or nonguarders (Balon, 1975a), the embryo within the egg and the developing young

fish will experience markedly different photic and social environments during ontogenesis. Consequently, considerable inter- and intraspecific variability exists in the development of visually mediated
behaviors in fishes and of coinciding changes in their visual system.

1. BRAIN STRUCTURES DIRECTLY ASSOCIATED WITH THE VISUAL SYSTEM

In visually oriented fishes, sensory information originating in the
retina travels along the afferent fibers of the optic nerves to the optic
tectum, the highest integration center for visual stimuli located in the
midbrain or mesencephalon (Bond, 1979; Jeserich and Rahmann,
1979). Structural and biochemical changes occur in the optic tectum
during ontogeny, and certain of these neural changes coincide with
developmental changes in behavior.

In brood hiders such as the brown trout *Salmo trutta* (Sharma,
1975) and the rainbow trout *S. gairdneri* (Rahmann and Jeserich,
1978), mitotic cell division and neuronal differentiation occur before
hatching. In rainbow trout, the major period of mitotic growth and
neuronal differentiation occurs during the eleutheroembryo phase
[*sensu* Balon (1975b)] when the young fish is still buried in the gravel
nest (Rahmann and Jeserich, 1978). This developmental phase is further characterized by an outgrowth of nerve fibers and synaptogenesis
in the optic tectum, with the densities of synapses and synaptic vesicles increasing over the course of this phase and reaching adult values
at the time that the alevin emerges from the gravel nest and becomes
free-swimming (Fig. 1a). Moreover, marked increases in the concentration of neuronal gangliosides, compounds presumed to be involved
in synaptogenesis and synaptic transmission (Breer and Rahmann,
1977; Seybold and Rahmann, 1985), and in the activity of their catabolic enzyme neuraminidase (Schiller *et al.*, 1979), occur during this
period of synaptic differentiation and maturation of the optic tectum
in teleosts. Coincident with these developmental changes in the optic
tectum of the trout eleutheroembryo is a very rapid improvement in
their visual acuity during a period when they are normally buried in a
gravel nest (Fig. 1B).

These structural and biochemical changes in the optic tectum are,
however, preceded by developmental changes in the retina, such as
synaptogenesis of photoreceptor cells and the onset of neural transmission (Rahmann and Jeserich, 1978). The rates of development of
neural structures and of synaptogenesis in the optic tectum and retina
of the eleutheroembryo of brood-hiding fishes are dependent at least

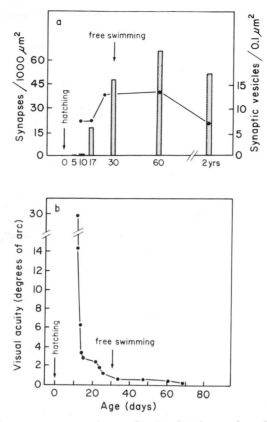

Fig. 1. (a) Changes in synaptic density (bars) and in the number of synaptic vesicles per unit area (line) in the optic tectum of rainbow trout (*Salmo gairdneri*) during ontogeny. (b) Changes in visual acuity of rainbow trout during early ontogeny. [After Rahmann and Jeserich (1978).]

on the light level reaching the young fish (Grün, 1979; Jeserich and Rahmann, 1979; Zeutzius *et al.*, 1984) and on water temperature through its general effects on metabolism and growth (Blaxter, 1969a). Consequently, at the time of emergence from the gravel nest (salmonid fishes) or release from the parent's buccal cavity (cichlid fishes) into the water column, the free-swimming young fish possesses acute vision and a mature and functional optic tectum capable of integrating visual stimuli from its new habitat (e.g., Fig. 1). Similarly, neuronal differentiation of the optic tectum and retina occurs in the eleutheroembryo of broadcast spawners such as the anchovy (*Engraulis mordax*) and sardine (*Sardinops caerulea*), and optic fibers connect the retina to the optic tectum by the onset of visually mediated external feeding in the larva (Schwassmann, 1965).

Other brain structures associated with vision also change during ontogenesis. For example, in cichlid fishes the number of afferent neuronal fibers from the nucleus olfacto–retinalis (NOR) in the telencephalon innervating the retina gradually increases during the eleutheroembryo period and reaches or approaches adult values near the onset of the free-swimming juvenile period, when the alevins or larvae leave their parent's buccal cavity or the pit in the substrate, depending on the species (Crapon de Caprona and Fritzsch, 1983).

2. DEVELOPMENTAL CHANGES IN THE EYE AND CORRESPONDING CHANGES IN BEHAVIOR

The eye is the major organ in fishes that detects photic stimuli and forms images, although the pineal body (Ralph, 1975) and putative unspecialized photoreceptors in the dermis or brain (Wales, 1975) have been implicated in photodetection and thus may play a role in mediating behavior. The anatomy, physiology, and ecology of vision in fishes have been previously reviewed (Ingle, 1971; Munz, 1971; Tomita, 1971; Ali, 1975a; Lythgoe, 1979; Ali and Klyne, 1985; Blaxter, 1986; Guthrie, 1986), and consequently these topics will not be covered here in detail.

The fish eye conforms to the vertebrate type in that it consists of a fluid-filled chamber that contains an inverted retina and a spherical, focusable lens (Guthrie, 1986). Since the pupillary diameter of the teleost eye generally does not change very much in response to varying ambient light intensity, adaptation is achieved by movement of the retinal photoreceptors relative to the retinal pigment layer, which simultaneously changes in thickness (i.e., retinomotor response) (Munz, 1971; Ali, 1975b; Guthrie, 1986). The fish retina is relatively complex, consisting of three or more cone photoreceptors (for color vision and acuity), rod photoreceptors (for contrast discrimination), two types of bipolar cell, three to five kinds of horizontal cell, and six or more types of ganglion cell (Munz, 1971; Guthrie, 1986). Most elasmobranchs are believed to have pure-rod retinas, whereas *adult* teleosts typically have duplex retinas (both rods and cones) (Guthrie, 1986). Typical of teleosts, the photoreceptor cells are not distributed uniformly over the retina, but rather are often concentrated in a high density area, the so-called *area centralis* or *area temporalis* (Munz, 1971; Guthrie, 1986). However, a true pit-like fovea is rare among teleosts.

The eyes of young fish differ markedly from the adult eye in several respects other than being smaller. During early fish ontogeny relatively rapid changes occur in the surface area of the retina, the diameter of the lens, the number and kinds of photoreceptor present, and their pattern of distribution within the retina, among other structural alterations. Such developmental changes in eye structure can alter visual function and thus result in changes in visually mediated behavior during fish ontogeny (see below). Since most teleosts have indeterminate growth, the relative rate of eye growth is generally dependent on the rate of body growth and the particular allometric relationship between body size and eye size (Fernald, 1985). Consequently, the ontogenetic timing and rate of structural and functional changes of the eye can be modified by ecological factors, such as temperature, food availability, ambient light regime, and social environment, that affect fish growth rate (Ali, 1975c; Brett, 1979; Grün, 1979; Rahmann et al., 1979; Zeutzius and Rahmann, 1984; Fernald, 1985).

In many cases, there exists a clear correlation between the ontogenetic timing of structural and functional changes in the eye and of changes in the behavior of developing fish, which in turn coincide with ontogenetic shifts in their ecology (e.g., Blaxter, 1975; Ahlbert, 1976; Noakes, 1978a,b). We review below certain of the major alterations in eye structure and function that occur during fish ontogeny and attempt to relate them to concurrent changes in the behavioral ecology of the developing fish.

As the young fish grows, its retina increases in surface area by stretching of the existing neural tissue and by the mitotic addition of visual cells in a germinal zone located at the retinal margin (Johns, 1981; Fernald, 1985). With continued retinal growth in teleosts, the absolute density of all retinal cells decreases, with the exception of rods, which increase numerically (Blaxter and Jones, 1967; Ahlbert, 1975, 1976; Boehlert, 1979; Guma'a, 1982; Neave, 1984; Fernald, 1985). Therefore, the ratio of rods to cones throughout the retina increases as the latter expands (Sandy and Blaxter, 1980; Johns, 1981; Guma'a, 1982; Pankhurst, 1984). Cone size as well as the proportion of double (twin) cones also increase during ontogeny (Blaxter and Jones, 1967; Ahlbert, 1975; Boehlert, 1979; Sandy and Blaxter, 1980; Johns, 1981; Guma'a, 1982). Moreover, the geometric arrangement of the visual cells within the retina is altered during ontogeny. In particular, the cones are organized in square mosaic units, which are evenly distributed throughout the retina in early developmental stages in many teleosts but gradually become more irregularly spaced into row mosaics as the individual grows (Ahlbert, 1975, 1976). The functional significance of such an alteration in cone geometry is unclear. In the salmonids at least, cone mosaics become relatively more concentrated in the ventro–temporal area (*area temporalis*) of the retina as body size increases, presumably resulting in a specific retinal area of high acuity that would be adaptive to feeding on aquatic insects drifting overhead in these stream-dwelling fish (Ahlbert, 1975, 1976). In herring (*Clupea harengus*) this high-acuity retinal area differentiates only at or shortly after metamorphosis (Blaxter and Jones, 1967), following which the juvenile fish feeds more actively on zooplankton (Rosenthal and Hempel, 1970).

Accompanying ontogenetic changes in the number, size, and distribution of photoreceptor cells are corresponding changes in the number and spatial organization of higher-order processing (e.g., ganglionic) cells in the retina and their synaptic contacts with the photoreceptors (Boehlert, 1979; Johns, 1981; Fernald, 1985). In addition to these morphometric changes in the retina, the diameter of the lens and, consequently, its focal length increase linearly as the young fish grows (Blaxter and Jones, 1967; Guma'a, 1982; Neave, 1984). However, the optical quality of the lens is preserved (Fernald, 1985) and the optical and retinal fields remain spherically symmetric and constant in size (Easter *et al.*, 1977) as the fish grows.

The aforementioned ontogenetic changes in the retina and lens result in alterations in visual function and thus have implications for visually mediated behavior. Photopic visual acuity, commonly defined as the minimum angle that a stimulus can subtend at the eye and still be resolved, is determined in part by the size and density of cones in the retina and their convergence onto higher-order retinal processing cells and in part by the focal length of the lens (Blaxter and Jones, 1967; Johns, 1981; Guma'a, 1982; Fernald, 1985). On the basis of ontogenetic changes in the eye and the optic tectum described above, visual activity in teleosts can theoretically increase with increasing body size.

Teleost fishes are capable of motion detection at or soon after hatching. This ability improves with ontogeny, as revealed by several studies showing visual acuity, measured using either morphological (e.g., intercone spacing, lens diameter) or behavioral (e.g., optomotor response) criteria, and increases at a decelerating rate to an asymptote with age or body size (Rahmann *et al.*, 1979; Blaxter, 1980; Guma'a, 1982; Breck and Gitter, 1983; Neave, 1984; Zeutzius and Rahmann, 1984; Li *et al.*, 1985). Commonly, the most rapid improvement in visual acuity coincides with eleutheroembryo emergence from its un-

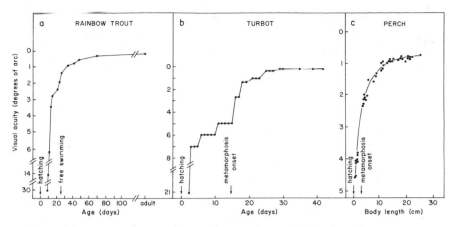

Fig. 2. Variation in the visual acuity (minimal separable angle in degrees of arc) of (a) rainbow trout [after Rahmann *et al.* (1979)], (b) turbot [after Neave (1984)], and (c) perch [after Guma'a (1982)] with age or body length.

dergravel nest (e.g., salmonids, Fig. 2a) or release from its parent's buccal cavity (e.g., cichlids), which corresponds to the period of transition from feeding on yolk reserves to external food sources (Dill, 1977; Noakes, 1978b; Brown, 1985), the metamorphosis of the pelagic larva into a juvenile (e.g., clupeids) (Fig. 2b,c), or the ontogenetic onset of predator avoidance responses (Taylor and McPhail, 1985), schooling (Shaw and Sachs, 1967; Rosenthal, 1968), and other social behaviors (Dill, 1977; Noakes, 1978b). A rapid improvement in visual acuity at a stage in development when the young fish is beginning to feed on external food, interact socially with parents, siblings, and other conspecifics, and actively avoid predators, among other activities, is of obvious ecological importance. With particular reference to foraging behavioral ecology, a greater visual acuity allows individual fish to detect prey items (and potential predators as well) further away, to consequently feed faster since more prey can theoretically be encountered in its larger visual reactive field per unit time (Hairston *et al.*, 1982; Breck and Gitter, 1983), to capture prey more successfully (Godin, 1978), and to more accurately assess absolute prey size (Li *et al.*, 1985), all of which have implications for size-selective foraging and thus for the fish's rate of net energy gain. Therefore, ontogenetic improvements in motion detection and stimulus discrimination are likely important proximate mechanisms determining fish growth and survivorship in early ontogeny (see Section IV,C).

In teleost species in which adults possess a duplex retina, the eleutheroembryos and larvae usually have only cones in their retina (Ali, 1959; Blaxter, 1968a,b, 1969b, 1975; Blaxter and Jones, 1967; Sandy and Blaxter, 1980; Guma'a, 1982) or, less commonly, rods only (Pankhurst, 1984). As a consequence, most newly hatched fish with either pure-cone or pure-rod retinae are incapable of retinomotor responses, that is, of light–dark adaptation. Therefore, at some point during the course of ontogeny the second type of visual cell (usually rods) appears in the growing retina. This occurs at or near the time of transition from the eleutheroembryo phase to the alevin phase (i.e., emergence or release from nest or cover) in brood hiders (Ali, 1959; Armstrong, 1964) or from the larval phase to juvenile phase (i.e., metamorphosis) in brood nonguarders (Guma'a, 1982) and broadcast spawners with planktonic larvae (Blaxter and Jones, 1967; Blaxter, 1968a,b; Blaxter and Staines, 1970; Sandy and Blaxter, 1980; O'Connell, 1981; Pankhurst, 1984).

The appearance of rods in the retina at time coinciding with these ontogenetic transitions enhances the young fish's scotopic sensitivity threshold to ambient light intensity, which gradually increases with age, paralleling ontogenetic increases in the number of rods in the retina (Armstrong, 1964; Blaxter, 1968a,b, 1969b, 1975; Blaxter and Staines, 1970; Boehlert, 1979; O'Connell, 1981). This means that the newly emerged or newly metamorphosed fish can feed, school, and respond to a variety of visual stimuli at increasingly lower light intensity levels, which are associated with an ontogenetic shift from pelagic to benthic habitats in certain species, thereby increasing the amount of time available daily to forage and interact socially. A duplex retina capable of retinomotor responses may also permit juvenile pelagic fish to discriminate between underwater light intensities (Blaxter, 1972, 1976) and thus facilitate vertical migration in the water column in response to diel changes in the vertical light gradient (Blaxter, 1976; O'Connell, 1981). Diel vertical migrations in fishes may serve several functions, including energy savings, predator avoidance, and increased feeding opportunities (e.g., Brett, 1971; Eggers, 1978). Such diel vertical migrations, however, do not require a duplex retina, since the larvae of several pelagic marine fishes have been reported to undergo diel vertical movements in the laboratory and field (Blaxter, 1968a; Wales, 1975; Hunter and Sanchez, 1976).

Prior to emergence from their gravel nest or pit in the substrate or release from their parent's mouth, the eleutheroembryos (possessing only a pure-cone retina) are strongly photophobic (negatively phototactic), which serves to keep them under darkened cover where they are protected from predators and from being displaced downstream at a stage when they are not effective swimmers (Armstrong, 1964; Noakes, 1981; Godin, 1982; Carey, 1985; Nunan and Noakes, 1985a). The rapid development of the duplex retina and retinomotor responses toward the end of the eleutheroembryo phase is accompanied by either a marked or gradual shift to photopositive behavior, which coincides with emergence from the gravel nest or pit or release from the parent's mouth, a major shift in habitat (Armstrong, 1964; Godin, 1982; Carey, 1985; Nunan and Noakes, 1985a). In comparison, the reversal of phototaxis in some species with pelagic larvae is in the opposite direction, and it coincides also with a shift in habitat, for example, from the pelagic to the littoral zone (Bulkowski and Meade, 1983). Although these ontogenetic shifts in photoresponse commonly coincide with the ontogenetic appearance of the duplex retina and retinomotor responses, the causal factors regulating the photoresponse reversal are unknown (cf. Godin, 1982).

For the salmonids at least, possession of a duplex retina and a shift to photopositive behavior are not likely the primary underlying causal mechanisms for upward movement through the gravel nest (i.e. emergence), since emergence from simulated nests can occur in complete darkness (Nunan and Noakes, 1985a). Although light influences the movements of the eleutheroembryos within the gravel nest in a species-typical manner (Godin, 1982; Carey, 1985; Nunan and Noakes, 1985a), gravity, not light nor water current direction, appears to be the major orientational cue used by the fish during emergence (Nunan and Noakes, 1985b). It is tempting to causally relate the reversal to photopositive behavior at about the time of emergence in salmonids to the corresponding development of retinomotor responses in the eye and the capability of the eye to adapt to varying ambient light intensities. The following observations suggest that this is too simple an explanation. First, the final state of photopositivity at emergence varies considerably among salmonid species under controlled conditions, and these interspecific differences appear to be genetically based (Godin, 1982; Carey, 1985; Nunan and Noakes, 1985a). Second, blind cavefish show ontogenetic changes in photobehavior, which of course cannot be related to concurrent physiological changes in the eye (Romero, 1985). Lastly, larval marine fishes, with pure-cone retinas that cannot undergo retinomotor changes, show photopositive behavior (Blaxter, 1975). Nevertheless, the duplex retina of the newly emerged, free-swimming alevins of brood hiders and bearers enable them to respond appropriately to the different photic conditions experienced in their new habitat outside the substrate or parent's mouth (Ali, 1959, 1975b; Armstrong, 1964). Such responses include emerging and dispersing from the gravel nest at night, thereby reducing predation risk (Godin, 1982), feeding over a wide range of ambient light intensities (Brett and Groot, 1963; Blaxter, 1980), and responding to relevant configurational (Coss, 1978) and colored stimuli in the environment (Noakes, 1978a,b, 1980; Russock, 1986).

B. Development of Chemosensory Systems and Chemoresponses

Fish detect water-borne chemical stimuli through at least two different sensory channels, olfaction (smell) and gustation (taste) (Kleerekoper, 1969; Hara, 1971, 1986; Bardach and Villars, 1974). A third chemosensory modality in fishes is the so-called common or general chemical sense (Hara, 1971). This latter channel of chemoreception is less sensitive than the olfactory and gustatory systems. Free nerve endings on the exposed body surface, which are supplied by spinal nerves, are believed to form the structural basis on the general chemical sense (Hara, 1971). In addition, free neuromasts associated with the lateral-line system in some teleosts are sensitive to various cations in the external medium (Katsuki and Yanagisawa, 1982). However, this latter chemosensitivity may simply be an incidental by-product of receptor physiology (Bleckmann, 1986). Because fish chemoreceptors respond physiologically to molecules dissolved in water, the functional distinction between chemosensory systems, particularly between olfaction and gustation, is not always as clear as in terrestrial, air-breathing vertebrates (Hara, 1971, 1986). Nevertheless, experimental evidence points to distinct functional roles for the olfactory and gustatory systems in teleosts (Hara, 1971, 1986; Bardach and Villars, 1974).

The olfactory organs of fishes show considerable interspecific diversity, reflecting in part habitat diversity (Kleerekoper, 1969; Hara, 1971, 1986). In teleosts the olfactory receptor cells are located within a sensory epithelium, which is usually folded into lamellae, on the floor of the paired olfactory pits or nasal cavities on the anterio–dorsal part of the head (Kleerekoper, 1969; Hara, 1971, 1986). The main function of the olfactory receptor cells is to detect, encode, and transmit information about the external chemical environment of the fish to the olfactory bulb and higher brain centers (Hara, 1986). The olfactory epithelium in teleosts is highly sensitive to chemicals, with response thresholds to a variety of chemicals commonly in the range of 0.1–10.0 nmol/l (Kleerekoper, 1969; Hara, 1971, 1986; Smith, 1985). Transmission of chemical information occurs along the olfactory nerve fibers, axons of the receptor cells, which converge onto the olfactory bulb where they make synaptic contact with second-order bulbar neurons (mitral cells). The axons of the mitral cells form the majority of the fibers of the olfactory tract, along which neural signals are conveyed from the olfactory bulb to the telencephalic hemispheres (Hara, 1986).

Gustation in teleosts is mediated by taste buds, which are located in the mouth and pharynx, as well as in the gill cavity and on gill arches, appendages (barbels, fins), and the external body surface (Hara, 1971; Bardach and Villars, 1974; Smith, 1985). The chemosensory cells of taste buds are innervated by branches of cranial nerves VII, IX, and X, which transmit the information to the dorsal medulla. In certain species, taste buds on modified fin rays are innervated by spinal nerves (Hara, 1971; Smith, 1985). Gustatory receptor cells are also very sensitive, with detection thresholds for certain chemicals commonly in the range of 0.2–1.0 mmol/l (Hara, 1971; Smith, 1985).

Considerable evidence points to olfaction as a general mediator of chemical signals involved in various teleost behaviors, including habitat selection, migration, mating, parental care, and predator avoidance (Kleerekoper, 1969; Hara, 1971, 1986; Bardach and Villars, 1974; Barnett, 1977a; Solomon, 1977; Liley, 1982; Pfeiffer, 1982; Smith, 1985). Gustation has been implicated in the searching and ingestion phases of feeding behavior (Bardach and Villars, 1974; Hara, 1986). Compared with the visual system, relatively little is known about structural and, particularly, functional changes that occur in the olfactory and gustatory systems during ontogeny in teleosts.

1. Ontogenetic Changes in the Olfactory and Gustatory Systems

The paired olfactory organs in teleosts develop from paired anlagen or placodes of ectodermal origin. The placodes appear ventrally on the head region of the embryo a few days after fertilization. During the embryonic period each placode increases in size, elongates anterio–posteriorly, invaginates to form an olfactory groove or pit, and migrates to a dorso–lateral and terminal position on the head. Concurrently, the olfactory nerves appear and increase in diameter with the addition of neural fibers and lead posteriorly to the olfactory bulb as the embryo grows (Watling and Hilleman, 1964; Vernier, 1969; Jahn, 1972; Zeiske et al., 1976; Breucker et al., 1979; Evans et al., 1982). Ciliated and microvillous receptor cells appear prior to hatching during differentiation of the olfactory epithelium (Evans et al., 1982).

Shortly after hatching, the anterior and posterior nares, which open into an enclosed olfactory chamber, are formed (Watling and Hilleman, 1964; Jahn, 1972; Verraes, 1976; Zeiske *et al.*, 1976; Breucker *et al.*, 1979; Evans *et al.*, 1982). In most adult teleosts the olfactory epithelium within the olfactory chamber is folded into lamellae (organized into a rosette in some species), which increase the epithelial surface area. The first lamella appears on the posterior floor of the olfactory chamber early during the embryonic or larval periods, the precise timing depending on the species. As the young fish grows, the lamellae increase in number and size up to a certain juvenile or adult body size, after which they remain constant (Watling and Hilleman, 1964; Jahn, 1972; Branson, 1975; Verraes, 1976; Iwai, 1980; Evans *et al.*, 1982; Hara, 1986). Lamellar growth is achieved largely by the addition of indifferent epithelium (supporting and glandular cells) (Branson, 1975). Therefore, the density of olfactory cells on the epithelial lamellae may not increase with ontogeny, although their absolute numbers do. No simple correlation, either intra- or interspecifically, has been established between the number of olfactory lamellae and the acuity of the olfactory system (Hara, 1986).

Other ontogenetic changes occur in the cellular components of the olfactory epithelium. For example in the atherinid fish *Nematocentris maccullochi*, only ciliated receptor cells appear in the olfactory epithelium early in the larval period. During the latter portion of this period, the microvillous receptor cell type develops and numerically matches the ciliated receptor cells by the juvenile period (Breucker *et al.*, 1979). Continuous cell turnover occurs in the developing olfactory epithelium of salmonids (Evans *et al.*, 1982). Whether changes in the sensitivity of the sensory epithelium to external chemical stimuli parallel the above-mentioned ontogenetic changes in its structure is unknown.

Less appears to be known about the development of taste buds in teleosts. Gustatory receptor cells appear as pear-shaped taste buds in the oro–pharyngeal cavity of the eleutheroembryo and larva within a couple of weeks posthatching, depending on the species (Iwai, 1980). Thereafter, the taste buds increase in number and size with fish growth. Differentiation of taste buds appears to be slower in marine fishes than in freshwater ones. In the former, the taste buds fully differentiate only after the onset of feeding on external food (Iwai, 1980).

On the basis of the above anatomical criteria, as well as behavioral criteria (White, 1915; Peterson, 1975; Dempsey, 1978; Iwai, 1980), the olfactory and gustatory systems of teleosts are likely functional early in ontogeny, during the embryonic and larval periods. The ability of young fish to detect chemical stimuli during these early ontogenetic periods, when they are still hidden in their nests or parent's mouth or free-swimming and still largely dependent on their yolk reserves, has implications for the development of chemically mediated behaviors in later ontogenetic periods. Detection of certain chemical stimuli in the environment by fish in early ontogeny can have a variety of effects on its behavior, which can either be immediate and short-term (e.g., avoidance of predators), immediate and long-term (e.g., species recognition), or delayed until later developmental stages (e.g., recognition of natal habitat).

2. Habitat Selection

Many teleosts, particularly anadromous salmonids and clupeids, show a strong tendency as adults to return to their natal habitat (usually a stream or lake) to breed after spending a period of variable duration (months to years) in one or more geographically distant habitats (e.g., ocean) where they grew into adults (Harden Jones, 1968; McKeown, 1984; Smith, 1985). This phenomenon is generally referred to as "homing." A wealth of experimental evidence from the laboratory and field indicates that the migrating adult fish can discriminate using olfaction between waters originating from its natal habitat and those originating from other sources, and thereby it is able to locate its natal tributary by exhibiting positive rheotaxis in chemically familiar water (reviewed in Cooper and Hirsch, 1982; Hasler and Scholz, 1983; Smith, 1985; Hara, 1986). The ability of the adult fish to select its natal spawning habitat from many other potential habitats nearby, mainly through olfactory means, appears to be based on its previous exposure during a specific period in early ontogeny to organic and inorganic chemicals particular to the waters of its natal (and commonly juvenile rearing) habitat. The olfactory hypothesis of homing proposes that the young fish becomes chemically or olfactorally "imprinted" during a sensitive phase in early ontogeny to the distinctive chemical composition of its natal stream. It then retains this chemical information after dispersing to another habitat, where it is no longer exposed to this composition of chemicals, and retains the ability to respond to these chemical stimuli at a later stage of the life cycle (as an adult) when searching for its natal habitat to reproduce (Cooper and Hirsch, 1982; Hasler and Scholz, 1983; Smith, 1985; Hara, 1986).

This latter ability is thus not inherited per se, but is acquired during the eleutheroembryo phase or juvenile period.

For salmonid species that undergo a physiological and behavioral transition, called smoltification (Hoar, 1976; also this volume, Chapter 4) just prior to seaward migration, the smoltification period appears to be the sensitive period during which the juvenile fish "learns" (imprints on) the chemical cues particular to its natal habitat (Cooper and Hirsch, 1982; Hasler and Scholz, 1983; Smith, 1985; Hara, 1986). Although this process of information acquisition remains poorly understood, it appears that it can be facilitated by thyroid hormones, whose plasma concentrations rise during smoltification (Hoar, 1976; Leatherland, 1982), partly through their sensitization of the olfactory system (Cooper and Hirsch, 1982; Hasler and Scholz, 1983). For species such as the pink (*Oncorhynchus gorbuscha*), chum (*O. keta*), and sockeye (*O. nerka*) salmon, whose alevins (fry) migrate seaward or lakeward within hours or days of emergence from the gravel nest, olfactory imprinting must necessarily occur earlier, probably during the eleutheroembryo phase or during the stream migration period itself (Smith, 1985). For populations of sockeye salmon whose newly emerged alevins migrate upstream in lake outlets to eventually enter the lake, the olfactory discrimination and positive rheotaxis to the waters of their nursery lake has an inherited (genetic) component in addition to being influenced by previous olfactory experience (Brannon, 1972; Godin, 1982; Smith, 1985).

An alternative, but not necessarily mutually exclusive, hypothesis for homing in salmonids is Nordeng's (1977) pheromone hypothesis. This hypothesis proposes that adults returning to their natal stream could use chemical stimuli (e.g., bile acids, mucus), released by smolts from their own population migrating downstream in the opposite direction to the ocean, to locate the natal spawning habitat. In support of the hypothesis, juvenile and adult salmonids have been shown to discriminate, on the basis of electrophysiological and behavioral criteria, population-specific odors (Smith, 1985; Groot *et al.*, 1986; Quinn and Tolson, 1986). Moreover, juvenile coho salmon (*O. kisutch*) are preferentially attracted to the chemical traces of siblings over nonsiblings of their own population (Quinn and Busack, 1985; Quinn and Hara, 1986). With regards to homing, the sibling recognition demonstrated in the above laboratory studies may be a special case of population recognition. In other contexts, recognition of siblings or conspecifics using pheromones learned in early ontogeny may influence certain behaviors, such as mate selection, in later stages of life history (see below and Section IV,D).

3. MATE CHOICE

In several groups of fishes, particularly in the Cichlidae, a close social association between parents and their mobile offspring continues for an extended period after the eggs have hatched (reviewed in Keenleyside, 1979). In brood-guarding cichlid fishes, a bidirectional communication system between parents and offspring maintains the integrity of the family unit (reviewed in Barnett, 1977a; Keenleyside, 1979; Pfeiffer, 1982). On the one hand, parental fish can distinguish visually and olfactorally (and perhaps gustatorially) their own offspring (eggs, eleutheroembryos, juveniles) from those of other conspecific or heterospecific fish (Myrberg, 1975; Barnett, 1977a; Keenleyside, 1979; Pfeiffer, 1982). This presumably prevents them from attacking and eating their own offspring and from directing (wastefully) parental care to the offspring of other fish. On the other hand, free-swimming cichlid juveniles can recognize their own siblings (e.g., Kühme, 1963) and their mother (e.g., Barnett, 1977b) using at least visual and chemical cues. Moreover, Barnett's (1977b) study on *Cichlasoma citrinellum* suggests that chemoreception may be more important than vision in the recognition of parents by juvenile cichlids when they become free-swimming, and that vision becomes increasingly the dominant modality for discrimination with age. Such sibling and parental recognition in cichlids could be important in keeping the juvenile fish together in schools (shoals), which are known to reduce individual risk of predation (Godin, 1986), and close to the protective mother, particularly at night and in turbid waters.

Because these young cichlid fish typically live in a family unit for several weeks, the potential exists for the chemosensory and visual experience gained by the young fish in early ontogeny to influence their choice of mate later on at maturity. The evidence for "filial or sexual imprinting" (cf. Immelmann, 1972) in fishes is, however, weak and contradictory. This is partly due to problems of small replicate sizes, lack of statistical data analysis, and experimental designs not allowing the fish to freely pair up and (or) to spawn with a partner of choice in many studies published on the subject. Nevertheless, in certain studies young cichlid fish have been raised from hatching (or shortly thereafter) with conspecific siblings and (or) parents or with foster parents and (or) young of another species or another color morph of their own species (Kop and Heuts, 1973; Sjölander and Fernö, 1973; Fernö and Sjölander, 1976; Crapon de Caprona, 1982; Siepen and Crapon de Caprona, 1986). The results of these studies suggest that juvenile fish learn the visual and chemical features of their broodmates or parents (genetic or foster) and are preferentially attracted to, show courtship behavior toward, and pair up with the adult fish of the species or color morph with which they were raised as juveniles. This learned preference, based on visual and chemical cues, wanes with time if not reinforced, but is retained for several months following the initial period of experience (Sjölander and Fernö, 1973; Crapon de Caprona, 1982; Siepen and Crapon de Caprona, 1986; see also Section IV,D).

Whether the learning processes observed in these studies are similar to classical imprinting processes (defined as stable over time and restricted to a sensitive ontogenetic period) remains questionable. Nonetheless, the evidence does indicate that sensory experience acquired in early ontogeny does influence the discrimination of sexual species-typical optical and chemical cues during a later life history period. The learned preferences formed can potentially affect mate choice at maturity and thus may play a role in the maintenance of stable color polymorphisms in natural populations and in speciation.

4. ALARM SUBSTANCE–FRIGHT REACTION SYSTEM

The epidermis of ostariophysian and gonorhynchiform fishes contains secretory cells that, when physically damaged—for example, by a predator—release a specific chemical substance, the so-called alarm substance (Smith, 1982). The alarm substance diffuses into the surrounding water, where it may be detected by other ostariophysians or gonorhynchiforms, which then exhibit a fright reaction as a result of chemically detecting the pheromone or of seeing other nearby fish responding to it (Smith, 1982). The fright reaction varies in form among species and is generally characterized by increased shoal cohesiveness, hiding, remaining still, or leaping at the surface, all of which are appropriate responses to predators (Smith, 1982). Although the major putative function of this chemical alarm system is antipredation (Smith, 1982), there exists no direct empirical evidence that such chemical alarm signals reduce the risk of mortality to predation of the receivers.

In cyprinids, epidermal alarm cells are present in the skin and are capable of producing the alarm substance within 1–2 weeks of hatching (Schutz, 1956; Pfeiffer, 1963). However, the young fish do not develop the ability to respond to (and presumably detect) the alarm substance until much later (at 1–2 months of age), although they are capable of schooling prior to the ontogenetic appearance of their fright reaction to the alarm substance. The form of the fright reaction may also change during the course of development (Smith, 1982). The ontogenetic timing of the fright reaction varies among species and depends partly on the relative rate of development, which of course can be influenced by temperature (Schutz, 1956; Pfeiffer, 1963). The underlying changes that occur in the olfactory system of the young fish, coincident with the ontogenetic appearance of its sensitivity and responsiveness to the alarm substance, remain unknown.

Once acquired, the fright reaction is generally assumed to remain in the behavioral repertoire of the fish throughout its life. However, in certain cyprinids the male typically loses its epidermal alarm cells temporarily during the breeding season, but retains the ability to respond to conspecific alarm signals (Smith, 1982). This seasonal loss of alarm substance is under the control of gonadal androgens, whose plasma levels of course rise during the reproductive season (Smith, 1982). The functional significance of this phenomenon is reviewed by Smith (1982).

C. Development of the Inner Ear and Lateral-Line System and Mechanoreceptor-Mediated Behavior

Mechanoreceptive neuromast cells associated with the lateral-line system and inner ear in fishes are the major receptors of external vibrational and gravitational stimuli (Lowenstein, 1971; Tavolga, 1971; Hawkins, 1973, 1986; Sand, 1981; Bleckmann, 1986).

The lateral-line system of teleost fishes typically consists of a row of pores along the trunk and head, leading into an underlying fluid-filled lateral-line canal (Bleckmann, 1986). Lateral-line organs may be categorized as ordinary (mechanoreceptive) or specialized (electroreceptive). Discussion will be restricted here to the former type, since comparatively little is known about the ontogeny of electroceptive cells (cf. Bullock, 1982).

Ordinary lateral-line organs (neuromasts) can be located inside the lateral-line canal (canal neuromasts) or superficially in shallow pits or grooves in the skin (free neuromasts) (Bleckmann, 1986). The stereocilia and kinociulium of the sensory hair cells of neuromasts are covered by a jelly-like cupula and are directionally sensitive to mechanical deflections (Sand, 1981; Bleckmann, 1986). Neural impulses from these cells are transmitted along the anterior and posterior lateral line nerves to the *octavolateralis area* of the medulla. Afferent neurons from this area project anteriorly to higher-order telencephalic and diencephalic centers (Northcutt, 1981). Moreover, branches of the auditory nerve innervate the lateral line organs (Tavolga, 1971; Hawkins, 1973). On the basis of anatomical, electrophysiological, and behavioral criteria, the lateral line functions in the detection of local water movements and surface waves in the vicinity of the fish's body and of low-frequency sound waves, but indirectly through particle displacements in the near field associated with the sound waves (Sand, 1981; Bleckmann, 1986). The ordinary lateral-line organs have been implicated in schooling behavior, predator avoidance, feeding, and social communication in adult teleost fishes (Sand, 1981; Bleckmann, 1986).

The inner ear (labyrinth) is the major organ for the detection of underwater sounds, linear acceleration, and gravity in fishes (Lowenstein, 1971; Tavolga, 1971; Hawkins, 1973, 1986). Fluid-filled semicircular canals interconnected with three sacs (ampullae) comprise the inner ear. Within each of these sacs, otoliths move above a sensory membrane (macula) containing numerous mechanoreceptive hair cells (neuromasts) that are also directionally sensitive to mechanical deflections (Hawkins, 1986). The auditory nerve innervates the inner ear, with branches projecting into the ampullary organs and maculae (Hawkins, 1986). Acoustic and gravistatic information are transmitted along this afferent nerve to the anterior octaval nucleus in the medulla, which projects afferents into a higher-order midbrain auditory center, the *torus semicircularis*. Fibers from this latter center in turn ascend bilaterally into the diencephalon and, in certain species, into telencephalic auditory areas (Northcutt, 1981).

Teleosts are sensitive to a rather limited range of sound wave frequencies compared with higher vertebrates, but many species possess an acute sensitivity to sound pressure (Hawkins, 1973, 1986). The sensitivity of fish to underwater sounds appears to depend a great deal on the degree of association between the inner ear and the gas-filled swimbladder, which can serve as a sound resonator/transducer (Hawkins, 1986). In many teleosts, the inner ear is not physically connected to the swim bladder. However, in the Cypriniformes (ostariophysian fishes), for example, the anterior portion of the swimbladder is coupled to the ear by a chain of small movable bones, the Weberian ossicles (Hawkins, 1986). In the Clupeiformes, the swimbladder extends into the cranial cavity and is linked to the ear through a pair of pro-otic bullae, which are partially filled with gas and which serve as pressure displacement converters. Further, there exist hydrodynamical connections between the ear and the lateral-line canal (Blaxter *et al.*, 1981a; Hawkins, 1986). This modified auditory system in clupeid fishes is referred to as the acousticolateralis system (Blaxter *et al.*, 1981a).

The inner ear and acousticolateralis system play important sensory roles in several behavioral phenomena in adult teleosts, including vertical migrations and associated depth–pressure adaptations, predator avoidance, equilibrium righting reflex, and acoustic communication in a variety of contexts (Lowenstein, 1971; Blaxter et al., 1981a; Myrberg, 1981; Hawkins, 1986).

1. ONTOGENY OF THE LATERAL-LINE SYSTEM AND INNER EAR

Both the lateral-line system and inner ear arise ontogenetically from ectodermal placodes early in ontogeny (Northcutt, 1981). Free neuromasts are apparent on the sides of the head and trunk of newly hatched fish in diverse species (Iwai, 1967, 1980; O'Connell, 1981; Blaxter, 1986). Cupulae cover these sensory cells at hatching or shortly afterward (Iwai, 1967, 1980; Cahn et al., 1968). The lateral-line canals and canal organs typically develop a few days after hatching, the exact timing being species-dependent (Cahn et al., 1968; Blaxter, 1986). It appears therefore that the mechanoreceptors of the lateral line system are functional at hatching or shortly thereafter in teleosts. In species whose eleutheroembryos develop within a darkened nest (e.g., salmonids) or whose larvae are pelagic, but with poorly developed vision initially (e.g., clupeids), functional free neuromasts may play a role in predator avoidance (Iwai, 1980), rheotaxis (Godin, 1982), and communication with nearby conspecifics or guarding parents (Myrberg, 1981; Bleckmann, 1986). Whether lateral-line mechanoreceptors are used in locating prey in early ontogeny, as they are in the adults of certain surface-feeding fishes (Bleckmann, 1986), remains uncertain.

Less is known about the development of the inner ear in fishes. The labyrinth and otoliths are present at the time of hatching (Armstrong and Higgins, 1971; Blaxter, 1986). On the basis of a limited number of species studied to date, sensory maculae also appear well developed by the time of hatching or shortly thereafter (Blaxter, 1986). The labyrinth therefore appears sufficiently well developed near the time of hatching to detect gravistatic stimuli and thus give the eleutheroembryos equilibrium perception (Dill, 1977; Blaxter, 1986) and to assist them in directing vertical movements within darkened gravel nests, for example (Nunan and Noakes, 1985b). It is uncertain, however, whether newly hatched fish can detect underwater sounds through the inner ear (Blaxter, 1986). This ability appears to develop somewhat later. For example, Atlantic herring larvae do not exhibit startle escape responses to sudden sound stimuli until the pro-otic bullae of the acousticolateralis system fills with gas at a body length of

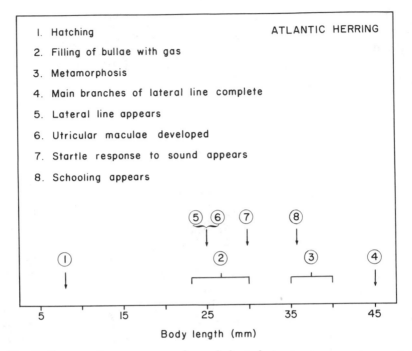

Fig. 3. Ontogenetic appearance of morphological structures or events associated with the inner ear and lateral-line system, and of certain behaviors in Atlantic herring (*Clupea harengus*) at varying body length. [After Allen *et al.* (1976), by permission of Cambridge University Press, Blaxter *et al.* (1981b), by permission of Cambridge University Press, and Blaxter (1986).]

23–30 mm, corresponding to several days posthatching (Fig. 3). The ontogenetic appearance of far-field sound detection in teleosts in general may well depend on the timing of the filling of the swimbladder with gas (Blaxter, 1986). This hypothesis requires further testing.

2. Parallel Development of Mechanoreception and Behavior

Only a limited number of examples illustrating ontogenetic coincidence in the development of the lateral line–inner ear systems and of specific behaviors in fishes have been well documented. Two such examples are briefly reviewed here.

The first example deals with schooling behavior. Lateral-line mechanoreceptors are involved in the maintenance of interindividual spacing within moving, polarized fish schools (Partridge and Pitcher,

1980) and probably in the social transmission of flight responses within the school (Godin, 1986). In silverside fishes (*Menidia* spp.), the ontogenetic appearance of coordinated, polarized schooling, which gradually develops from simple approaching and parallel orientation behaviors, coincides with the innervation of the free neuromasts on the trunk and head and with the formation of the lateral line canals. This occurs when the young fish reach a length of 9–14 mm, corresponding to several days posthatching (Cahn *et al.*, 1968). Similarly, schooling does not appear ontogenetically in herring until the lateral-line system has developed to a considerable degree (Fig. 3) (see also Section IV,D).

Startle responses comprise the second example. The lateral-line and inner-ear systems are major sensory pathways involved in the initiation of startle ("tail flip") responses in teleosts (Eaton and Nissanov, 1985; Eaton and DiDomenico, 1986). Typically, such startle responses are short-latency, C-type, fast-start accelerations to a variety of transient external stimuli, and they function mainly in predator avoidance and perhaps in hatching from the egg capsule (Eaton and Nissanov, 1985; Eaton and DiDomenico, 1986). Most startle responses in teleosts are triggered by the motor output from large Mauthner-type neurons located in the brainstem, which in turn receive sensory input from the visual and octavolateralis systems; the latter includes sensory neurons from free neuromasts, inner ear, and lateral line, and from Rohon–Beard tactile receptors on the body surface (Eaton and Nissanov, 1985; Eaton and DiDomenico, 1986).

The Mauthner cells appear early in embryogenesis, differentiate rapidly, and reach mature morphological form at about the time of hatching in teleosts. In addition, the Mauthner neurons have synapsed with sensory neurons from the visual, octavolateralis, and Rohon–Beard systems prior to hatching (Eaton and Nissanov, 1985; Eaton and DiDomenico, 1986). Therefore, the embryo possesses the full, functional neural circuitry for the initiation of coordinated startle responses while in the egg capsule. Research on zebra danios (*Brachydanio rerio*) and herring, in particular, have revealed that distinct developmental changes occur in the sensory systems associated with the startle response. Ontogenetically, the embryos respond earliest to tactile stimuli while in the egg capsule. The first startle responses to sound stimuli are observed in the free-swimming eleutheroembryos and larvae, and this coincides with the development of the calcified otoliths (in the danios) and the filling of the pro-otic bullae with gas (in the herring) (Blaxter *et al.*, 1981b; Eaton and Nissanov, 1985; Blaxter, 1986; Eaton and DiDomenico, 1986). In addition, the

Mauthner cells appear to fire spontaneously up to the time of hatching, but not afterward unless stimulated (Eaton and Nissanov, 1985; Eaton and DiDomenico, 1986). In brood-hiding salmonids, embryos similarly exhibit startle responses to tactile stimuli prior to hatching, and the proportions of fish that exhibit startle responses to acousticolateralis or visual stimulation increase progressively during the eleutheroembryo phase, reaching maximal values at the time of gravel emergence (Taylor and McPhail, 1985). At this free-swimming stage, the juveniles enter a more visual world, and their acute responsiveness to visual and vibrational stimuli, which can originate from predators, is thus adaptive. Furthermore, the kinematic features of the startle response also change as the young fish grows (Eaton and Nissanov, 1985; Taylor and McPhail, 1985; Eaton and DiDomenico, 1986).

The early development of the startle response and its associated motor and sensory systems in teleosts, as described above, reflects therefore the ecological importance of predator escape and hatching in early life history, and consequently deserves more attention than hitherto given (Eaton and Nissanov, 1985; Eaton and DiDomenico, 1986).

IV. DEVELOPMENT OF BEHAVIOR

A. Developmental Intervals

An animal must not only develop from a single cell to a large, complex adult organism, but also successfully interact with other organisms and the physical environment during that entire time. We tend to think of fish as definitive (i.e., adult) phenotypes and to look for explanations of behavior and other attributes in this light. However, embryos, larvae, and juveniles cannot be "generalized precursors," anymore than fossil forms could be "generalized ancestors" of any living species. The recognition of this can provide useful insights into our consideration of ontogeny.

In conjunction with an appreciation of life-history styles, we can focus on specific cases to illustrate this approach. The times during development when there are relatively rapid transitions in the relationship between the developing individual and its environment are of special interest. Examples of such transitions include the change from endogenous to exogenous nutrition, hatching from the egg envelope (or release from the body of the parent in species that bear the young), movement from the embryonic to larval or juvenile habitat, and settling of planktonic larvae from open water to the substrate. Whether by conscious design or coincidence, many studies of early behavior and physiology have dealt with such phenomena. This consideration is incxtricably wrapped up with the whole issue of recognizing and defining stages in development (Noakes, 1978a,b). Unfortunately, we have too often been impressed by what we think are the major features of our own development, and assume that the same general pattern must apply to all other species, including fishes—hence our fixation with birth (or hatching, as an equivalent) as the most significant transitional event in early development. Even in humans, birth does not necessarily correlate with developmental events, as evidenced by the relatively common occurrence of "premature" births. Similary, hatching (or release) of young fish is important, but probably not such an absolute landmark as many have assumed (Noakes, 1981).

There are significant differences among species in the relative timing and duration of these developmental events and intervals (Figs. 4 and 5). Every species must have at least the embryonic and adult intervals, by definition, but either or both of the "larval" and "juvenile" intervals may be eliminated. Many fish species lack the larval interval (Fig. 5), and some others probably lack the juvenile interval only (i.e., they go directly from larva to adult). The tiny, neotenic (pedomorphic) goby and cyprinid described by Winterbottom and Emery (1981) and Roberts (1986) are sexually mature while still retaining typically larval features (e.g., incomplete scalation, limited pigmentation, and partial skeletal ossification). There is even some evidence that both the larval and juvenile intervals can be absent in a species. Shaw et al. (1974) reported that male dwarf surfperch (*Micrometrus minimus*) are sexually mature and able to inseminate females shortly after release (= "birth") from their mother's body. Thus, they go directly from being embryos to adults (Noakes, 1981), without either a larval or juvenile period.

The apparent insistence by some that every fish must have a larval period during its ontogeny is not the only, but probably the most common, misconception perpetuated in the fish literature. Terms such as "yolk-sac larva" and "free-swimming larva" are frequently encountered examples of this attitude. No doubt such individuals would be perplexed by the case of a species of mite (*Acarophenax tribolii*) that dies before it is born (Gould, 1982, pp. 69–75; see Hamilton, 1967, for further details). Males of this species become sexually mature and mate with their sisters while still inside their mother's

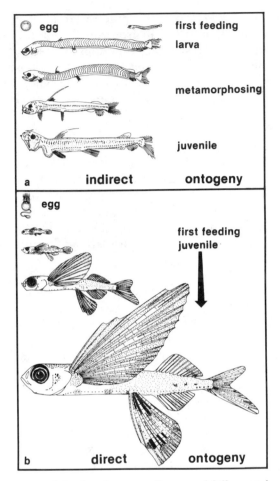

Fig. 4. Comparison of the development of young of different teleost species. (a) A mesopelagic viperfish *Chauliodus sloani* (a salmoniform protacanthopterygian) has indirect, metamorphic ontogeny with larvae and decreases in size during metamorphosis. (b) An epipelagic flying-fish *Hirundichthys rondeleltii* (an atheriniform exocoetoidei) develops from eggs about the same size as the viperfish but with a much greater yolk density, through a direct, ametamorphic ontogeny without larvae. [From Balon (1985).]

body. Males die before they, and their now inseminated sisters, are released from their mother's body (= born).

   We are not aware of any similar examples from fish species, but the possibility should not surprise us if we are thinking in developmental terms. It is just one of a range of possible life-history styles we could predict from first principles. Anyone fixated on birth, or hatching,

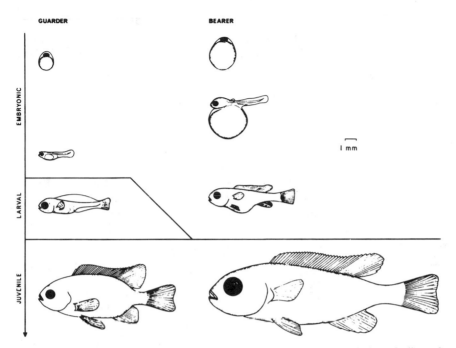

Fig. 5. Early ontogeny of representative cichlid bearer ("mouth brooder") and guarder ("substrate brooder") species. [From Noakes and Balon (1982).]

however, must be perplexed by this. We must extend and elaborate the general argument to fish, if we are to establish the common reasoning and universal nature of a truly comparative approach to studies of ontogeny.

There has been a lot of detailed consideration of developmental rates in fish ontogeny, but most often without regard to developmental intervals. Such studies range from general models of early growth (Zweifel and Lasker, 1976) to fine-grained genetic analyses of differences in development (Thorpe *et al.*, 1983; Danzmann *et al.*, 1986). An extensive literature exists for studies of the timing of activity of genes during development (Schmidtke *et al.*, 1976; Frankel and Hart, 1977; Barlow, 1981; Frankel, 1981; Stanley, 1983). Single-gene (allele) differences can significantly alter such diverse behavioral attributes as swimming performance (Klar *et al.*, 1979; DiMichele and Powers, 1982a) and homing propensity (Bams, 1976). Similar effects have also been demonstrated for broader categories of behavior influenced by sets or complexes of genes. For example, the early social behavior of chars (*Salvelinus* spp.) varies markedly between species, and these differences are largely genetically determined (Ferguson and Noakes, 1982, 1983ab; Ferguson *et al.*, 1983). Further details of other examples are discussed in Noakes (1986).

B. Hatching

There is no question that different fish species hatch at different times in ontogeny (Fig. 4). The remarkable thing is that this relationship is typically viewed in reverse—that is, hatching is assumed to be the common reference point for comparisons between species, and the species are described as differing in development at the time of hatching. Statements based on this viewpoint typically describe fish in terms of their age (hours or days) since hatching. This is a convenient and useful measure, but only within a species, under specified conditions (Balon, 1985).

Since hatching occurs at such different states of development in different species, it is quite misleading to even consider this measure in comparisons between species. It also happens that hatching is not a fixed developmental event even within a species. In salmonids, coregonids, centrarchids, and ictalurids, for example, hatching can be significantly accelerated or retarded by factors such as dissolved oxygen and temperature (Carlson and Siefert, 1974; Carlson *et al.*, 1974; Brooke and Colby, 1980; Peterson *et al.*, 1980). Single-gene (allele) differences can produce significant differences in hatching time, in both rainbow trout (*Salmo gairneri*) and mummichog (*Fundulus heteroclitus*) (DiMichele and Powers, 1982b; Danzmann *et al.*, 1986).

Certain killifish (family Cyprinodontidae), known as annual fishes, are perhaps the most extreme examples of deferred (or accelerated) hatching (Cunningham and Balon, 1986). In genera such as *Nothobranchius* and *Pterolebias*, developing embryos have one or more diapause intervals, when development is arrested for varying times. In their native habitats, these species live in seasonally flooded pools. Hatching from the egg membranes is triggered by stimuli that normally indicate the initial flooding to create new pools where the developing embryos lie. The young fish (juveniles) hatch and undergo a very brief period of rapid growth and sexual maturation. Courtship, mating, and deposition of fertilized eggs all take place before the seasonal pools desiccate, with the embyros awaiting the seasonal trigger to hatch for the next annual cycle.

Changes in dissolved gases and in ionic and osmotic factors can all act to trigger hatching. The timing of hatching in these killifish, whether measured in absolute time (hours, days, weeks, or months) or temperature units (degree-hours), would have no meaning. Perhaps a less spectacular example is the grunion (*Leuresthes tenuis*) of the Pacific Coast of California. These fish spawn just above the tide line shortly after a high spring tide. Hatching of the young from the egg membranes is triggered by the mechanical and osmotic action of waves washing over the sand at the next high spring tide (Walker, 1952; Ehrlich and Farris, 1971).

C. Feeding

Probably no aspect of early behavior in fishes has received so much study as feeding (e.g., Blaxter, 1974, and this volume). The critical importance of first exogenous feeding has been repeatedly stressed, particularly for pelagic young. Failure to begin feeding or to acquire sufficient food, has often been suggested as a major cause of early mortality (see review by Hunter, 1981).

The transition from endogenous nutrition (whether from yolk or via specialized "placental" tissues from the parent) to exogenous feeding is obviously one of the most significant events in the early life history of fishes (Wourms, 1981; and this volume, Chapter 1). This transition is the major feature used to define the end of the embryonic period (Balon, 1986). The consequences of this change in nutrition are substantial, but have sometimes been overshadowed by what is probably an unrealistic emphasis on hatching as the event of primary significance in early life history (Noakes, 1978a,b, Noakes, 1981; Noakes and Balon, 1982).

The presence of suitable prey items prior to the stage when exogenous feeding is normally initiated has no effect on first feeding (e.g., Galat and Eipper, 1975), although a period of mixed feeding (some exogenous food taken before yolk reserves are completely depleted) may occur in some species (e.g., Crane, 1981; Balon, 1985). Oviphagy (young feeding on sibling eggs and/or embryos) while "in utero" is know for some species (e.g., Fujita, 1981). It is really a special form of endogenous nutrition, since the source of nutrients is entirely from the mother. Facultative (or obligate) mucus-feeding by young fish from the bodies of the parents should be interpreted similarly (Noakes, 1976).

Prey items taken by young fish at their first exogenous feeding appear to be a consequence of their abundance, size, visibility, and evasive responses. Rajasilta and Vuorinen (1983), for example, studied the food habits of planktivorous young (less than 40 mm in length) of several fish species. When food was abundant, the young fish specialized and took the largest prey items available. If food was scarce, however, the fish generalized their diet and fed equally on all sizes of prey items. Prey abundance thus appeared to be the major factor influencing prey selection. Prey size (= visibility) and prey behavior (= mobility, evasion) probably also had significant effects on prey choice.

Similarly, larval Atlantic mackerel (*Scomber scombrus*) appear to feed selectively on their planktonic prey based on prey visibility (Peterson and Ausubel, 1984). In this species, young must ingest 25–75% of their body weight in prey each day to survive and grow. Such figures clearly demonstrate the critical nature of early feeding, especially the availability of suitable prey and the appropriate responses of the fish to the prey.

Mason (1974) demonstrated the limiting nature of food and feeding to young coho salmon in freshwater summer nursery streams by adding supplemental food (marine euphausiids) to the streams. This counteracted the normal negative density effects on fish survival, growth and biomass. The growth of fish increased significantly, as did the biomass. The frequency distribution of sizes of fish was much more normally distributed when food was added (compared to the skewed Poisson distribution in "typical" streams). This came about by increased growth rates in smaller fish. Fish had greater lipid reserves, and so were probably more resistant to subsequent winter starvation after supplemental feeding.

Feeding preferences and specializations can develop with experience (e.g., Dill, 1983; Ringler, 1983). This can come about either as a result of experience itself (= learning), or through maturational changes in the fish, such as increasing visual acuity (see Section III,A). A comparative study of the ontogeny of feeding behavior in centrarchids, for example, showed that behavior was initially similar in the four species studied and became more specialized with growth of the young (Brown and Colgan, 1984). Interestingly, these changes in behavior appeared to be more a consequence of size (= developmental age) than chronological age. Development of locomotor ability in herring, which is an important component of feeding behavior, proceeds in coordination with growth and differentiation of other body parts (Batty, 1984). Swimming mode changes after the median fins have developed, and again in coordination with differentiation of red muscle fibers and the switch to branchial respiration. Young paddlefish (*Polyodon spathula*) switch from particulate to filter feeding only when their gill rakers have developed sufficiently to retain zooplankton (Michaletz *et al.*, 1982). In a comparable fashion, the swimming and feeding modes of both mackerel (*S. japonicus*) and red sea bream (*Pagrus major*) have been reported to develop in three stages, as the caudal fin, jaws, and pharyngeal teeth develop (Kohno *et al.*, 1983, 1984).

There have been some suggestions that the timing of first exogenous feeding and the availability of suitable prey items at that time are quite critical. Young fish, if deprived of exogenous food, will reach a so-called "point of no return," when the effects of food deprivation are manifested as irreversible starvation (e.g., Dabrowski, 1976; Hunter, 1981). However, this should not be taken too literally in all cases. Some experimental investigations have shown that the interval over which exogenous feeding may be successfully initiated can be from a few days up to 6 weeks, depending very much on the species (Twongo and MacCrimmon, 1976; Hunter, 1981).

Among freshwater fishes, the salmonids have been studied most frequently because of their importance in aquaculture. It is important not to add food to rearing tanks before the young fish begin feeding so as to prevent food wastage and water contamination. On the other hand, food must not be withheld for too long, otherwise fish may suffer impaired growth or starvation. Twongo and MacCrimmon (1976) found that young rainbow trout did not begin exogenous feeding prior to the initiation of free-swimming (= "swim-up"), shortly before the yolk was completely absorbed. Histological study showed that the cavity of the gut was not open before that time, so exogenous feeding would not be possible. They reported that exogenous food could be withheld for up to 42 days after initiation of free-swimming without lasting deleterious effects on the fish. Initial exogenous feeding appeared to be equally successful, at whatever age, within this time span. Fish given exogenous food for the first time after such a delay showed increased growth rates and more uniform growth rates within groups than did fish fed for the first time immediately after initation of free swimming. The physiological basis for this effect was not investigated.

Studies of marine fish larvae have stressed the significance of first exogenous feeding and starvation (Hunter, 1981). Considerable effort has been expended in the search for histological or other field criteria that could be used to assess the nutritional status of larvae sampled from the environment (e.g., O'Connell, 1976; Buckley, 1979). Generally these criteria have been based on series of larvae raised with known food concentrations/availability in controlled laboratory conditions. With increasing food deprivation, larvae suffer from progressive emaciation and depletion of liver and muscle energy reserves. Obviously, the behavior of these fish will also be progressively impaired, so that beyond some point they would no longer be able to find and ingest prey, or avoid predators; so death would be inevitable.

These criteria have been also used to define and recognize suitable feeding areas (= patches) for larvae in nature (e.g., Lasker, 1975). Combined with detailed estimates of locomotor ability, food searching, ingestion rates, and food assimilation rates (e.g., Hunter, 1981), these have been a major component of studies of commercially important marine species (e.g., Werner and Blaxter, 1980; Townsend, 1983).

In general, the alimentary canal of teleosts develops by a series of relatively rapid changes, coincident with the transitions between embryonic and larval and then between larval and juvenile periods (Govani et al., 1986). These marked changes in structure of the gut correspond to changes in the diet and feeding mode. This differs from the more gradual and continuous development of other major organ systems of the body (O'Connell, 1981).

Precocial young begin exogenous feeding when they are relatively large, with well-developed sensory systems, jaws, and feeding structures (Noakes and Balon, 1982). Their alimentary systems are correspondingly well-developed, and are often coiled or otherwise elaborated. In comparison, altricial young have relatively less developed alimentary canals, typically consisting of a simple, straight tube (Govani et al., 1986), and begin exogenous feeding when they are smaller and less differentiated and developed (Noakes and Balon, 1982).

The production of digestive enzymes also appears to be synchronized with other developmental changes in the digestive system and feeding behavior. The rate at which food items pass through the gut varies directly with food intake, whereas the assimilation efficiency varies inversely with passage time of food in the gut (Govani et al., 1986). Superimposed on these relationships are the effects of the structure of the alimentary tract itself. The relatively straight digestive tracts of altricial young pass food relatively more quickly than do the more elaborate, coiled tracts of precocial young. The typical development of the gut from a simple to a more complex, coiled, elaborate structure parallels this change with the ontogeny of each species (Table I). Hence, the feeding strategies of young fish might be correlated to the structure and function of the digestive system (Govani et al., 1986). For example, the altricial young of clupeoids appear adapted to feed on contagiously distributed planktonic prey in patches of high concentration. Precocial young could be expected to feed more efficiently on more widely scattered, individual prey items.

Table I
General Pattern of Differentiation of the Alimentary Canal of Teleost Fishes[a]

Embryo	Larva	Juvenile/adult
		Esophagus
	Foregut	
		Stomach
Incipient gut (simple, straight tube)		
	Midgut	Anterior intestine
	Hindgut	Posterior intestine

[a] From Govani et al. (1986). The condition of the canal is given for the embryonic, larval, and juvenile/adult periods.

These differences in early development, that is, altricial versus precocial, no doubt have a number of important correlates and consequences. The altricial pattern of development, for example, requires that young fish have to seek out food, capture, ingest, and digest food items while still relatively undeveloped, compared to precocial young. The latter, as a consequence of a large endogenous food supply and/or transfer of nutrients via absorptive structure from the parent, do not have to begin foraging and feeding themselves on exogenous food items until they are (typically) differentiated juvenile phenotypes (Balon, 1986; Hart and Werner, 1987).

D. Social Responses

Only members of a relatively small fraction of all fish species have reproductive habits that lead to situations where young fish interact socially with adults (typically their parents) (Breder and Rosen, 1966). However, the young of many species without any parental care still develop social responses to conspecifics (sometimes, but not necessarily, siblings). The development of schooling behavior is certainly the best studied of these and perhaps the most widespread.

Fully developed schooling behavior typically involves conspecifics, of similar age and size, engaged in the same activities at a given time, with a high degree of spatial and temporal synchrony (Keenleyside, 1979). A distinction has been suggested between this and shoaling behavior (Pitcher, 1986). The latter term is applied more generally to social aggregations that lack the tight coordination and synchrony of schools. Such aggregating responses (whether schooling or shoaling has not always been specified) have been studied in most detail in altricial species (Shaw, 1970).

Initially, shortly after the young fish have hatched from the egg membranes, they show no particular social reactions. Not surprisingly, any responses tend to be avoidance movements, if they approach each other closely. With increasing size and improved swimming ability, they spend longer periods of time swimming close to each other when they happen to encounter another individual. This is seen as short bouts of parallel swimming before the fish break off and travel independently again. Eventually, by the time the fish are juveniles, they spend the majority of their time swimming in groups with several other fish, turning and moving in coordination with each other. The correspondence between the development of these social responses and the development of the optomotor reaction of young fish is obvious, both in terms of timing and the responses involved (since most schooling is considered to be visually mediated). In both cases, the response is completely developed by the time the young have completed their metamorphosis from larva to juvenile (e.g., Kawamura and Hara, 1980; see also Section III,C).

The question of species recognition—that is, the ability of an individual to distinguish between conspecifics and heterospecifics, and to respond positively to (usually by physical association) conspecifics— might be involved in the development of these social responses. It cannot necessarily be concluded that an active recognition and discrimination has taken place, just because fish end up associating with conspecifics. First, there must be some alternative available. Localized spawning by adults of a species, water and wind currents, and other physical factors can all lead to single-species aggregations, without any particular action on the part of the young fish. Second, responses of young fish to environmental cues (e.g., light, temperature, food) can also produce single-species aggregations without the need for social recognition, or even social responses. Finally, physical limitations and/or characteristics of locomotion (e.g., swimming speed, depth, etc.) will tend to result in fish of the same species and similar size swimming together, or at least ensure that fish with different swimming characteristics will not associate without some overriding social attractiveness and responses.

There is even evidence from some studies that fish may be able to recognize genetic kin (Ferguson and Noakes, 1981; Quinn and Busack, 1985; Quinn and Hara, 1986; Quinn and Tolson, 1986). In such cases, fish could (at least theoretically) preferentially respond to conspecifics as a function of their genetic relatedness. The extent of such capabilities remains to be established, as do the consequences (Colgan, 1983). The sensory basis for this discrimination are known for some examples, and they would lend themselves readily to experimental studies of development.

Any social responses of young fish will require that the appropriate sensory system(s) is sufficiently developed to allow the fish to detect, localize, and respond appropriately to the relevant stimuli (see Section III). In addition, the fish must have the motor capacity to perform whatever response(s) is appropriate to manifest this social recognition. At the least, this will usually require that this fish be capable of sustained locomotion, and be able to coordinate that locomotion with relevant social stimuli. The well-known alarm reaction of ostariophysian fishes is a good example (Smith, 1982). Young zebrafish, for example, first show the alarm reaction to skin extracts from conspecifics shortly after they begin producing the alarm substance in their own skin. The exact chronological age is not critical, but rather the developmental state of the individual fish (Waldman, 1982). The developmental age (rather than chronological age) has also been shown to be important in the approach and withdrawal social responses of young tilapia (*Tilapia mossambica*) (Goude et al., 1972). Similarly, Chen et al. (1983) reported that the timing of dispersal of young jewel cichlids (*Hemichromis bimaculatus*) was regulated by developmental processes, rather than factors such as crowding.

Early social experience not only depends on anatomical and physiological development, it can itself affect those parameters. For example, it has been shown that early social experience affects the development of interneurons in the optic tectum of jewel cichlids. The normal pattern of early development of tectal neurons in jewelfish is a marked increase in number and an increased regularity of their spacing (Burgess et al., 1982). Fish reared with deprived social (conspecific) stimulation have fewer dendritic spines and branches on these neurons, and the spines and branches present are shorter (Coss and Globus, 1978, 1979). Crowding (i.e., rearing fish at higher densities in laboratory aquaria) results in a decrease in dendritic spine number in the optic tectum and changes in dendritic spine lengths (Burgess and Coss, 1982). There have even been reports that rearing with parents is essential for the survival, growth, and normal differentiation of some cichlids (Noakes, 1976). Social interactions of juveniles have also been shown to affect such diverse measures as growth rate, size, and age at maturity, and social behavior in several species (Farr, 1980; Fraley and Fernald, 1982). Parent–young interactions have been reported as essential for physiological development, differentiation, and survival of some species (Ward and Barlow, 1967), but the extent or physiological basis for this effect has not been clearly established (Minnier and Cole, 1973; Noakes, 1976).

Early experience may also have much longer lasting, or delayed, effects. It has been reported (Crapon de Caprona, 1982) that the social preferences of adult male and female cichlids (*Haplochromis burtoni*) are affected by learning processes associated with exposure to both chemical and visual social stimuli. This could be of particular importance as related to species recognition in general, or mate choice, more specifically. The latter possibility is much more likely to be significant in species in which there is prolonged social contact and interaction between parents and their offspring (e.g., Barlow, 1983). For example, Stearns (1983) reported that mosquitofish (*Gambusia affinis*) reared in social isolation from 10 days of age until maturity did not reproduce as adults. However, it must be noted that the evidence for social imprinting (filial or sexual) is still inconclusive for any fish species. The evidence is in part contradictory, and equivocal at best (Sjölander and Fernö, 1973; Fernö and Sjölander, 1976; Weber and Weber, 1976), given methodological and procedural concerns (Barlow, 1983; see also Section III,B3).

The development of social responses to parents has been best studied in the Cichlidae. This family is characterized by elaborate social behavior, with parental care of young, in all species (Keenleyside, 1979). The parental care of these species has been characterized as being of two types, guarding and bearing (Fig. 5; Noakes and Balon, 1982). These types have variously been described as substrate brooding (or substrate breeding, or substrate spawning) and mouth brooding (or mouth breeding, or mouth bearing) in an extensive literature (Breder and Rosen, 1966). Guarding species typically form extended, monogamous pair bonds, and both parents tend the fertilized eggs and developing young over a period of days, weeks, or even months (Baylis, 1981). Bearing species are characterized by relatively brief pair bonds, often only for the duration of the actual spawning act, and subsequent care of the young by only one parent, typically the female. The fertilized eggs and developing young are carried within the buccal cavity of the parent until their subsequent release some weeks later.

The young of guarding species are relatively more altricial, and those of bearing species more precocial, in their ontogeny (Noakes and Balon, 1982). The former have smaller amounts of yolk material at fertilization, and so must pass through a larval interval and begin exogenous feeding while still incompletely developed (Fig. 5). Young bearers, on the other hand, develop from eggs with larger yolk reserves, and pass through the early development and differentiation while protected inside the parent's buccal cavity. There have been many studies of reproductive and parental behavior of cichlids, but relatively few detailed investigations of early behavior in different species, and even fewer direct comparisons of guarding and bearing species (Noakes, 1978a,b).

As in other comparisons, the tendency has been to relate the behavior of the young to their age since hatching or release from the parent's mouth. Of course these events are comparable, inasmuch as they represent the times at which the young must face the world on their own, and fend for themselves. However, this comparison overlooks, and in fact misrepresents, the ontogeny of the animals, since at the time of hatching a young guarder is still an embryo, while at the time of release from the parent's mouth a young bearer is a juvenile (Fig. 5). Not surprisingly, the young of these species differ remarkably at this time in almost any behavioral measure. A more reasonable comparison of developmental stages must take into account the fact that the corresponding early stages of young bearers pass while the young fish is carried inside the parent's mouth.

When the comparison is made with this realization in mind, the contrast is less striking, although no less significant. Ignoring species-typical differences for the moment, it could be that all species are on essentially the same developmental trajectory. It is the relative timing of events on this trajectory, particularly the ecologically significant event of first exogenous feeding, that is most critical. The few detailed studies of early ontogeny of particular species clearly illustrate this point. Young bearers, after release by the parent, behave essentially as one would expect, that is, as independent juveniles. They may aggregate or disperse, depending on the species, but they are fully developed and competent in terms of morphology, coloration, locomotion, and feeding performance (e.g., Balon, 1977).

Young guarders, by contrast, undergo a prolonged development, from a very undifferentiated condition at the time of hatching (Ward and Barlow, 1967; Noakes and Barlow, 1973; Wyman and Ward, 1973). The performance of almost every behavioral action is initially either nonexistent or at best rudimentary. It is not until these young have grown and become differentiated as juveniles that they leave the custodial care of the parents, and behave in a comparable fashion to the newly released bearer young.

Study of the behavioral ontogeny of young bearers would be particularly important, not just to add to our store of factual knowledge. If it is true that bearer species evolved from guarder ancestors, then it follows that behavioral ontogeny of the young should in some sense reflect that origin and evolutionary history. Young bearers pass through an extended embryonic interval, and lack a larval interval, in comparison to guarders. In addition, this all occurs while the young are being carried inside the parent's mouth. Several possibilities could be predicted as to the behavioral development of a bearer species, in comparison to its closest guarder relative. If there has not been any strong selection acting on young bearers (because of the protection from their parents), then one might expect remnants of the early behavior typical of guarders to persist in their early ontogeny. Clearly, there should be some behavioral adaptations associated with the oral bearing habit, such as responses to retain young in the mouth, or to ensure they return to the mouth or are picked up by the parent if released prematurely. But these should be relatively easy to recognize. The question is particularly attractive, since young of both types (guarders and bearers) can readily be produced from captive breedings, and young of both are accessible for observation, large enough to permit individual observations and manipulations, and the behavior has already been described in sufficient detail in representative species (Noakes, 1978a,b).

Young guarders, of course, must undergo considerable behavioral (and other corresponding) development from the time of hatching until they reach a comparable state. For example, young guarders develop the ability to recognize and discriminate among parents, other conspecifics, and heterospecifics, using visual and chemical (and possibly other) sensory modalities (e.g., Noakes and Barlow, 1973; Barnett, 1977b; Coss, 1979). Social interactions with siblings progress from simple physical encounters and direct avoidance, to more elaborate lateral posturing, more complex stimulus–response situations, and eventually recognizable dominance and sexual interactions (Wyman and Ward, 1973). Young bearers, upon release by the parent, behave as independent juveniles in all respects, as we would predict.

V. RECAPITULATION, PERSPECTIVES, AND PROSPECTS

Of the relatively few major objectives in this chapter, perhaps the principal one was to emphasize the necessity, and utility, of applying a consistent scheme of terminology to studies of ontogeny. This terminology must have some generality, not just to particular groups of fish, but ideally across animal species. It is only with such a broad perspective that general patterns and processes can be recognized. A uniform terminology also brings details of individual ontogenies into focus, and raises a series of important issues.

One of these is whether the functional development of behavior proceeds in a smooth, progressive fashion, or in a more stepwise manner. Another is the ecological and evolutionary significance of differences in ontogeny, especially the timing of exogenous feeding in relation to development. Yet another is the question of the functional significance of metamorphosis during early development (Werner, 1986).

It seems clear that there is a significant, consistent correlation between the physiological substrate and corresponding behavioral development in fishes. This is hardly unexpected, and in fact it would be very surprising were it not the case. Nonetheless, it need not be the case a priori. So it is somewhat satisfying to accumulate the empirical evidence for this association.

Possibly the most exciting area for future studies would be to unravel the causal links between the physiological substrates and the behavioral actions. Some examples of the association between genes and behavior hold promise in this regard. They would not only be amenable to experimental study, but are most likely to be cases where an immediate application or management procedure could be incorporated. The examples of single-gene differences associated with differences in time of hatching, early growth rate, and locomotor ability are cases in point.

Behavior, as the integrated expression of the animal's response to its environment (both internal and external), is the key to this approach. For ethologists, questions as to both proximate and ultimate causation must constantly be asked of behavior. Immediate, proximate mechanisms of behavior are essentially synonymous with the physiological basis of behavior. Ultimate causation, or function, of behavior will be expressed in ecological and evolutionary terms. The recognition of this distinction and an appreciation of the different approaches to be used, and answers to be accepted to resolve these two kinds of questions, should always be kept in mind. Ecological and evolutionary studies in ethology, often referred to as behavioral ecology (Noakes and Grant, 1987), deal with ultimate questions about fish behavior. The distinction between ethology and physiology may not always be clear, and in the final analysis depends mostly on the answer being sought by the investigator. To physiologists, we would suggest that it is worth considering, if only as a possibility, the view that physiology can be a consequence of behavior (rather than the converse, as is usually the case) (McFarland, 1986).

ACKNOWLEDGMENTS

This chapter was prepared while both authors were recipients of research operating grants from N.S.E.R.C. of Canada. We are grateful for this support.

REFERENCES

Abu-Gideiri, Y. B. (1966). The behaviour and neuro-anatomy of some developing teleost fishes. *J. Zool.* **149**, 215–241.

Abu-Gideiri, Y. B. (1969). The development of behaviour in *Tilapia nilotica* L. *Behaviour* **34**, 17–28.

Ahlbert, I. B. (1975). Organization of the cone cells in the retinae of some teleosts in relation to their feeding habits. Ph.D. Thesis, University of Stockholm, Stockholm.

Ahlbert, I.-B. (1976). Organization of the cone cells in the retinae of salmon (*Salmo salar*) and trout (*Salmo trutta trutta*) in relation to their feeding habits. *Acta Zool. (Stockholm)* **57**, 13–35.

Ali, M. A. (1959). The ocular structure, retinomotor and photo-behavioral responses of juvenile Pacific salmon. *Can. J. Zool.* **37**, 965–996.

Ali, M. A., ed. (1975a). "Vision in Fishes: New Approaches in Research." Plenum, New York.

Ali, M. A. (1975b). Retinomotor responses. *In* "Vision in Fishes: New Approaches in Research" (M. A. Ali, ed.), pp. 313–355. Plenum, New York.

Ali, M. A. (1975c). Temperature and vision. *Rev. Can. Biol.* **34**, 131–186.

Ali, M. A., and Klyne, M. A. (1985). "Vision in Vertebrates." Plenum, New York.

Allen, J. M., Blaxter, J. H. S., and Denton, E. J. (1976). The functional anatomy and development of the swimbladder–inner ear–lateral line system in herring and sprat. *J. Mar. Biol. Assoc. U.K.* **56**, 471–486.

Arey, L. B. (1962). "Developmental Anatomy," 6th ed. Saunders, Philadelphia, Pennsylvania.

Armstrong, P. B. (1964). Photic responses in developing bullhead embryos. *J. Comp. Neurol.* **123**, 147–160.

Armstrong, P. B., and Higgins, D. C. (1971). Behavioral encephalization in the bullhead embryo and its neuroanatomical correlates. *J. Comp. Neurol.* **143**, 371–384.

Baerends, G. P. (1971). The ethological analysis of fish behavior. *In* "Fish Physiology" (W. S. Hoar and D. J. Randall, eds.), Vol. V, pp. 279–370. Academic Press, New York.

Balon, E. K. (1975a). Reproductive guilds of fishes: A proposal and definition. *J. Fish. Res. Board Can.* **32**, 821–864.

Balon, E. K. (1975b). Terminology of intervals in fish development. *J. Fish. Res. Board Can.* **32**, 1663–1670.

Balon, E. K. (1977). Early ontogeny of *Labeotropheus* Ahl, 1927 (Mbuna, Cichlidae, Lake Malawi), with a discussion on advanced protective styles in fish reproduction and development. *Environ. Biol. Fishes* **2**, 147–176.

Balon, E. K., ed. (1985). "Early Life Histories of Fishes. New Developmental, Ecological and Evolutionary Perspectives." Martinus Nijhoff/Dr. W. Junk Publishers, Dordrecht, The Netherlands.

Balon, E. K. (1986). Types of feeding in the ontogeny of fishes and the life history model. *Environ Biol. Fishes* **16**, 11–24.

Bams, R. A. (1976). Survival and propensity for homing as affected by presence or absence of locally adapted paternal genes in two transplanted populations of pink salmon (*Oncorhynchus gorbuscha*). *J. Fish. Res. Board Can.* **33**, 2716–2725.

Bardach, J. E., and Villars, T. (1974). The chemical senses of fishes. *In* "Chemoreception in Marine Organisms" (P. T. Grant and A. M. Mackie, eds.), pp. 49–104. Academic Press, London.

Barlow, G. W. (1981). Genetics and development of behavior, with special reference to patterned motor output. *In* "Behavioral Development" (K. Immelmann, G. W. Barlow, L. Petrinovitch, and M. Main, eds.), pp. 191–251. Cambridge Univ. Press, London and New York.

Barlow, G. W. (1983). The benefits of being gold: Behavioral consequences of polychromatism in the midas cichlid, *Cichlasoma citrinellum*. *Environ. Biol. Fishes* **8**, 73–86.

Barnett, C. (1977a). Aspects of chemical communication with special reference to fish. *Biosci. Commun.* **3**, 331–392.

Barnett, C. (1977b). Chemical recognition of the mother by the young of the cichlid fish, *Cichlasoma citrinellum*. *J. Chem. Ecol.* **3**, 461–466.

Batty, R. S. (1984). Development of swimming movements, and musculature of larval herring (*Clupea harengus*). *J. Exp. Biol.* **110**, 217–229.

Baylis, J. R. (1981). The evolution of parental care in fishes, with reference to Darwin's rule of male sexual selection. *Environ. Biol. Fishes* **6**, 223–251.

Blaxter, J. H. S. (1968a). Visual thresholds and spectral sensitivity of herring larvae. *J. Exp. Biol.* **48**, 39–53.

Blaxter, J. H. S. (1968b). Light intensity, vision, and feeding in young plaice. *J. Exp. Mar. Biol. Ecol.* **2**, 293–307.

Blaxter, J. H. S. (1969a). Development: Eggs and larvae. *In* "Fish Physiology" (W. S. Hoar and D. J. Randall, eds.), Vol. 3, pp. 178–252. Academic Press, New York.

Blaxter, J. H. S. (1969b). Visual thresholds and spectral sensitivity of flatfish larvae. *J. Exp. Biol.* **51**, 221–230.

Blaxter, J. H. S. (1972). Brightness discrimination in larvae of plaice and sole. *J. Exp. Biol.* **57**, 693–700.

Blaxter, J. H. S. ed. (1974). "The Early Life History of Fish." Springer-Verlag, Berlin and New York.

Blaxter, J. H. S. (1975). The eyes of larval fish. *In* "Vision in Fishes. New Approaches in Research" (M. A. Ali, ed.), pp. 427–443. Plenum, New York.

Blaxter, J. H. S. (1976). The role of light in the vertical migration of fish—A review. *In* "Light as an Ecological Factor: II" (G. C. Evans, R. Bainbridge, and O. Rackham, eds.), pp. 189–210. Blackwell, Oxford.

Blaxter, J. H. S. (1980). Vision and feeding of fishes. *In* "Fish Behavior and Its Use in the Capture and Culture of Fishes" (J. E. Bardach, J. J. Magnuson, R. C. May, and J. M. Reinhart, eds.), pp. 32–56. ICLARM Resour. Manage. Manila, Philippines.

Blaxter, J. H. S. (1986). Development of sense organs and behaviour of teleost larvae with special reference to feeding and predator avoidance. *Trans. Am. Fish. Soc.* **115**, 98–114.

Blaxter, J. H. S., and Jones, M. P. (1967). The development of the retina and retinomotor responses in the herring. *J. Mar. Biol. Assoc. U.K.* **47**, 677–697.

Blaxter, J. H. S., and Staines, M. (1970). Pure-cone retinae and retinomotor responses in larval teleosts. *J. Mar. Biol. Assoc. U.K.* **50**, 449–460.

Blaxter, J. H. S., Denton, E. J., and Gray, J. A. B. (1981a). Acousticolateralis system of clupeid fishes. *In* "Hearing and Sound Communication in Fishes" (W. N. Tavolga, A. N. Popper, and R. R. Fay, eds.), pp. 39–59. Springer-Verlag, Berlin and New York.

Blaxter, J. H. S., Denton, E. J., and Gray, J. A. B. (1981b). The auditory bullae-swimbladder system in late stage herring larvae. *J. Mar. Biol. Assoc. U.K.* **61**, 315–326.

Bleckmann, H. (1986). Role of the lateral line in fish behaviour. *In* "The Behaviour of Teleost Fishes" (T. J. Pitcher, ed.), pp. 177–202. Croom Helm, London.

Boehlert, G. W. (1979). Retinal development in postlarval through juvenile *Sebastes diploproa:* Adaptations to a changing photic environment. *Rev. Can. Biol.* **38**, 265–280.

Bond, C. E. (1979). "Biology of Fishes." Saunders, Philadelphia, Pennsylvania.

Bonner, J. T., ed. (1982). "Evolution and Development." Springer-Verlag, Berlin and New York.

Brannon, E. L. (1972). Mechanisms controlling migration of sockeye salmon fry. *Int. Pac. Salmon Fish., Comm. Bull.* **21**, 1–86.

Branson, D. A. (1075). Post hatching sequence of olfactory lamella formation and pigmentation in *Hybopsis aestivalis* (Pisces: Cyprinidae). *Copeia,* pp. 109–112.

Breck, J. E., and Gitter, M. J. (1983). Effect of fish size on the reactive distance of bluegill (*Lepomis macrochirus*) sunfish. *Can. J. Fish. Aquat. Sci.* **40**, 162–167.

Breder, C. M., Jr., and Rosen, D. E. (1966). "Modes of Reproduction in Fishes." Natural History Press, Garden City, New York.

Breer, H., and Rahmann, H. (1977). Cholinesterase-Aktivitat and Hirnganglioside wahrend der Fisch-Entwicklund. *Wilhelm Roux's Arch. Dev. Biol.* **181**, 65–72.

Brett, J. R. (1971). Energetic response of salmon to temperature. A study of some thermal relations in the physiology and freshwater ecology of sockeye salmon (*Oncorhynchus nerka*). *Am. Zool.* **11**, 99–113.

Brett, J. R. (1979). Environmental factors and growth. *In* "Fish Physiology" (W. S. Hoar, D. J. Randall, and J. R. Brett, eds.), Vol. 8, pp. 599–675. Academic Press, New York.

Brett, J. R., and Groot, C. (1963). Some aspects of olfactory and visual responses in Pacific salmon. *J. Fish. Res. Board Can.* **20**, 287–303.

Breucker, H., Zeiske, E., and Melinkat, R. (1979). Development of the olfactory organ in the rainbow fish *Nematocentris maccullochi* (Atheriniformes, Melanotaeniidae). *Cell Tissue Res.* **200**, 53–68.

Brooke, L. T., and Colby, P. J. (1980). Development and survival of embryos of lake herring at different constant oxygen concentrations and temperature. *Prog. Fish-Cult.* **42**, 3–9.

Brown, J. A. (1985). The adaptive significance of behavioural ontogeny in some centrarchid fishes. *Environ. Biol. Fishes* **13**, 25–34.

Brown, J. A., and Colgan, P. W. (1984). The ontogeny of feeding behaviour in four species of centrarchid fish. *Behav. Process.* **9**, 395–411.

Buckley, L. J. (1979). Relationships between RNA–DNA ratio, prey density and growth rate in Atlantic cod (*Gadus morhua*) larvae. *J. Fish. Res. Board. Can.* **36**, 1497–1502.

Bulkowski, L., and Meade, J. W. (1983). Changes in phototaxis during early development of walleye. *Trans. Am. Fish. Soc.* **112**, 445–447.

Bullock, T. H. (1982). Electroreception. *Annu. Rev. Neurosci.* **5**, 121–170.

Burgess, J. W., and Coss, R. G. (1982). Effects of chronic crowding stress on midbrain development: Changes in dendritic spine density and morphology in jewel fish optic tectum. *Dev. Psychobiol.* **15**, 461–470.

Burgess, J. W., Monachello, M. P., and McGinn, M. D. (1982). Early development of spiny neurons in fish and mouse: Morphometric measures of dendritic spine formation pattern. *Dev. Brain Res.* **4**, 465–472.

Cahn, P. H., Shaw, E., and Atz, E. H. (1968). Lateral-line histology as related to the development of schooling in the atherinid fish. *Menidia. Bull. Mar. Sci.* **18**, 660–670.

Carey, W. E. (1985). Comparative ontogeny of photobehavioural responses of charrs (*Salvelinus* species). *Environ. Biol. Fishes* **12**, 189–200.

Carlson, A. R., and Siefert, R. E. (1974). Effects of reduced oxygen on the embryos and larvae of lake trout (*Salvelinus namaycush*) and largemouth bass (*Microptrus salmoides*). *J. Fish. Res. Board Can.* **31**, 1393–1396.

Carlson, A. R., Siefert, R. E., and Herman, L. J. (1974). Effects of lowered dissolved oxygen concentrations on channel catfish (*Ictalurus punctatus*) embryos and larvae. *Trans. Am. Fish. Soc.* **103**, 623–626.

Chen, M. J., Goss, R. G., and Goldthwaite, R. O. (1983). Timing of dispersal in juvenile jewel fish during development is unaffected by available space. *Dev. Psychobiol.* **16**, 303–310.

Colgan, P. W. (1983). "Comparative Social Recognition." Wiley, New York.

Cooper. J. C., and Hirsch, P. J. (1982). The role of chemoreception in salmonid homing. *In* "Chemoreception in Fishes" (T. J. Hara, ed.), pp. 343–362. Elsevier, Amsterdam.

Coss, R. G. (1978). Development of face aversion by the jewel fish (*Hemichromis bimaculatus*, Gill 1862). *Z. Tierpsychol.* **48**, 28–46.

Coss, R. G. (1979). Delayed plasticity of an instinct recognition and avoidance of 2 facing eyes by the jewel fish. *Dev. Psychobiol.* **12**, 335–345.

Coss, R. G., and Globus, A. (1978). Spine stems on tectal interneurons in jewel fish are shortened by social stimulation. *Science* **200**, 787–790.

Coss, R. G., and Globus, A. (1979). Social experience affects the development of dendritic spines and branches on tectal interneurons in the jewel fish. *Dev. Psychobiol.* **12**, 347–358.

Crane, J. M., Jr. (1981). Feeding and growth by the sessile larvae of the teleost *Porichthys notatus. Copeia*, pp. 895–897.

Crapon de Caprona, M.-D. (1982). The influence of early experience on preferences for optical and chemical cues produced by both sexes in the cichlid fish *Haplochromis burtoni* (*Astatotilapia burtoni*, Greenwood 1979). *Z. Tierpsychol.* **58**, 329–361.

Crapon de Caprona, M.-D., and Fritzsch, B. (1983). The development of the retinopetal nucleus olfacto-retinalis of two cichlid fish as revealed by horseradish peroxidase. *Dev. Brain Res.* **11**, 281–301.

Cunningham, J. E. R., and Balon, E. K. (1986). Early ontogeny of *Adinia xenica* (Pisces, Cyprinodontidae). 3. Comparison and evolutionary significance of some patterns in epigenesis of egg-scattering, hiding and bearing cyprinodontiforms. *Environ. Biol. Fishes* **15**, 91–105.

Dabrowski, K. (1976). An attempt to determine the survival time for starving fish larvae. *Aquaculture* **8**, 189–192.

Danzmann, R. G., Ferguson, M. M., Allendorf, F. W., and Knudsen, K. L. (1986). Heterozygosity and developmental rate in a strain of rainbow trout (*Salmo gairdneri*). *Evolution (Lawrence, Kans.)* **40**, 86–93.

Dawkins, R. (1982). "The Extended Phenotype." Freeman, San Francisco, California.

Dempsey, C. H. (1978). Chemical stimuli as a factor in feeding and intraspecific behaviour of herring larvae. *J. Mar. Biol. Assoc. U.K.* **58**, 739–747.

Dill, L. M. (1983). Adaptive flexibility in the foraging behavior of fishes. *Can. J. Fish. Aquat. Sci.* **40**, 398–408.

Dill, P. A. (1977). Development of behaviour in alevins of Atlantic salmon, *Salmo salar*, and rainbow trout, *S. gairdneri*. *Anim. Behav.* **25**, 116–121.

DiMichele, L., and Powers, D. A. (1982a). Physiological basis for swimming endurance differences between LDH-B phenotypes of *Fundulus heteroclitus*. *Science* **216**, 1014–1016.

DiMichele, L., and Powers, D. A. (1982b). LDH-B genotype-specific hatching times of *Fundulus heteroclitus* embryos. *Nature (London)* **296**, 563–564.

Easter, S. S., Jr., Johns, P. R., and Baumann, L. R. (1977). Growth of the adult goldfish eye. I. Optics. *Vision Res.* **17**, 469–477.

Eaton, R. C., and DiDomenico, R. (1986). Role of the teleost escape response during development. *Trans. Am. Fish. Soc.* **115**, 128–142.

Eaton, R. C., and Nissanov, J. (1985). A review of Mauthner-initiated escape behavior and its possible role in hatching in the immature zebrafish, *Brachydanio rerio*. *Environ. Biol. Fishes* **12**, 265–279.

Eaton, R. C., Bombarieri, R. A., and Meyer, D. L. (1978). The Mauthner-initiated startle response in teleost fish. *J. Exp. Biol.* **66**, 65–81.

Eggers, D. M. (1978). Limnetic feeding behavior of juvenile sockeye salmon in Lake Washington and predator avoidance. *Limnol. Oceanogr.* **23**, 1114–1125.

Ehrlich, K. F., and Farris, D. A. (1971). Some influences of temperature on the development of the grunion *Leuresthes tenuis* (Ayres). *Calif. Fish Game* **57**, 58–68.

Evans, R. E., Zielinski, B., and Hara, T. J. (1982). Development and regeneration of the olfactory organ in rainbow trout. *In* "Chemoreception in Fishes" (T. J. Hara, ed.), pp. 15–37. Elsevier, Amsterdam.

Farr, J. A. (1980). The effects of juvenile social interaction on growth rate, size and age at maturity, and adult social behavior in *Girardinus metallicus* Poey. *Z. Tierpsychol.* **52**, 247–268.

Ferguson, M. M., and Noakes, D. L. G. (1981). Social grouping and genetic variation in common shiner, *Notropis cornutus* (Pisces, Cyprinidae). *Environ. Biol. Fishes* **6**, 357–360.

Ferguson, M. M., and Noakes, D. L. G. (1982). Genetics of social behaviour in charrs (*Salvelinus* species). *Anim. Behav.* **30**, 128–134.

Ferguson, M. M., and Noakes, D. L. G. (1983a). Behaviour-genetics of lake charr (*Salvelinus namaycush*) and brook charr (*S. fontinalis*): Observation of backcross and F2 generations. *Z. Tierpsychol.* **62**, 72–86.

Ferguson, M. M., and Noakes, D. L. G. (1983b). Movers and stayers: Genetic analysis of mobility and positioning in hybrids of lake charr *Salvelinus namaycush* and brook charr *Salvelinus fontinalis*. *Behav. Genet.* **13**, 213–222.

Ferguson, M. M., and Noakes, D. L. G. (1983c). Behavioral plasticity of lake charr (*Salvelinus namaycush*) × brook charr (*S. fontinalis*) F1 hybrids in response to varying social environment. *Behav. Process.* **8**, 147–156.

Ferguson, M. M., Noakes, D. L. G., and Romani, D. (1983). Restricted behavioural plasticity of juvenile lake charr, *Salvelinus namaycush*. *Environ. Biol. Fishes* **8**, 151–156.

Fernald, R. D. (1985). Growth of the teleost eye: Novel solutions to complex constraints. *Environ. Biol. Fishes* **13**, 113–123.

Fernö, A., and Sjölander, S. (1976). Influence of previous experience on the mate selection of two colour morphs of the convict cichlid, *Cichlasoma nigrofasciatum* (Pisces, Cichlidae). *Behav. Process.* **1**, 3–14.

Fraley, N. B., and Fernald, R. D. (1982). Social control of developmental rate in the African cichlid, *Haplochromis burtoni*. *Z. Tierpsychol.* **60**, 66–82.

Frankel, J. S. (1981). Alcohol dehydrogenase ontogeny and liver maturation in the leopard danio, *Brachydanio nigrofasciatus. Comp. Biochem. Physiol. B* **70B**, 643–644.

Frankel, J. S., and Hart, N. H. (1977). Lactate dehydrogenase ontogeny in the genus *Brachydanio* (Cyprinidae). *J. Hered.* **68**, 81–86.

Fujita, K. (1981). Oviphagous embryos of the pseudo-carchariid shark *Pseudocarcharias kamoharai*, from the central Pacific. *Jpn. J. Ichthyol.* **28**, 37–44.

Galat, D. L., and Eipper, A. W. (1975). Presence of food organisms in the prolarval environment as a factor in the growth and mortality of larval muskellunge (*Esox masquinongy*). *Trans. Am. Fish. Soc.* **104**, 338–341.

Godin, J.-G. J. (1978). Behavior of juvenile pink salmon (*Oncorhynchus gorbuscha* Walbaum) toward novel prey: Influence of ontogeny and experience. *Environ. Biol. Fishes* **3**, 261–266.

Godin, J.-G. J. (1982). Migrations of salmonid fishes during early life history phases: Daily and annual timing. *In* "Salmon and Trout Migratory Behavior Symposium" (E. L. Brannon and E. O. Salo, eds.), pp. 22–50. Univ. of Washington Press, Seattle.

Godin, J.-G. J. (1986). Antipredator function of shoaling in teleost fishes: A selective review. *Naturaliste can. Rev. Ecol. Syst.* **113**, 241–250.

Goude, G., Edlund, B., Engqvist-Edlund, U., and Andersson, M. (1972). Approach and withdrawal in young of *Tilapia mossambica* (Cichlidae, Pisces) as a function of age and onset of stimulation. *Z. Tierpsychol.* **31**, 60–77.

Gould, S. J. (1982). "The Panda's Thumb." Norton, New York.

Govani, J. J., Boehlert, G. W., and Wantanabe, Y. (1986). The physiology of digestion in fish larvae. *Environ. Biol. Fishes* **16**, 59–77.

Groot, C., Quinn, T. P., and Hara, T. J. (1986). Responses of migrating adult sockeye salmon (*Oncorhynchus nerka*) to population-specific odours. *Can. J. Zool.* **64**, 926–932.

Grün, G. (1979). Light-induced acceleration of retina development in a mouth-brooding teleost. *J. Exp. Zool.* **208**, 291–302.

Guma'a, S. A. (1982). Retinal development and retinomotor responses in perch, *Perca fluviatilis* L. *J. Fish. Biol.* **20**, 611–618.

Guthrie, D. M. (1986). Role of vision in fish behaviour. *In* "The Behaviour of Teleost Fishes" (T. J. Pitcher, ed.), pp. 75–113. Croom Helm, London.

Hairston, N. G., Jr., Li, K. T., and Easter, S. S., Jr. (1982). Fish vision and the detection of planktonic prey. *Science* **218**, 1240–1242.

Hamilton, W. D. (1967). Extraordinary sex ratios. *Science* **156**, 477–488.

Hara, T. J. (1971). Chemoreception. *In* "Fish Physiology" (W. S. Hoar and D. J. Randall, ed.), Vol. 5, pp. 79–120. Academic Press, New York.

Hara, T. J. (1986). Role of olfaction in fish behaviour. *In* "The Behaviour of Teleost Fishes" (T. J. Pitcher, ed.), pp. 152–176. Croom Helm, London.

Harden Jones, F. R. (1968). "Fish Migration." Arnold, London.

Hart, T. F., Jr., and Werner, R. G. (1987). Effects of prey density on growth and survival of white sucker, *Catostomus commersoni*, and pumkinseed, *Lepomis gibbosus*, larvae. *Environ. Biol. Fishes* **18**, 41–50.

Hasler, A. D., and Scholz, A. T. (1983). "Olfactory Imprinting and Homing in Salmon." Springer-Verlag, Berlin and New York.

Hawkins, A. D. (1973). The sensitivity of fish to sounds. *Oceanogr. Mar. Biol.* **11**, 291–340.

Hawkins, A. D. (1986). Underwater sound and fish behaviour. *In* "The Behaviour of Teleost Fishes" (T. J. Pitcher, ed.), pp. 114–151. Croom Helm, London.

Hoar, W. S. (1976). Smolt transformation: Evolution, behaviour, and physiology. *J. Fish. Res. Board Can.* **33**, 1234–1252.

Hubbs, C. L. (1943). Terminology of early stages of fishes. *Copiea*, p. 260,

Hunter, J. R. (1981). Feeding ecology and predation of marine fish larvae. *In* "Marine Fish Larvae: Morphology, Ecology and Relation to Fisheries" (R. Lasker, ed.), Washington Sea Grant Program, pp. 33–77. Univ. of Washington Press, Seattle.

Hunter, J. R., and Sanchez, C. (1976). Diel changes in swim bladder inflation of the larvae of the northern anchovy, *Engraulis mordax. U.S. Fish Wildl. Serv. Bull.* **74**, 847–855.

Huntingford, F. A. (1986). Development of behaviour in fishes. *In* "The Behaviour of Teleost Fishes" (T. J. Pitcher, ed.), pp. 47–68. Croom Helm, London.

Immelmann, K. (1972). Sexual and other long-term aspects of imprinting in birds and other species. *Adv. Study Behav.* **4**, 147–174.

Ingle, D. (1071). Vision: The experimental analysis of visual behavior. *In* "Fish Physiology" (W. S. Hoar and D. J. Randall, eds.), Vol. 5, pp. 59–77. Academic Press, New York.

Iwai, T. (1967). Structure and development of lateral line cupulae in teleost larvae. *In* "Lateral Line Detectors" (P. H. Cahn, ed.), pp. 27–44. Indiana Univ. Press, Bloomington.

Iawi, T. (1980). Sensory anatomy and feeding of fish larvae. *In* "Fish Behavior and its Use in the Capture and Culture of Fishes" (J. E. Bardach, J. J. Magnuson, R. C. May, and J. M. Reinhart, eds.), pp. 124–145. ICLARM Resour. Manage., Manila, Philippines.

Jahn, L. A. (1972). Development of the olfactory apparatus of the cutthroat trout. *Trans. Am. Fish. Soc.* **101**, 284–289.

Jamieson, I. G. (1986). The functional approach to behavior: Is it useful? *Am. Nat.* **127**, 195–208.

Jeserich, G., and Rahmann, H. (1979). Effect of light deprivation on fine structural changes in the optic tectum of the rainbow trout (*Salmo gairdneri*, Rich.) during ontogenesis. *Dev. Neurosci.* **2**, 19–24.

Johns, P. R. (1981). Growth of fish retinas. *Am. Zool.* **21**, 447–458.

Katsuki, Y., and Yanagisawa, K. (1982). Chemoreception in the lateral-line organ. *In* "Chemoreception in Fishes" (T. J. Hara, ed.), pp. 227–242. Elsevier, Amsterdam.

Kawamura, G., and Hara, S. (1980). The optomotor reaction of milkfish larvae and juveniles. *Bull. Jpn. Soc. Sci. Fish.* **46**, 929–932.

Keenleyside, M. H. A. (1979). "Diversity and Adaptation in Fish Behaviour." Springer-Verlag, Berlin and New York.

Klar, G. T., Stalnaker, C. B., and Farley, T. M. (1979). Comparative physical and physiological performance of rainbow trout, *Salmo gairdneri*, of distinct lactate dehydrogenase B phenotypes. *Comp. Biochem. Physiol. A* **63A**, 229–235.

Kleerekoper, H. (1969). "Olfaction in Fishes." Indiana Univ. Press, Bloomington.

Kohno, H., Taki, Y., Ogasawara, T., Shirojo, Y., Masakazu, T., and Inoue, Y. (1983). Development of swimming and feeding functions in larval *Pagurus major. Jpn. J. Ichthyol.* **30**, 47–59.

Kohno, H., Shimizu, M., and Nose, Y. (1984). Morphological aspects of the development of swimming and feeding functions in larval *Scomber japonicus. Bull. Jpn. Soc. Sci. Fish.* **50**, 1125–1137.

Kop, P.P.A.M., and Heuts, B. A. (1973). An experiment on sibling imprinting in the jewel fish *Hemichromis bimaculatus* (Gill 1862, Cichlidae). *Rev. Comp. Anim.* **7**, 63–76.

Kühme, W. (1963). Verhaltensstudien am maulbrutenden (*Betta anabatoides* Bleeker) und am nestbauenden Kampffisch (*B. splendens* Regan). *Z. Tierpsychol.* **18**, 33–55.

Lasker, R. (1975). Field criteria for survival of anchovy larvae: The relation between inshore chlorophyll maximum layers and successful first feeding. *Fish. Bull.* **73**, 453–462.

Leatherland, J. F. (1982). Environmental physiology of the teleostean thyroid gland: A review. *Environ. Biol. Fishes* **7**, 83–110.

Li, K. T., Wetterer, J. K., and Hairston, N. G., Jr. (1985). Fish size, visual resolution and prey selectivity. *Ecology* **66**, 1729–1735.

Liley, N. R. (1982). Chemical communication in fish. *Can. J. Fish. Aquat. Sci.* **39**, 22–35.

Lowenstein, O. (1971). The labyrinth. *In* "Fish Physiology" (W. S. Hoar and D. J. Randall, eds.), Vol. 5, pp. 207–240. Academic Press, New York.

Lythgoe, J. N. (1979). "The Ecology of Vision." Oxford Univ. Press, London and New York.

McFarland, D. J. (1986). "Animal Behaviour." Pitman, New York.

McKeown, B. A. (1984). "Fish Migration." Croom Helm, London.

Magurran, A. E. (1986). Individual differences in fish behaviour. *In* "The Behaviour of Teleost Fishes" (T. J. Pitcher, ed.), pp. 338–365. Croom Helm, London.

Mason, J. C. (1974). A first appraisal of the response of juvenile coho salmon (*Oncorhynchus kisutch*) to supplemental feeding in an exprimental rearing stream. *Tech. Rep—Fish. Mar. Serv. (Can.)* **469**.

Michaletz, P. H., Rabeni, C.F., Taylor, W. W., and Russell, T. R. (1982). Feeding ecology and growth of young-of-the-year paddlefish in hatchery ponds. *Trans. Am. Fish. Soc.* **111**, 700–709.

Minnier, D. E., and Cole, J. E. (1973). Parental influence on the larval and early postlarval development of young *Etroplus maculatus* (Osteichthyes: Cichlidae). *Proc. Pa. Acad. Sci.* **47**, 117–121.

Munz, F. W. (1971). Vision: Visual pigments. *In* "Fish Physiology" (W. S. Hoar and D. J. Randall, eds.), Vol. 5, pp. 1–32. Academic Press, New York.

Myrberg, A. A., Jr. (1975). The role of chemical and visual stimuli in the preferential discrimination of young by the cichlid fish *Cichlasoma nigrofasciatum* (Günther). *Z. Tierpsychol.* **37**, 274–297.

Myrberg, A. A., Jr. (1981). Sound communication and interception in fishes. *In* "Hearing and Sound Communication in Fishes" (W. N. Tavolga, A. N. Popper, and R. R. Fay, eds.), pp. 395–426. Springer-Verlag, Berlin and New York.

Neave, D. A. (1984). The development of visual acuity in larval plaice (*Pleuronectes platessa* L.) and turbot (*Scophthalmus maximus*). *J. Exp. Mar. Biol. Ecol.* **78**, 167–175.

Noakes, D. L. G. (1976). Parent-touching behaviour in fishes: Incidence, function and causation. *Environ. Biol. Fishes* **2**, 89–100.

Noakes, D. L. G. (1978a). Ontogeny of behavior in fishes: A survey and suggestions. *In* "The Development of Behavior: Comparative and Evolutionary Aspects" (G. M. Burghardt and M. Bekoff, eds.), pp. 103–125. Garland STPM Press, New York.

Noakes, D. L. G. (1978b). Early behaviour in fishes. *Environ. Biol. Fishes* **3**, 321–326.

Noakes, D. L. G. (1980). Social behaviour in young charrs. *In* "Charrs. Salmonid Fishes of the Genus *Salvelinus*" (E. K. Balon, ed.), pp. 683–701. Junk, The Hague.

Noakes, D. L. G. (1981). Comparative aspects of behavioral development: A philosophy from fishes. *In* "Behavioral Development" (K. Immelmann, G. W. Barlow, L. Petrinovitch, and M. Main, eds.), pp. 491–508. Cambridge Univ. Press, London and New York.

Noakes, D. L. G. (1986). Genetic basis of fish behaviour. In "The Behaviour of Teleost Fishes" (T. J. Pitcher, ed.), pp. 3–22. Croom Helm, London.

Noakes, D. L. G., and Balon, E. K. (1982). Life histories of tilapias: An evolutionary perspective. In "The Biology and Culture of Tilapias" (R. S. V. Pullin and R. H. Lowe-McConnel, eds.), pp. 61–82. ICLARM, Manila, The Philippines.

Noakes, D. L. G., and Barlow, G. W. (1973). Ontogeny of parent-contacting in young Cichlasoma citrinellum (Pisces: Cichlidae). Behaviour 46, 221–255.

Noakes, D. L. G., and Grant, J. W. A. (1987). Behavioural ecology and production of riverine fishes. Arch. Pol. Hydrobiol. 19.

Nordeng, H. (1977). A pheromone hypothesis for homeward migration in anadromous salmonids. Oikos 28, 155–159.

Northcutt, R. G. (1981). Audition and the central nervous system of fishes. In "Hearing and Sound Communication in Fishes" (W. N. Tavolga, A. N. Popper, and R. R. Fay, eds.), pp. 331–355. Springer-Verlag, Berlin and New York.

Nunan, C. P., and Noakes, D. L. G. (1985a). Light sensitivity and substrate penetration by eleutheroembryos of brook (Salvelinus fontinalis) and lake charr (Salvelinus namaycush) and their F1 hybrid, splake. Exp. Biol. 44, 221–228.

Nunan, C. P., and Noakes, D. L. G. (1985b). Response of rainbow trout (Salmo gairdneri) embryos to current flow in simulated substrates. Can. J. Zool. 63, 1813–1815.

O'Connell, C. P. (1976). Histologica criteria for diagnosing the starving condition in early post yolk sac larvae of the northern anchovy, Engraulis mordax Girard. J. Exp. Mar. Biol. Ecol. 25, 285–312.

O'Connell, C. P. (1981). Development of organ systems in the northern anchovy, Engraulis mordax, and other teleosts. Am. Zool. 21, 429–446.

Pankhurst, N. W. (1984). Retinal development in larval and juvenile European eel, Anguilla anguilla (L.). Can. J. Zool. 62, 335–343.

Partridge, B. L., and Pitcher, T. J. (1980). The sensory basis of fish schools: Relative roles of lateral line and vision. J. Comp. Physiol. A 135A, 315–325.

Peterson, R. H. (1975). Pectoral fin and opercular movements of Atlantic salmon (Salmo salar) alevins. J. Fish. Res. Board Can. 32, 643–647.

Peterson, R. H., Daye, P. G., and Metcalf, J. L. (1980). Inhibition of Atlantic salmon (Salmo salar) hatching at low pH. Can. J. Fish. Aquat. Sci. 37, 770–774.

Peterson, W. T., and Ausubel, S. J. (1984). Diets and selective feeding by larvae of Atlantic mackerel Scomber scombrus on zooplankton. Mar. Ecol. Prog. Ser. 14, 65–75.

Pfeiffer, W. (1963). The fright reaction in North American fish. Can. J. Zool. 41, 69–77.

Pfeiffer, W. (1982). Chemical signals in communication. In "Chemoreception in Fishes" (T. J. Hara, ed.), pp. 307–326. Elsevier, Amsterdam.

Pitcher, T. J., ed. (1986). "The Behaviour of Teleost Fishes." Croom-Helm, London.

Quinn, T. P., and Busack, C. A. (1985). Chemosensory recognition of siblings in juvenile coho salmon (Oncorhynchus kisutch). Anim. Behav. 33, 51–56.

Quinn, T. P., and Hara, T. J. (1986). Sibling recognition and olfactory sensitivity in juvenile coho salmon (Oncorhynchus kisutch). Can. J. Zool. 64, 921–925.

Quinn, T. P., and Tolson, G. M. (1986). Evidence of chemically mediated population recognition in coho salmon (Oncorhynchus kisutch. Can. J. Zool. 64, 84–87.

Rahmann, H., and Jeserich, G. (1978). Quantitative morphogenetic investigations on fine structural changes in the optic tectum of the rainbow trout (Salmo gairdneri) during ontogenesis. Wilhelm Roux's Arch. Dev. Biol. 184, 83–94.

Rahmann, H., Jeserich, G., and Zeutzius, I. (1979). Ontogeny of visual acuity of rainbow trout under normal conditions and light deprivation. *Behaviour* 68, 315–322.

Rajasilta, M., and Vuorinen, I. (1983). A field study of prey selection in planktivorous fish larvae. *Oecologia* (Berlin) 59, 65–68.

Ralph, C. L. (1975). The pineal complex: A retrospective view. *Am. Zool.* 15, Suppl. 1, 105–116.

Richards, W. J. (1976). Some comments on Balon's terminology of fish developmental intervals. *J. Fish. Res. Board Can.* 33, 1253–1254.

Ringler, N. H. (1983). Variation in foraging tactics of fishes. *In* "Predators and Prey in Fishes" (D. L. G. Noaekes, D. G. Lindquist, G. S. Helfman, and J. A. Ward, eds.), pp. 159–171. Junk, The Hague.

Roberts, T. R. (1986). *Danionella translucida*, a new genus and species of cyprinid fish from Burma, one of the smallest living vertebrates. *Environ. Biol. Fishes* 16, 231–241.

Romero, A. (1985). Ontogenetic change in phototactic responses of surface and cave populations of *Astyanax fasciatus* (Pisces: Characidae). *Copeia*, pp. 1004–1011.

Rosenthal, H. (1968). Beobachtungen über die Entwicklung des Schwarmverhaltens bei den Larven des Herings *Clupea harengus. Mar. Biol.* 2, 73–76.

Rosenthal, H., and Hempel, G. (1970). Experimental studies in feeding and food requirements of herring larvae (*Clupea harengus* L.). *In* "Marine Food Chains" (J. H. Steele, ed.), pp. 344–364. Oliver & Boyd, Edinburgh.

Russock, H. I. (1986). Preferential behaviour of *Sarotherodon* (*Oreochromis*) *mossambicus* (Pisces: Cichlidae) fry to maternal models and its relevance to the concept of imprinting. *Behaviour* 96, 304–321.

Sand, O. (1981). The lateral line and sound reception. *In* "Hearing and Sound Communication in Fishes" (W. N. Tavolga, A. N. Popper, and R. R. Fay, eds.), pp. 459–480. Springer-Verlag, Berlin and New York.

Sandy, J. M., and Blaxter, J. H. S. (1980). A study of retinal development in larval herring and sole. *J. Mar. Biol. Assoc. U.K.* 60, 59–71.

Schiller, H., Segler, K., Jeserich, G., Rosner, H., and Rahmann, H. (1979). Properties of membrane-bound neuraminidase in the developing trout brain. *Life Sci.* 25, 2029–2033.

Schmidtke, J., Kuhl, P., and Engel, W. (1976). Transitory hemizygosity of paternally derived allele in hybrid trout embryos. *Nature* (*London*) 260, 319–320.

Schutz, F. (1956). Vergleichende Untersuchungen über dei Schreckreaktion bei Fischen und Deren Verbreitung. *Z. Vergl. Physiol.* 38, 84–135.

Schwassmann, H. O. (1965). Functional development of visual pathways in larval sardines and anchovies. *Rep. Calif. Coop. Oceanogr. Fish. Invest.* 10, 64–70.

Seybold, V., and Rahmann, H. (1985). Changes in developmental profiles of brain gangliosides during ontogeny of a teleost fish (*Sarotherodon mossambicus*, Cichlidae). *Roux's Arch. Dev. Biol.* 194, 166–172.

Sharma, S. C. (1975). Development of the optic tectum in brown trout. *In* "Vision in Fishes: New Approaches in Research" (M. A. Ali, ed.), pp. 411–417. Plenum, New York.

Shaw, E. (1970). Schooling in fishes: Critique and review. *In* "Development and Evolution of Behavior" (L. R. Aronson, E. Tobach, D. S. Lehrman, and J. S. Rosenblatt, eds.), pp. 452–480. Freeman, San Francisco, California.

Shaw, E., and Sachs, B. D. (1967). Development of the optomotor response in the schooling fish *Menidia menidia. J. Comp. Physiol. Psychol.* 63, 385–388.

Shaw, E., Allen, J., and Stone, R. (1974). Notes on a collection of shiner perch, *Cymatogaster aggregata*, in Bodega Harbour, California. *Calif. Fish Game* **60**, 15–22.

Siepen, G., and Crapon de Carprona, M.-D. (1986). The influence of parental color morph on mate choice in the cichlid fish *Cichlasoma nigrofasciatum*. *Ethology* **71**, 187–200.

Sjölander, S., and Fernö, A. (1973). Sexual imprinting on another species in a cichlid fish, *Haplochromis burtoni*. *Rev. Comp. Anim.* **7**, 77–81.

Smith, R. J. F. (1982). The adaptive significance of the alarm substance–fright reaction system. *In* "Chemoreception in Fishes" (T. J. Hara, ed.), pp. 327–342. Elsevier, Amsterdam.

Smith, R. J. F. (1985). "The Control of Fish Migration." Springer-Verlag, Berlin and New York.

Solomon, D. J. (1977). A review of chemical communication in freshwater fish. *J. Fish Biol.* **11**, 363–376.

Stanley, J. G. (1983). Gene expression in haploid embryos of Atlantic salmon. *J. Hered.* **74**, 19–22.

Stearns, S. C. (1983). A tractable model system in which social deprivation early in life leads to behaviour-mediated functional sterility: The mosquitofish, *Gambusia affinis*. *Anim. Behav.* **31**, 950–951.

Tavolga, W. N. (1971). Sound production and detection. *In* "Fish Physiology" (W. S. Hoar and D. J. Randall, eds.), Vol. 5, pp. 135–205. Academic Press, New York.

Taylor, E. B., and McPhail, J. D. (1985). Ontogeny of the startle response in young coho salmon *Oncorhynchus kisutch. Trans. Am. Fish. Soc.* **114**, 552–557.

Thorpe, J. E., Morgan, R. I. G., Talbot, C., and Miles, M. S. (1983). Inheritance of developmental rates in Atlantic salmon, *Salmo salar* L. *Aquaculture* **33**, 119–128.

Tomita, T. (1971). Vision: Electrophysiology of the retina. *In* "Fish Physiology" (W. S. Hoar and D. J. Randall, eds.), Vol. 5, pp. 33–57. Academic Press, New York.

Townsend, D. W. (1983). The relations between larval fishes and zooplankton in two inshore areas of the Gulf of Maine. *J. Plankton Res.* **5**, 145–173.

Twongo, K. T., and MacCrimmon, H. R. (1976). Significance of the timing of initial feeding in hatchery rainbow trout, *Salmo gairdneri. J. Fish. Res. Board Can.* **33**, 1914–1921.

Vernier, J.-M. (1969). Table chronologique due développement embryonaire de la truite arc-en-ciel, *Salmo gairdneri* Richardson 1836. *Ann. Embryol. Morphol.* **2**, 495–520.

Verraes, W. (1976). Postembryonic development in the nasal organs, sacs and surrounding skeletal elements in *Salmo gairdneri* (Teleostei: Salmonidae), with some functional interrelationships. *Copeia*, pp. 71–75.

Waldman, B. (1982). Quantitative and developmental analyses of the alarm reaction in the zebra danio, *Brachydanio rerio. Copeia*, pp. 1–9.

Wales, W. (1975). Extraretinal photosensitivity in fish larvae. *In* "Vision in Fishes: New Approaches in Research" (M. A. Ali, ed.), pp. 445–450. Plenum, New York.

Walker, B. (1952). A guide to the grunion. *Calif. Fish Game* **38**, 409–420.

Ward, J. A., and Barlow, G. W. (1967). The maturation and regulation of glancing off the parents by young orange chromides (*Etroplus maculatus*; Pisces, Cichlidae). *Behaviour* **29**, 1–56.

Watling, H., and Hilleman, H. H. (1964). The development of the olfactory apparatus of the grayling (*Thymallus arcticus*). *J. Fish. Res. Board Can.* **21**, 373–396.

Weber, P. G., and Weber, S. P. (1976). The effects of female color, size, dominance, and early experience upon mate selection in male convict cichlids, *Cichlasoma nigrofasciatum* Günther (Pisces, Cichlidae). *Behaviour* **54**, 116–135.

Werner, E. E. (1986). Amphibian metamorphosis: Growth rate, predation risk, and the optimal size at transformation. *Am. Nat.* **128**, 319–341.

Werner, R. G., and Blaxter, J. H. S. (1980). Growth and survival of larval herring (*Clupea harengus*) in relation to prey density. *Can. J. Fish. Aquat. Sci.* **37**, 1063–1069.

White, G. M. (1915). The behavior of brook trout embryos from the time of hatching to the absorption of the yolk sac. *J. Anim. Behav.* **5**, 44–60.

Winterbottom, R., and Emery, A. R. (1981). A new genus and two new species of gobiid fishes (Perciformes) from the Chagos Archipelago, Central Indian Ocean. *Environ. Biol. Fishes* **6**, 139–149.

Wourms, J. P. (1981). Viviparity: The maternal–fetal relationship in fishes. *Am. Zool.* **21**, 473–515.

Wyman, R. L., and Ward, J. A. (1973). The development of behavior in the cichlid fish *Etroplus maculatus* (Bloch). *Z. Tierpsychol.* **33**, 461–491.

Zeiske, E., Kux, J., and Melinkat, R. (1976). Development of the olfactory organ of oviparous and viviparous cyprinodonts (Teleostei). *Z. Zool. Syst. Evolutions forsch.* **14**, 34–40.

Zeutzius, I., and Rahmann, H. (1984). Influence of dark-rearing on the ontogenetic development of *Sarotherodon mossambicus* (Cichlidae, Teleostei): I: Effects on body weight, body growth pattern, swimming activity, and visual activity. *Exp. Biol.* **43**, 77–85.

Zeutzius, T., Probst, W., and Rahmann, H. (1984). Influence of dark-rearing on the ontogenetic development of *Sarotherodon mossambicus* (Cichlidae, Teleostei): II: Effects on allomerical growth relations and differentiation of the optic tectum. *Exp. Biol.* **43**, 87–96.

Zweifel, J. R., and Lasker, R. (1976). Prehatch and posthatch growth of fishes—a general model. *U.S. Fish Wildl. Serv. Fish. Bull.* **74**, 609–621.

AUTHOR INDEX

Numbers in italics refer to the pages on which the complete references are listed.

SYSTEMATIC INDEX

Note: Names listed are those used by the authors of the various chapters. No attempt has been made to provide the current nomenclature where taxonomic changes have occurred. Boldface letters refer to Parts A and B of Volume XI.

SUBJECT INDEX

Note: Boldface **A** refers to entries in Volume XIA; **B** refers to entries in Volume XIB.

A

Absolute aerobic scope, **A**, 99–106
Acid rain, *see* pH
Acoustic system, development, **A**, 40–41, 44
Activation
  defined, **A**, 167
  ion fluxes and electrical events, **A**, 178–183, 217–220, 238
Adelphophagy, **B**, 37, 43–48, 81–82
Adrenal gland, *see* Interrenal gland, and smolting
Aerobic scope, *see* Absolute aerobic scope
Air incubation, **A**, 482
Alarm substance, **B**, 362–363, 379
Alimentary canal, development, **B**, 376–377
Allometric growth, **A**, 12, 39
Anabolic steroids, *see* Testosterone
Androgens, in vitellogenesis, **A**, 352, 360
Annual fish, **B**, 372
Adrenocorticotropic hormone (ACTH), *see* Interrenal gland, and smolting
Apstein's stages, **A**, 8
Atricial young, **B**, 376–377
Astaxanthin, *see* Carotenoids
Atroposic model, **B**, 244–251, 256, 260, 261

B

Behavior
  development of, **B**, 346–395
  feeding, **B**, 373–377

myogenic vs. neurogenic origin, **B**, 346
pollutants and, **A**, 273–275, 313–315
Boundary layer, **A**, 61–64
Branchiostegal rays, development, **B**, 242
Buoyancy
  lipids and, **A**, 420–421
  salinity and, **A**, 121
  of smolts, **B**, 281
  wax esters and, **A**, 382

C

Calorimetry, **A**, 83–87
Calyces nutriciae, **B**, 101
Cannibalism, **B**, 28, 43
Carbonic anhydrase, in smolting, **B**, 284
Cardiac rate, pollutant effects, **A**, 271–272, 317
Carotenoids
  in eggs, **A**, 23, 384, 392–393, 421
  in respiration, **A**, 76–77
Chemosensory system
  alarm substance and fright reaction, **B**, 362–363
  development of, **A**, 41, **B**, 356–363
  in habitat selection, **B**, 359–360
  in mate choice, **B**, 361–362
Chloride cells
  "acid rain" and, **A**, 229
  in deionized water, **A**, 229
  in embryos and larvae, **A**, 214, 224–234, **B**, 24
  in smolts, **B**, 288–289, 301, 306, 322
  structure of, **A**, 225–228

427